MW01329252

- Logic Synthesis for Low Power
- Low Power Arithmetic Components
- Low Power Memory Design
- Low Power Microprocessor Design
- Applications for Low Power Design

COURSE LOAD:

2 EXAMS — 35% each

HW/Labs — 30%

LOW POWER DESIGN METHODOLOGIES

LOW POWER DESIGN METHODOLOGIES

edited by

Jan M. Rabaey
University California

and

Massoud Pedram
University of Southern California

KLUWER ACADEMIC PUBLISHERS
Boston / Dordrecht / London

Distributors for North America:
Kluwer Academic Publishers
101 Philip Drive
Assinippi Park
Norwell, Massachusetts 02061 USA

Distributors for all other countries:
Kluwer Academic Publishers Group
Distribution Centre
Post Office Box 322
3300 AH Dordrecht, THE NETHERLANDS

Consulting Editor: Jonathan Allen, Massachusetts Institute of Technology

Library of Congress Cataloging-in-Publication Data

A C.I.P. Catalogue record for this book is available
from the Library of Congress.

Copyright © 1996 by Kluwer Academic Publishers. Fifth Printing 2002.

All rights reserved. No part of this publication may be reproduced, stored in a retrieval system or transmitted in any form or by any means, mechanical, photo-copying, recording, or otherwise, without the prior written permission of the publisher, Kluwer Academic Publishers, 101 Philip Drive, Assinippi Park, Norwell, Massachusetts 02061

Printed on acid-free paper.

Printed in the United States of America

This printing is a digital duplication of the original edition.

Table of Contents

PART II Logic and Module Design Levels

PART III Architecture and System Design Levels

Preface

Most of the research and development efforts in the area of digital electronics have been oriented towards increasing the speed and the complexity of single chip digital systems. This has resulted in a powerful, but power-hungry, design technology, which enabled the development of personal workstations, sophisticated computer graphics, and multi-media capabilities such as real-time speech recognition and real-time video. While focusing the attention on speed and area, power consumption has long been ignored.

This picture is, however, undergoing some radical changes. Power consumption of individual components is reaching the limits of what can be dealt with by economic packaging technologies, resulting in reduced device reliability. Dealing with power is rapidly becoming one of the most demanding issues in digital system design. This situation is aggravated by the increasing demand for portable systems in the areas of communication, computation and consumer electronics. Improvements in battery technology are easily offset by the increasing complexity and performance of those applications. To guarantee a reasonable battery operation time, a dramatic (e.g., 100×) reduction of the power consumption is essential.

These realizations spurred a renewed interest in low power design over the last five years. Researchers learned quickly that there is no single solution to the power dissipation problem. In fact, to be meaningful and have a real impact, power optimization should occur at all levels of the design hierarchy, including the technology, circuit, layout, logic, architectural and algorithmic levels. It is our experience that combining optimizations at all those levels easily results in orders of magnitude of power reduction.

Realizing these potential savings requires a thorough understanding of where power is dissipated in a digital integrated circuit. Once the dominant sources of dissipation are identified, a whole battery of low power design techniques can be brought in action. A significant portion of the overall challenge of making low power design techniques and methodologies practical involves going through the existing synopsis of power conscious techniques and finding the right combination of methods and tools for a particular application domain. It is a fair statement to say that low power design is foremost an educational problem. For too long, digital integrated circuit courses focused solely on performance and area optimization. To make low power design a reality, it is essential to make the power dimension an integral part of the design process, even in the early phases of the design conception.

This book has been conceived as an effort to bring all aspects of low power design together in a single document. It covers all layers of the design hierarchy from the technology, circuit, logic and architectural levels up to the system layer. Besides offering an in-depth insight into the mechanisms of power dissipation in digital circuits, it also presents the state-of-the-art approaches to power estimation and reduction. Finally, it introduces a global view on low power design methodologies and how these are being captured in the latest design automation environments.

The different chapters from this manuscript were developed by the leading researchers in their respective areas. Contributions are from both academia and industry. The many contributors of this book have extensively documented the various approaches for designing and implementing power efficient circuits and systems, and have presented them in a way that is understandable and useful to the designers as well as developers. The book can also be used as a textbook for teaching an advanced course on low power design methodologies and approaches. Instructors can select various combinations of chapters and augment them with some of the many references provided at the end of each chapter to tailor the book to their educational needs.

We are convinced that this document will serve as a broad and thorough introduction for anyone interested in the low-power dimension of design and hope that it will spur further research in all aspects of low power design. Again, mastering the power problem is mandatory if integrated circuits are to maintain their growth curve of the last decades.

Jan M. Rabaey, Tokyo
Massoud Pedram, Los Angeles

Author Index

Arthur Abnous
University of California, Berkeley
abnous@eecs.berkeley.edu

William Athas
Information Sciences Institute (ISI)
athas@isi.edu

Thomas K. Callaway
University of Texas, Austin
tkc@pine.ece.utexas.edu

Wayne W-M. Dai
University of California, Santa Cruz
dai@ce.ucsc.edu

Sonya Gary
Motorola Somerset
sonyag@ibmoto.com

Benjamin M. Gordon
Stanford University
bgordon@tilden.stanford.edu

Chenming Hu
University of California, Berkeley
hu@eecs.berkeley.edu

Kiyoh Itoh
Hitachi Ltd., Kokubunji-shi
81-423-27-7694 (fax)

Paul E. Landman
Texas Instruments, Dallas
landman@nikki.hc.ti.co

David Lidsky
University of California, Berkeley
lidsky@eecs.berkeley.edu

Dake Liu
University of Linkoping, Sweden
dake_l@ifm.liu.se

Renu Mehra
University of California, Berkeley
mehra@eecs.berkeley.edu

Teresa H. Meng
Stanford University
teresa@tilden.stanford.edu

Massoud Pedram
University of Southern California
massoud@zugros.usc.edu

Jan M. Rabaey
University of California, Berkeley
jan@eecs.berkeley.edu

Christer Svensson
University of Linkoping, Sweden
chs@ifm.liu.se

Earl E. Swartzlander
University of Texas, Austin
e.swartzlander@compmail.com

Ely K. Tsern
Stanford University
tsern@tilden.stanford.edu

Joe G. Xi
University of California, Santa Cruz
joex@ce.ucsc.edu

LOW POWER DESIGN METHODOLOGIES

Introduction

Jan M. Rabaey, Massoud Pedram and Paul E. Landman

1.1. Motivation

Historically, system *performance* has been synonymous with circuit speed or processing power. For example, in the microprocessor world, performance is often measured in Millions of Instructions Per Second (MIPS) or Millions of Floating point Operations Per Second (MFLOPS). In other words, the highest "performance" system is the one that can perform the most computations in a given amount of time. The question of *cost* really depends on the implementation strategy being considered. For integrated circuits there is a fairly direct correspondence between silicon area and cost. Increasing the implementation area tends to result in higher packaging costs as well as reduced fabrication yield with both effects translating immediately to increased product cost. Moreover, improvements in system performance generally come at the expense of silicon real estate. So, historically, the task of the VLSI designer has been to explore the Area-Time (AT) implementation space, attempting to strike a reasonable balance between these often conflicting objectives.

But area and time are not the only metrics by which we can measure implementation quality. *Power consumption* is yet another criterion. Until recently, power considerations were often of only secondary concern, taking the back seat to both area and speed. Of course, there are exceptions to this rule; for example, designers of portable devices such as wrist watches have always placed considerable emphasis on minimizing power in order to maximize battery life. For the most part, however, designers of mainstream electronic systems have considered power consumption only as an afterthought — designing for maximum performance regardless of the effects on power.

In recent years, however, this has begun to change and, increasingly, power is being given comparable weight to area and speed considerations. Several factors have contributed to this trend. Perhaps the most visible driving factor has been the remarkable success and growth of the portable consumer electronics market [16],[17],[18],[19],[29]. Lap-top computers, Personal Digital Assistants (PDA's), cellular phones, and pagers have enjoyed considerable success among consumers, and the market for these and other portable devices is only projected to increase in the future.

For these applications, *average* power consumption has become a critical design concern and has perhaps superceded speed and area as the overriding implementation constraint [5][7][11]. The reason for this is illustrated with the simple example of a future portable multi-media terminal, which supports high bandwidth wireless communication, bi-directional motion video, high quality audio, speech and pen-based input and full texts/graphics. The projected power budget for a such a terminal, when implemented using off-the-shelf components not designed for low-power operation [9], hovers around 40 W. With modern Nickel-Cadmium battery technologies offering around 20 Watt-hours/pound (Figure 1.1), this terminal would require 20 pounds of batteries for 10 hours of operation between recharges [13][15]. More advanced battery technologies such as Nickel-Metal-Hydride offer little relief at 30-35 Watt-hours/pound, bringing battery weights down to a still unacceptable 7 pounds. From Figure 1.1, it can be observed that battery capacity has only improved with a factor 2 to 4 over the last 30 years (while the computational power of digital IC's has increased with more than 4 orders of magnitude). Even with new battery technologies such as rechargeable lithium or polymers, it is anticipated that the expected battery lifetime will increase with no more than 30 to 40% over the next 5 years. In the absence of low-power design techniques then, current and future portable devices will suffer from either very short battery life or unreasonably heavy battery packs unless a low power design approach is adopted.

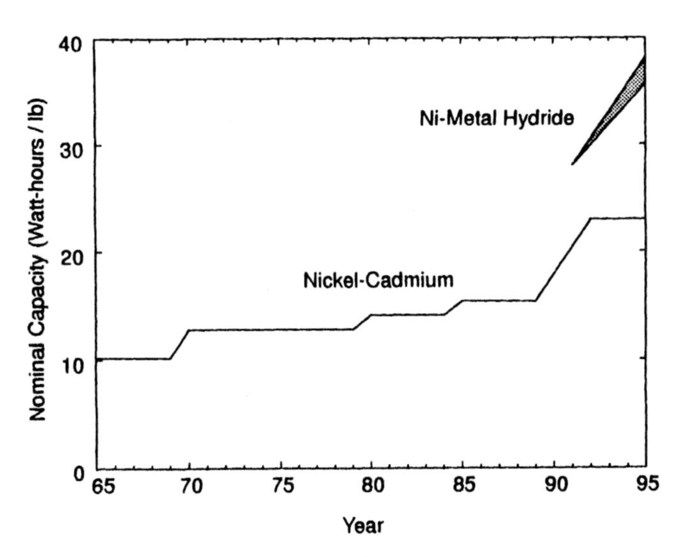

Figure 1.1 Energy storage capacity trends for common battery technologies [13]

On the other hand, *peak power* (maximum possible power dissipation) determines the electrical limits of a design, dictates battery type and power distribution network, and impacts the signal integrity through the $R*i$ and Ldi/dt problems. It is therefore essential to have the peak power under control.

Portability is by no means the sole driving force behind the push for low-power. There exists a strong pressure for producers of high-end products to reduce their power consumption as well. Figure 1.2 shows the trend in microprocessor power consumption. It plots the dissipation of a number of well known processors as a function of die area × clock frequency. The figure demonstrates that contemporary high performance processors dissipate as much as 30 W [12]! More importantly, it shows that the dissipation increases approximately linearly with the area-frequency product. The following expression seems to hold approximately

$$P = \alpha \cdot area \cdot f_{clock} \text{ with } \alpha = 0.063 \text{ W/cm}^2 \cdot \text{MHz} \qquad (1.1)$$

Newer, high performance processors do not display a considerably different behavior. For instance, the DEC 21164 (300 MHz on a die area of 3 cm^2) [4] consumes 50 Watts, compared to the 56.6 Watt predicted by Eq. (1.1). Assuming that the same trend continues in the future, it can be extrapolated that a 10 cm^2

microprocessor, clocked at 500 MHz (which is a not too aggressive estimate for the next decade) would consume 315 Watt. The cost associated with packaging and cooling such devices is becoming prohibitive. Since core power consumption must be dissipated through the packaging, increasingly expensive packaging and cooling strategies are required as chip power consumption increases [20]. Consequently, there is a clear financial advantage to reducing the power consumed by high performance systems.

In addition to cost, there is the issue of reliability. High power systems tend to run hot, and high temperature tends to exacerbate several silicon failure mechanisms. Every 10 °C increase in operating temperature roughly doubles a component's failure rate [22]. Figure 1.3 illustrates this very definite relationship between temperature and the various failure mechanisms such as electromigration, junction fatigue, and gate dielectric breakdown.

From the environmental viewpoint, the smaller the power dissipation of electronic systems, the lower the heat pumped into the rooms, the lower the electricity consumed and therefore, the less the impact on global environment.

Obviously, the motivations in reducing power consumption differ from application to applications. For instance, for a number of application domains such as pace makers and digital watches, minimizing power to an absolute minimum is the absolute prime requirement. In this class of *ultra low power* applications, overall power levels are normally held below 1 mW. In a large class of

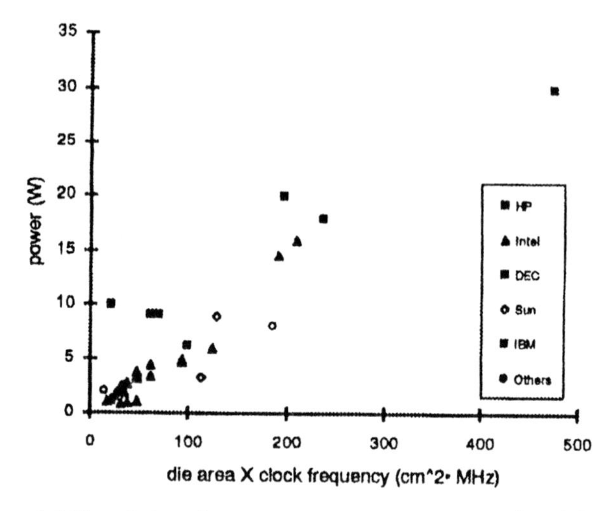

Figure 1.2 Trends in microprocessor power consumption (after [30]).

applications, such as cellular phones and portable computers (*low power*), the goal is to keep the battery lifetime reasonable and packaging cheap. Power levels below 2 Watt, for instance, enable the use of cheap plastic packages. Finally, for *high performance* systems such as workstations and set-top computers, the overall goal of power minimization is to reduce system cost (cooling, packaging, energy bill). These different requirements impact how power optimization is addressed and how much the designer is willing to sacrifice in cost or performance to obtain his power objectives.

1.2. Sources of Dissipation in Digital Integrated Circuits

To set the scene for the rest of this book, it is judicious at this point to briefly discuss the mechanisms for power consumption in CMOS circuits. Consider the CMOS inverter of Figure 1.4. The power consumed when this inverter is in use can be decomposed into two basic classes: static and dynamic. Each of these components will now be analyzed individually.

1.2.1. Static Power

Ideally, CMOS circuits dissipate no static (DC) power since in the steady state there is no direct path from V_{dd} to ground. Of course, this scenario can never be realized in practice since in reality the MOS transistor is not a perfect switch. Thus, there will always be leakage currents and substrate injection currents, which will give rise to a static component of CMOS power dissipation. For a sub-

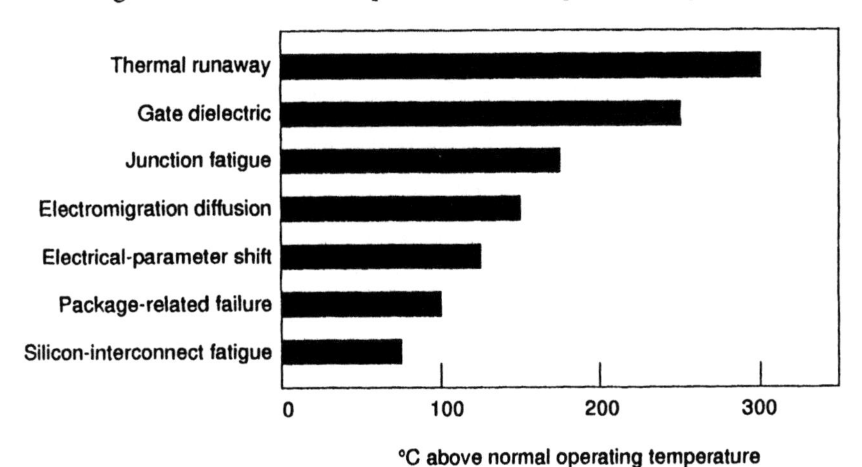

Figure 1.3 Onset temperatures of various failure mechanisms (after [22]).

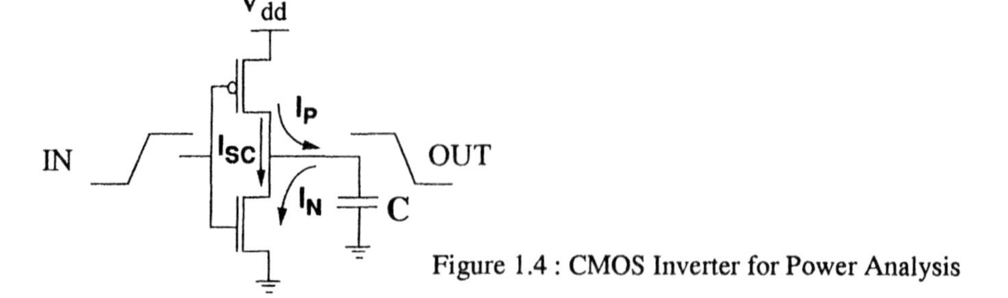

Figure 1.4 : CMOS Inverter for Power Analysis

micron NMOS device with an effective $W/L = 10/0.5$, the substrate injection current is on the order of 1-100μA for a V_{dd} of 5V [28]. Since the substrate current reaches its maximum for gate voltages near $0.4V_{dd}$ and since gate voltages are only transiently in this range as devices switch, the actual power contribution of the substrate injection current is several orders of magnitude below other contributors. Likewise, reverse-bias junction leakage currents associated with parasitic diodes in the CMOS device structure are on the order of nanoamps and will have little effect on overall power consumption.

Another form of static power dissipation occurs for so-called ratioed logic [21]. Pseudo-NMOS, as depicted in is an example of a ratioed CMOS logic family. In this example, the PMOS pull-up is always on and acts as a load device for the NMOS pull-down network. Therefore, when the gate output is in the low-state, there is a direct path from V_{dd} to ground and static currents flow. In this state, the exact value of the output voltage depends on the ratio of the strength of the PMOS and NMOS networks - hence the name. The static power consumed by these logic families can be considerable. For this reason, logic families such as this which experience static power consumption should be avoided for low-power design. With that in mind, the static component of power consumption in *low-power* CMOS circuits should be negligible, and the focus shifts primarily to dynamic power consumption.

1.2.2. Dynamic Power

The dynamic component of power dissipation arises from the transient switching behavior of the CMOS device. At some point during the switching transient, both the NMOS and PMOS devices in Figure 1.4 will be turned on. This occurs for gates voltages between V_{tn} and $V_{dd}-|V_{tp}|$. During this time, a short-circuit exists between V_{dd} and ground and currents are allowed to flow. A detailed

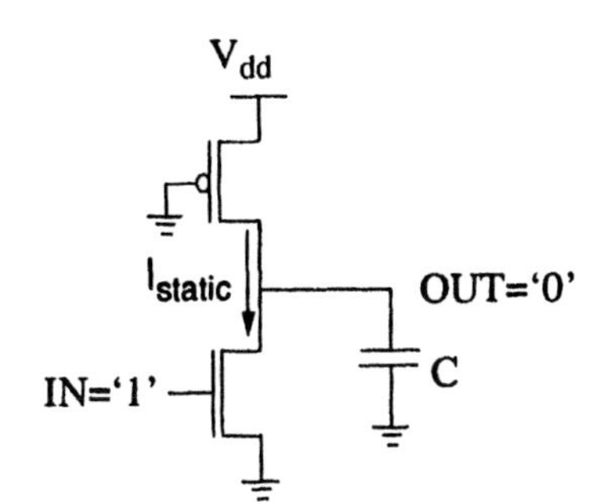

Figure 1.5 Static power dissipation in pseudo-NMOS inverter

analysis of this phenomenon by Veendrick reveals that with careful design of the transition edges this component can be kept below 10-15% of the total power [27]; this can be achieved by keeping the rise and fall times of all the signals throughout the design within a fixed range (preferably equal). Thus, although short circuit dissipation cannot always be completely ignored, it is certainly not the dominant component of power dissipation in well-designed CMOS circuits.

Instead, dynamic dissipation due to capacitance charging consumes most of the power used by CMOS circuits. This component of dynamic power dissipation is the result of charging and discharging parasitic capacitances in the circuit. The situation is modeled in Figure 1.4 where the parasitic capacitances are lumped at the output in the capacitor C. Consider the behavior of the circuit over one full cycle of operation with the input voltage going from V_{dd} to ground and back to V_{dd} again. As the input switches from high to low, the NMOS pull-down network is cut off and PMOS pull-up network is activated charging load capacitance C up to V_{dd}. This charging process draws an energy equal to CV_{dd}^2 from the power supply. Half of this is dissipated immediately in the PMOS transistors, while the other half is stored on the load capacitance. Then, when the input returns to V_{dd} the process is reversed and the capacitance is discharged, its energy being dissipated in the NMOS network. In summary, every time a capacitive node switches from ground to V_{dd} (and back to ground), an energy of CV_{dd}^2 is consumed.

This leads to the conclusion that CMOS power consumption depends on the switching *activity* of the signals involved. In this context, we can define activity, α, as the expected number of zero to one transitions per data cycle. If this is coupled with the average data-rate, f, which may be the clock frequency in a synchronous system, then the effective frequency of nodal charging is given by the product of the activity and the data rate: αf. This leads to the following formulation for average CMOS power consumption:

$$P_{dyn} = \alpha C V_{dd}^2 f \qquad (1.2)$$

This classical result illustrates that the dynamic power is proportional to switching activity, capacitive loading, and the square of the supply voltage. In CMOS circuits, this component of power dissipation is by far the most important, (typically) accounting for at least 90% of the total power dissipation [27]) as illustrated by the previous discussion.

1.3. Degrees of Freedom

The previous section revealed the three degrees of freedom inherent in the low-power design space: voltage, physical capacitance, and activity. Optimizing for power invariably involves an attempt to reduce one or more of these factors. Unfortunately, these parameters are not completely orthogonal and cannot be optimized independently. This section briefly discusses each of these factors, describing their relative importance, as well as the interactions that complicate the power optimization process.

1.3.1. Voltage

With its quadratic relationship to power, voltage reduction offers the most direct and dramatic means of minimizing energy consumption. Without requiring any special circuits or technologies, a factor of two reduction in supply voltage yields a factor of four decrease in energy (see Figure 1.6a) [10]. Furthermore, this power reduction is a global effect, experienced not only in one sub-circuit or block of the chip, but throughout the entire design. Because of this quadratic rela-

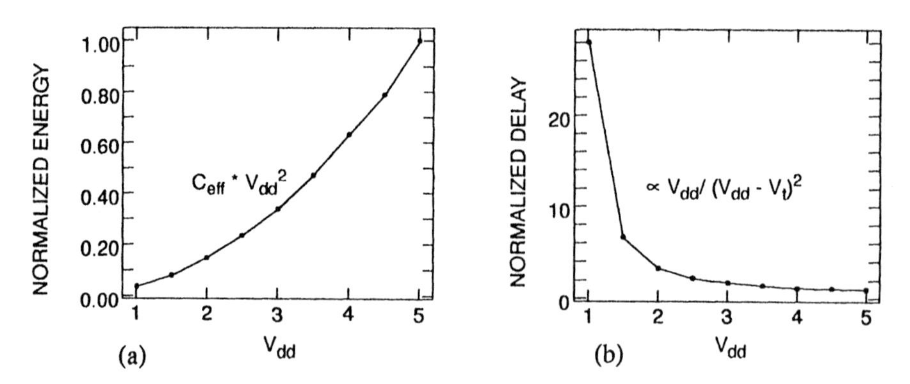

Figure 1.6 : Energy and delay as a function of supply voltage

tionship, designers are often willing to sacrifice increased physical capacitance or circuit activity for reduced voltage. Unfortunately, supply voltage cannot be decreased without bound. In fact, several factors other than power influence selection of a system supply voltage. The primary determining factors are performance requirements and compatibility issues.

As supply voltage is lowered, circuit delays increase (see Figure 1.6b) leading to reduced system performance. For $V_{dd} \gg V_t$ delays increase linearly with decreasing voltage. In order to meet system performance requirements, these delay increases cannot go unchecked. Some techniques must be applied, either technological or architectural to compensate for this effect. This works well until V_{dd} approaches the threshold voltage at which point delay penalties simply become unmanageable. This tends to limit the advantageous range of voltage supplies to a minimum of about $2V_t$.

Performance is not, however, the only limiting criterion. When going to non-standard voltage supplies, there is also the issue of compatibility and inter-operability. Most off-the-shelf components operate off either of 5 V supply or, more recently, a 3.3 V supply [6][24]. Unless the entire system is being designed completely from scratch it is likely that some amount of communications will be required with components operating at a standard voltage. The severity of this dilemma is lessened by the availability of highly efficient (> 90%) DC-DC level converters, but still there is some cost involved in supporting several different supply voltages [23]. This suggests that it might be advantageous for designers to support only a small number of distinct intra-system voltages. For example, custom chips in the system could be designed to operate off a single low voltage (e.g. $2V_t$) with level shifting only required for communication with the outside world.

To summarize, reducing supply voltage is paramount to lowering power consumption, and it often makes sense to increase physical capacitance and circuit activity in order to further reduce voltage. There are, however, limiting factors such as minimum performance and compatibility requirements that limit voltage scaling. These factors will likely lead designers to fix the voltage within a system. Once the supply has been fixed, it remains to address the issues of minimizing physical capacitance and activity at that operating voltage. The next two sections address these topics.

1.3.2. Physical Capacitance

Dynamic power consumption depends linearly on the physical capacitance being switched. So, in addition to operating at low voltages, minimizing capacitances offers another technique for minimizing power consumption. In order to consider this possibility we must first understand what factors contribute to the physical capacitance of a circuit. Only then can we consider how those factors can be manipulated to reduce power.

The physical capacitance in CMOS circuits stems from two primary sources: devices and interconnect. For past technologies, device capacitances dominated over interconnect parasitics. As technologies continue to scale down, however, this no longer holds true and we must consider the contribution of interconnect to the overall physical capacitance [2],[21].

With this understanding, we can now consider how to reduce physical capacitance. From the previous discussion, we recognize that capacitances can be kept at a minimum by using less logic, smaller devices, fewer and shorter wires. Example techniques for reducing the active area include resource sharing, logic minimization and gate sizing. Example techniques for reducing the interconnect include register sharing, common sub-function extraction, placement and routing. As with voltage, however, we are not free to optimize capacitance independently. For example, reducing device sizes reduces physical capacitance, but it also reduces the current drive of the transistors making the circuit operate more slowly. This loss in performance might prevent us from lowering V_{dd} as much as we might otherwise be able to do. In this scenario, we are giving up a possible quadratic reduction in power through voltage scaling for a linear reduction through capacitance scaling. So, if the designer is free to scale voltage it does not make sense to minimize physical capacitance without considering the side effects. Similar arguments can be applied to interconnect capacitance. If voltage and/or activity can be significantly reduced by allowing some increase in physical interconnect capacitance, then this may result in a net decrease in power. The key point to recognize is that low-power design is a joint optimization process in which the variables cannot be manipulated independently.

1.3.3. Activity

In addition to voltage and physical capacitance, switching activity also influences dynamic power consumption. A chip can contain a huge amount of physical capacitance, but if it does not switch then no dynamic power will be consumed. The activity determines how often this switching occurs. As mentioned above, there are two components to switching activity. The first is the data rate, *f*, which reflects how often on average new data arrives at each node. This data might or might not be different from the previous data value. In this sense, the data rate *f* describes how often on average switching, not will but *could* occur. For example, in synchronous systems *f* might correspond to the clock frequency (see Figure 1.7).

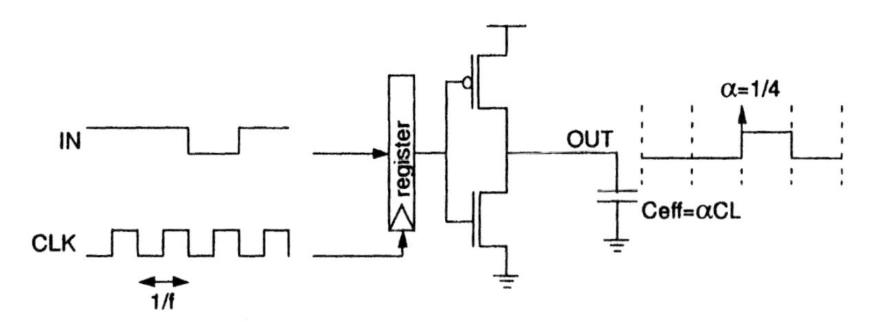

P = αCLVdd2f = CeffVdd2f

Figure 1.7 : Interpretation of switching activity in synchronous systems

The second component of activity is the data activity, α. This factor corresponds to the expected number of energy consuming transitions that will be triggered by the arrival of each new piece of data. So, while *f* determines the average periodicity of data arrivals, α determines how many transitions each arrival will spark. For circuits that do not experience *glitching*, α can be interpreted as the probability that an energy consuming (zero to one) transition will occur during a single data period. Even for these circuits, calculation of α is difficult as it depends not only on the switching activities of the circuit inputs and the logic function computed by the circuit, but also on the spatial and temporal correlations among the circuit inputs. The data activity inside a 16-bit multiplier may change by as much as one order of magnitude as a function of input correlations.

For certain logic styles, however, glitching can be an important source of signal activity and, therefore, deserves some mention here [3]. Glitching refers to spurious and unwanted transitions that occur before a node settles down to its final steady-state value. Glitching often arises when paths with unbalanced propagation delays converge at the same point in the circuit. Calculation of this spurious activity in a circuit is very difficult and requires careful logic and/or circuit level characterization of the gates in a library as well as detailed knowledge of the circuit structure [26]. Since glitching can cause a node to make several power consuming transitions instead of one (i.e. $\alpha > 1$) it should be avoided whenever possible.

The data activity α can be combined with the physical capacitance C to obtain an effective capacitance, $C_{eff}=\alpha C$, which describes the average capacitance charged during each $1/f$ data period. This reflects the fact that neither the physical capacitance nor the activity alone determine dynamic power consumption. Instead, it is the effective capacitance, which combines the two, that truly determines the power consumed by a CMOS circuit

$$P = \alpha C V_{dd}^2 f = C_{eff} V_{dd}^2 f \tag{1.3}$$

Evaluating the effective capacitance of a design is non-trivial as it requires a knowledge of both the physical aspects of the design (that is, technology parameters, circuit structure, delay model) as well as the signal statistics (that is, data activity and correlations). This explains why, lacking proper tools, power analysis is often deferred to the latest stages of the design process or is only obtained from measurements on the finished parts.

As with voltage and physical capacitance, we can consider techniques for reducing switching activity as a means of saving power. For example, switching activity of a finite state machine can be significantly reduced via power-conscious state encoding [25] and multi-level logic optimization [14]. As another example, certain data representations such as sign magnitude have an inherently lower activity than two's-complement [8]. Since sign-magnitude arithmetic is much more complex than two's-complement, however, there is a price to be paid for the reduced activity in terms of higher physical capacitance. This is yet another indication that low-power design is truly a joint optimization problem. In particular, optimization of activity cannot be undertaken independently without consideration for the effects on voltage and capacitance.

1.4. Recurring Themes in Low-Power

The previous sections have provided a strong foundation from which to consider low-power CMOS design. The different chapters in this book present a number of specific power reduction techniques applicable at various levels of abstraction. Many of these techniques follow a small number of common themes.

The four principle themes are trading area-performance for power, adapting designs to environmental conditions or data statistics, avoiding waste, and exploiting locality.

Unquestionably, the most important theme is *trading area-performance for power*. As mentioned earlier, power can be reduced by decreasing the system supply voltage and allowing the performance of the system to degrade. This is clear example of trading performance for power. If the system designer is not willing to give up the performance, she can consider applying techniques such as parallel processing to maintain performance at low voltage. Since many of these techniques incur an area penalty, we can think of this as trading area for power.

An effective power saving technique is to *dynamically change operation of the circuits* as the characteristics of the environment and/or the statistics of the input streams vary. Examples include choosing the most economic communication medium and changing error recovery and encoding to suit the channel noise and error tolerance. Another example is to selectively precompute the output logic values of the circuits one clock cycle before they are required, and then use the precomputed values to reduce internal switching activity in the succeeding clock cycle [1].

Another recurring low-power theme involves *avoiding waste*. For example, clocking modules when they are idle is a waste of power. Glitching is another good example of wasted power and can be avoided by path balancing and choice of logic family. Other strategies for avoiding waste include using dedicated rather than programmable hardware and reducing control overhead by using regular algorithms and architectures. Avoiding waste can also take the form of designing systems to meet rather than surpass performance requirements. If an application requires 25 MIPS of processing performance, there is no advantage gained by implementing a 50 MIPS processor at twice the power.

Exploiting locality is another recurring theme of low-power design. Global operations inherently consume a lot of power. Data must be transferred from one part of the chip to another at the expense of switching a large bus capacitance. Furthermore, the same data might need to be stored in many parts of the chip wasting still more power. In contrast, a design partitioned to exploit locality of reference can minimize the amount of expensive global communications employed in favor of much less costly local interconnect networks. Moreover, especially for DSP applications local data is more likely to be correlated and, therefore, to require much fewer power consuming transitions. So, in its various forms, locality is an important concept in low-power design.

While not all low-power techniques can be classified as trading-off area-performance for power, avoiding waste, and exploiting locality these basic themes do describe many of the strategies that will be presented in this book.

1.5. Emerging Low Power Approaches — An Overview

While long being neglected, the topic of low power digital design has vaulted to the forefront of the attention in the last couple of years and, suddenly, a myriad of low power design techniques has emerged from both universities, research laboratories and industry.

Barring a dramatic introduction of a novel low power manufacturing technology, it is now commonly agreed that low power digital design requires optimization at all levels of the design hierarchy, i.e. technology, devices, circuits, logic, architecture (structure), algorithm (behavior) and system levels, as is illustrated in Figure 1.8. Furthermore, an accompanying computer aided design methodology is required. The goal of this book is to give a comprehensive overview of the different approaches that are currently being conceived at the various levels of design abstraction.

The presented techniques and approaches ultimately all come down to a fundamental set of concepts: dissipation is reduced by lowering either *the supply voltage, the voltage swing, the physical capacitance, the switching activity* or a combination of the above (assuming that a reduction in performance is not allowable).

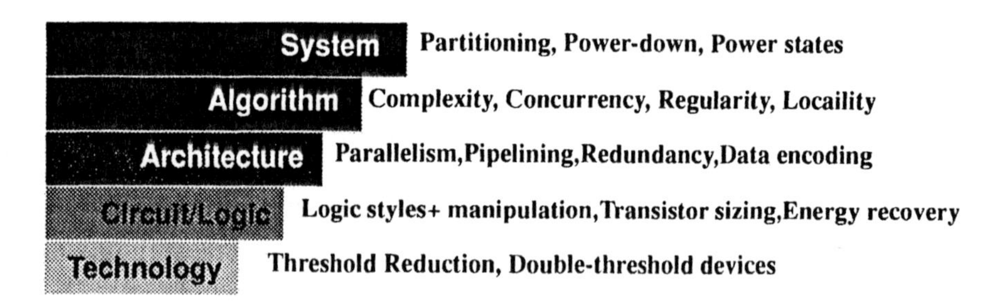

Figure 1.8 An integrated low-power methodology requires optimization at all design abstraction layers.

The first part of the book discusses power minimization approaches at the technology and circuit design levels. The various roads towards a low voltage technology are explored in Chapter 2. An important conclusion from this chapter is that various improvements and innovations in silicon technology can help us to keep power dissipation within bounds but that no huge break-through's should be expected. Chapter 3 presents an in depth analysis and comparison of the design options at the circuit level, including the impact of device sizing, circuit style selection and clocking techniques. An entirely new and alternative approach to low power circuits is presented in Chapter 4 that discusses the topic of energy-recovery CMOS. Since clock distribution is one of the important sources of power dissipation (up to 60% in some of the highest performing microprocessors), a careful analysis and optimization of the low power clock generation and distribution networks is recommendable. This is the topic of Chapter 5.

In the second part, we focus in power analysis and minimization at the logic and module levels. Chapter 6 presents a complete design methodology and computer-aided design approach for low-power design at the logic level. The next two chapters focus on the power minimization of complete modules. Chapter 7 gives an in-depth and experimental view on the design of low-power arithmetic modules, while Chapter 8 analyzes the options in low-power memory design.

The last three chapters of the book address the highest levels in the design hierarchy, being the architecture and system levels. Both the low-power options in the high-performance and the portable communities are represented. Chapter 9 discusses current and future trends in low-power microprocessor design, while Chapter 10 analyzes how to reduce power dissipation in portable, wireless communication devices. Finally, Chapter 11 presents an integrated design methodology for these higher levels of design abstraction.

1.6. Summary

To conclude this introduction, it is worthwhile to summarize the major challenges that, to our belief, have to be addressed if we want to keep power dissipation within bounds in the next generations of digital integrated circuits.

- A low voltage/low threshold technology and circuit design approach, targeting supply voltages around 1 Volt and operating with reduced thresholds. This may include power savings techniques that recycle the signal energies rather than dissipating them as heat.

- Low power interconnect, using advanced technology, reduced swing or reduced activity approaches.
- Introduction of low-power system synchronization approaches, using either self-timed, locally synchronous or other activity minimizing approaches.
- Dynamic power management techniques, varying supply voltage and execution speed according to activity measurements. This includes self-adjusting and adaptive circuit architectures that can quickly and efficiently respond to the environmental change as well as varying data statistics.
- Application specific processing. This might rely on the increased use of application specific circuits or application or domain specific processors.
- A conscientious drive towards parallel and distributed computing, even in the general purpose world. This might mean a rethinking of the programming paradigms, currently in use.
- A system-level approach towards power minimization. System performance can be improved by moving the work to less energy constrained parts of the system, for example, by performing the task on fixed stations rather than mobile sites, by using asymmetric communication protocols, or unbalanced data compression schemes.
- An integrated design methodology - including synthesis and compilation tools. This requires the development of power conscious techniques and tools for behavioral synthesis, logic synthesis and layout optimization. Prime requirements for this are accurate and efficient estimation of the power cost of alternative implementations and the ability to minimize the power dissipation subject to given performance constraints and supply voltage levels. This might very well require a progression to higher level programming and specification paradigms (e.g. data flow or object oriented programming).

Overall, low power design will require a change in mind-set. This can only be achieved by introducing the power element early in the educational process, not only with regards to circuit and logic design but also at the architectural level.

References

[1] M. Alidina, J. Monteiro, S. Devadas, A. Ghosh, and M. Papaefthymiou. " Precomputation-based Sequential Logic Optimization for Low Power. " In *Proceedings of the 1994 International Workshop on Low Power Design*, pages 57-62, April 1994.

[2] H. Bakoglu, *Circuits, Interconnections, and Packaging for VLSI*, Addison-Wesley, Menlo Park, CA, 1990.

[3] L. Benini, M. Favalli, and B. Ricco, "Analysis of Hazard Contributions to Power Dissipation in CMOS IC's," *1994 International Workshop on Low-Power Design*, Napa Valley, CA, pp. 27-32, April 1994.

[4] W Bowhill, et al., "A 300 MHz Quad-Issue CMOS RISC Microprocessor," *Technical Digest of the 1995 ISSCC Conference*, San Fransisco, 1995.

[5] R. W. Brodersen et al, "Technologies for Personal Communications," Proceedings of the VLSI Symposium '91 Conference, Japan, pp. 5-9, 1991.

[6] D. Bursky, "A Tidal Wave of 3-V IC's Opens Up Many Options; Logic Chips that Operate at 3V or Less Offer a Range of Low-Power Choices for Portable Systems," *Electronic Design*, vol. 40, no. 17, pp. 37-45, August 20, 1992.

[7] B. Case, "Low-Power Design, Pentium Dominate ISSCC," *Microprocessor Report*, vol. 8, no. 4, pp. 26-28, March 28, 1994.

[8] A. Chandrakasan, R. Allmon, A. Stratakos, and R. W. Brodersen, "Design of Portable Systems," Proceedings of CICC '94, San Diego, May 1994.

[9] A. Chandrakasan, S. Sheng, and R. W. Brodersen, "Low-Power Techniques for Portable Real-Time DSP Applications," VLSI Design '92, India, 1992.

[10] A. Chandrakasan, S. Sheng, and R. W. Brodersen, "Low-Power CMOS Design," *IEEE Journal of Solid-State Circuits*, pp. 472-484, April 1992.

[11] B. Cole, "At CICC: Designers Face Low-Power Future," *Electronic Engineering Times*, no. 745, pp. 1-2, May 10, 1993.

[12] D. Dobberpuhl et al, "A 200MHz, 64b, Dual Issue CMOS Microprocessor," *Digest of Technical Papers, ISSC '92*, pp. 106-107, 1992.

[13] J. Eager, "Advances in Rechargeable Batteries Spark Product Innovation," *Proceedings of the 1992 Silicon Valley Computer Conference*, Santa Clara, pp. 243-253, Aug. 1992.

[14] S. Iman and M. Pedram. " Multi-level network optimization for low power. " In *Proceedings of the* IEEE *International Conference on Computer Aided Design*, pages 372–377, November 1994.

[15] D. Maliniak, "Better Batteries for Low-Power Jobs," *Electronic Design*, vol. 40, no. 15, pp. 18, July 23, 1992.

[16] D. Manners, "Portables Prompt Low-Power Chips," *Electronics Weekly*, no. 1574, pp. 22, Nov. 13, 1991.

[17] J. Mayer, "Designers Heed the Portable Mandate," *EDN*, vol. 37, no. 20A, pp.65-68, November 5, 1992.

[18] J. Mello and P. Wayner, "Wireless Mobile Communications," Byte, vol. 18, no. 2, pp. 146-153, Feb. 1993.

[19] -, "Mobile Madness: CCD Becomes PC," *LAN Magazine*, vol. 9, no. 1, pp. 46-48, 1994.

[20] D. Pivin, "Pick the Right Package for Your Next ASIC Design," *EDN*, vol. 39, no. 3, pp. 91-108, Feb. 3, 1994.

[21] J. Rabaey, *Digital Integrated Circuits: A Design Perspective*, Prentice Hall, Englewood Cliffs, N.J. , Nov. 1995.

[22] C. Small, "Shrinking Devices Put the Squeeze on System Packaging," *EDN*, vol. 39,

no. 4, pp. 41-46, Feb. 17, 1994.

[23] A. Stratakos, R. W. Brodersen, and S. R.Sanders, "High-Efficiency Low-Voltage DC-DC Conversion for Portable Applications," *1994 International Workshop on Low-Power Design*, Napa Valley, CA, April 1994.

[24] J. Sweeney and K. Ulery, "3.3V DSPs for Multimedia and Communications Products; Designers Harness Low-Power, Application-Specific DSPs," *EDN*, vol. 38, no. 21A, pp. 29-30, Oct. 18, 1993.

[25] C-Y. Tsui, M. Pedram, C-H. Chen, and A. M. Despain. " Low power state assignment targeting two- and multi-level logic implementations. " In *Proceedings of the* IEEE *International Conference on Computer Aided Design*, pages 82–87, November 1994.

[26] C-Y. Tsui, M. Pedram, and A. M. Despain. " Efficient estimation of dynamic power dissipation under a real delay model. " In *Proceedings of the* IEEE *International Conference on Computer Aided Design*, pages 224–228, November 1993.

[27] H. J. M. Veendrick, "Short-Circuit Dissipation of Static CMOS Circuitry and its Impact on the Design of Buffer Circuits," *IEEE JSSCC*, pp. 468-473, August 1984.

[28] R. K. Watts, ed., *Submicron Integrated Circuits*, John Wiley & Sons, New York, 1989.

[29] R. Wilson, "Phones on the Move; Pocket Phone Sales are on Line for a Boom," Electronics Weekly, no. 1606, pp. 25, Aug. 12, 1992.

[30] Lemnios et al, "Issues in Low Power Design", internal document ARPA 1993.

PART I

Technology and Circuit Design Levels

2

Device and Technology Impact on Low Power Electronics

Chenming Hu

2.1. Introduction

In this chapter, we will explore the interplay between device technology and low power electronics. For device designers, this study may contain lessons for how to optimize the technology for low power. For circuit designers, a more accurate understanding of device performance limitations and new possibilities both for the present and the future should emerge from reading this chapter.

This chapter will focus on silicon CMOS technology for it is the dominant technology today and probably for a long, long time to come. Alternative materials and novel devices will be briefly mentioned.

2.2. Dynamic Dissipation in CMOS

Let us consider a CMOS logic gate. The power dissipation consists of a static and dynamic component. Dynamic power is usually the dominant component and is incurred only when the node voltage is switched.

$$P = P_{dynamic} + P_{static} \tag{2.1}$$

$$P_{dynamic} = CV_{dd}^2\alpha f + P_{sc} \tag{2.2a}$$

$$P_{sc} \approx V_{dd} I_{sc} \frac{\tau_{in}}{4} 2f \approx V_{dd}^2 f \frac{C}{10} \tag{2.2b}$$

The second term in Eq. (2.2a) is known as the short-circuit power. When the inverter input is around $V_{dd}/2$ during the turn-on and turn-off switching transients, both the PFET and the NFET are on and a short circuit current I_{sc} flows from V_{dd} to ground. The width of this short-circuit current pulse is about 1/4 of the input rise and fall time. This term is typically only 10% of the first term as shown in Eq. (2.2b). There, we assumed $\tau_{in} \approx \tau_{out} \approx V_{dd}C/I_{dsat} \approx V_{dd}C/5I_{sc}$. Combining Eq. (2.2a) and Eq. (2.2b).

$$P_{dynamic} = kCV_{dd}^2\alpha f \tag{2.3}$$

$$\text{Switching Energy, } E = kCV_{dd}^2 \tag{2.4}$$

$$\begin{aligned} C &= \text{oxide capacitance} + \text{junction capacitance} + \text{interconnect capacitance} \\ &= C_{ox} + C_j + C_{int} \end{aligned} \tag{2.5a}$$

$$= \frac{b}{T_{ox}} + C_j + C_{int} \tag{2.5b}$$

What can one do to reduce $P_{dynamic}$? k is approximately 1.1 with a lower bound of 1.0. No one seems to have a clever idea for reducing it except for raising the ratio V_t/V_{dd}. We will discuss ways of minimizing C later. f is the clock frequency. αf is the average rate of cycling this node experiences. For example, an idle block of the circuit may not experience switching because the clock signals to the function blocks are gated. In this case α may be much smaller than one.

2.3. Effects of V_{dd} and V_t on Speed

Reducing V_{dd} in Eq. (2.3) is an obvious and very effective way of reducing CMOS power. However, lower V_{dd}, for a given device technology, leads to lower gate speed. Parallelism can be applied to compensate for speed loss to realize a large net reduction in power [1]. On the other hand, optimizing the technology for a lower V_{dd} could minimize or eliminate the speed loss due to V_{dd} reduction. SPICE simulations confirm that the gate delay may be expressed as

$$\tau = \frac{CV_{dd}}{4}\left(\frac{1}{I_{dsat\,n}} + \frac{1}{I_{dsat\,p}}\right) \tag{2.6a}$$

$$\approx \frac{CV_{dd}}{4I_{dsat\,n}}\left(1 + 2.2\frac{W_n}{W_p}\right) \tag{2.6b}$$

τ may be interpreted as the average of the time for the NFET saturation current, I_{dsatn} to discharge C from V_{dd} to $V_{dd}/2$, when the switching may be considered to be complete, and the time for $I_{dsat\,p}$ to charge C from zero to $V_{dd}/2$. In Eq. (2.6b), we used the fact $I_{dsat\,n} \approx 2.2 I_{dsat\,p}$ if $W_n = W_p$ (see Table 2.2 in a later section).

The classical MOSFET model

$$I_{dsat\,o} = q\mu C_{ox} W (V_g - V_t)^2/L \tag{2.7}$$

over-states the benefit of L reduction and the disadvantage of V_g, i.e. V_{dd}, reduction. A very accurate model makes use of the fact that all inversion layer charge carriers move with the saturation velocity, v_{sat}, at the pinch-off point in the channel, where the potential is V_{dsat}. Therefore, [2]

$$(I_{dsat} = W \cdot \upsilon_{sat}) \cdot \text{inversion charge density at "pinch-off" point}$$

$$= W\upsilon_{sat} C_{ox} (V_g - V_t - V_{dsat}) \tag{2.8}$$

where W is the channel width, V_{dsat} is the drain saturation voltage. $\upsilon_{sat} = 8 \times 10^6 cm/s$ for electrons and $6.5 \times 10^5 cm/s$ for holes. It can be shown that V_{dsat} is a function of L [2], and therefore

$$I_{dsat} = \frac{I_{dsato}}{1 + \frac{(V_g - V_t)\mu}{2\upsilon_{sat} L}} \tag{2.9}$$

Eq. (2.9) agrees well with experimental data (Fig. 2.1). It reduces to Eq. (2.7) for large L or small $V_g - V_t$ as expected. In the limit of very small L,

$$I_{dsat} = W\upsilon_{sat} C_{ox} (V_g - V_t) \tag{2.10}$$

The surface mobility, μ, which appears in Eq. (2.9) in two places, is a function of T_{ox} and V_g as well [3]. The approximate L, V_g and T_{ox} dependence of I_{dsat} is

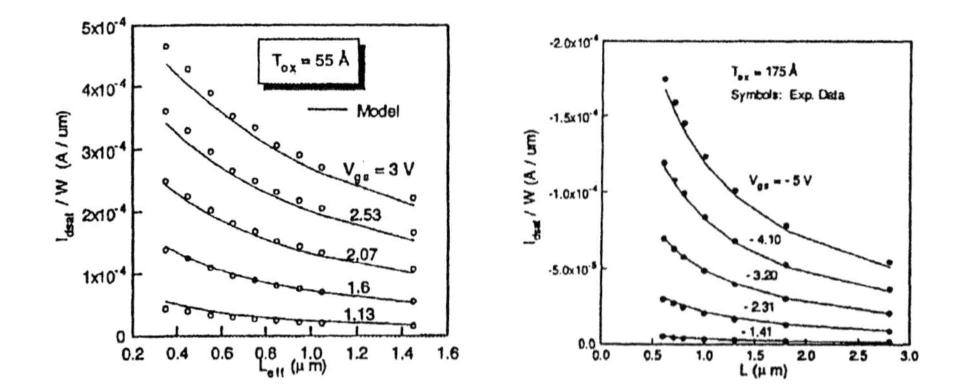

Figure 2.1 Measured data and Equation (2.9) agree within 10% for a large range of technologies.

$$I_{dsat} \propto L^{-0.5} T_{ox}^{-0.5} (V_g - V_t)^{1.3} \tag{2.11}$$

Eq. (2.10) and Eq. (2.11) indicate that the benefits of reducing T_{ox} and L are smaller than the classical model predicted due to velocity saturation. Fortunately, the negative impact of reducing V_{dd}, hence V_g, is also smaller as shown in Figure 2.2. Combining Eq. (2.6b) and Eq. (2.11) with $V_g \approx 0.9V_{dd}$ (it takes a long time for the input voltage to rise from $0.9V_{dd}$ to V_{dd}), we obtain

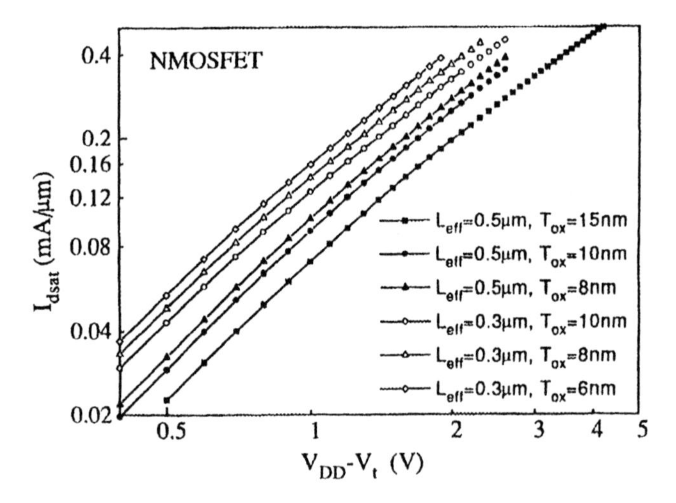

Figure 2.2 nMOSFET I_{dsat} vs V_{dd} and technology. $I_{dsat} \propto (V_{dd}-V_T)^{1.3}$.

$$\tau \propto \frac{CL^{0.5}T_{ox}^{0.5}}{V_{dd}^{0.3}(0.9 - V_t/V_{dd})^{1.3}}\left(\frac{1}{W_n} + \frac{2.2}{W_p}\right) \tag{2.12}$$

Eq. (2.12) shows that delay is a strong function of V_t/V_{dd}. For a fixed V_t/V_{dd} ratio, τ increases only slowly with decreasing V_{dd}, i.e. $\tau \propto V_{dd}^{-0.3}$. These facts are in excellent agreement with experimental and simulation those such as those shown in Figure 2.3 [4]. If V_t/V_{dd} is constant, $\tau \propto T_{ox}^{0.5}/V_{dd}^{0.3}$. Since T_{ox} can be expected to scale with V_{dd}, τ will not increase and may decrease with decreasing V_{dd}. Figure 2.4 shows Eq. (2.12) and the $E \cdot \tau$ product.

$$E \cdot \tau \propto \frac{C^2 V_{dd}^{1.7} L^{0.5} T_{ox}^{0.5}}{(0.9 - V_t/V_{dd})^{1.3}}\left(\frac{1}{W_n} + \frac{2.2}{W_p}\right) \tag{2.13}$$

To a first order, τ is a function of V_{dd}/V_t independent of V_t or V_{dd}.The $E \cdot \tau$ product is minimized when $V_{dd} = 2V_t$.

2.4. Constraints on V_t Reduction

Clearly, we want to minimize V_{dd} in order to minimize $P_{dynamic}$ (Eq. 2.3). V_{dd} should be larger than $4V_t$ if speed is not to suffer excessively (Eq. 2.12 and Fig. 2.4).

The lower bound of V_t is set by the subthreshold static current that the transistors conduct at $V_{gs} = 0V$. At 50°C,

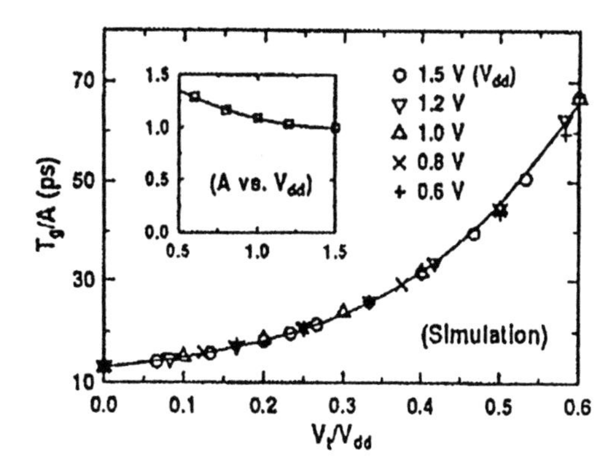

Figure 2.3 Simulated gate delay of a 0.1μm CMOS inverter as a function of V_t/V_{dd}. τ is normalized to a factor A (inset) which is a weak function of V_{dd}, i.e. $V_{dd}^{-0.3}$

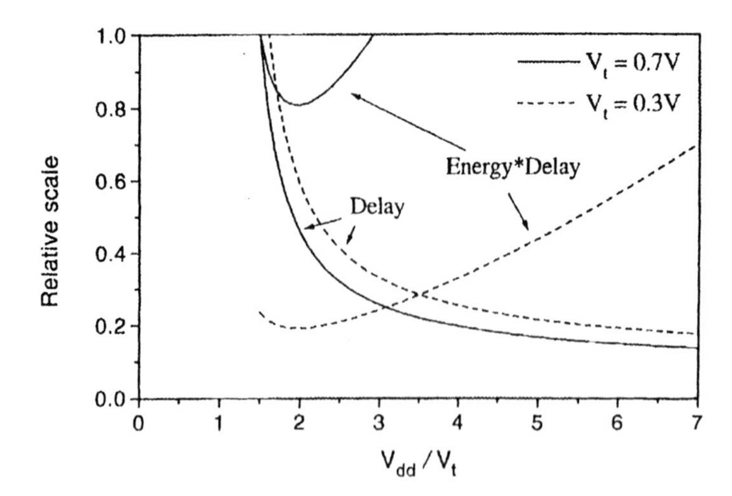

Figure 2.4 Speed is almost independent of V_{dd} as long as V_t is linearly scaled with V_d. τ E is a strong function of V_{dd}.

$$I_{static} \approx 10\frac{\mu A}{\mu m} \bullet W \bullet 10^{-V_t/95mV} \tag{2.14}$$

A large chip may contain 10^8µm of transistor width. The acceptable chip static current may be 100 µA for low power memory to 100 mA for high speed processors. This means V_t must be 0.68 V or 0.38 V respectively. Every 0.1 V reduction in V_t raises I_{static} by 10 times. In order to accommodate low V_{dd} and keep $V_{dd}/V_t > 4$, lower V_t's may be tolerated by accepting higher static current or use circuit techniques to raise V_s [5] or V_t (by raising the body potential) in idle circuit blocks. We may eventually see V_t's as low as 0.2 V for 1 V circuits. It is also likely that more than one (fixed) V_t may be used on the same chip, e.g. in memory arrays and in logic blocks or for different devices in a logic circuit.

2.5. Transistor Sizing and Optimal Gate Oxide Thickness

For a given $W_n + W_p$, there is a certain W_p/W_n ratio that minimizes τ (Eq. 2.12 and τE (Eq. 2.13) as shown in Figure 2.5. This value can be obtained by differentiating $1/(\kappa - W_p) + 2.2/W_p$ with respect to W_p ($\kappa = W_n + W_p$ = constant) and equating the differential to zero. This optimal ratio is independent of the load capacitance.

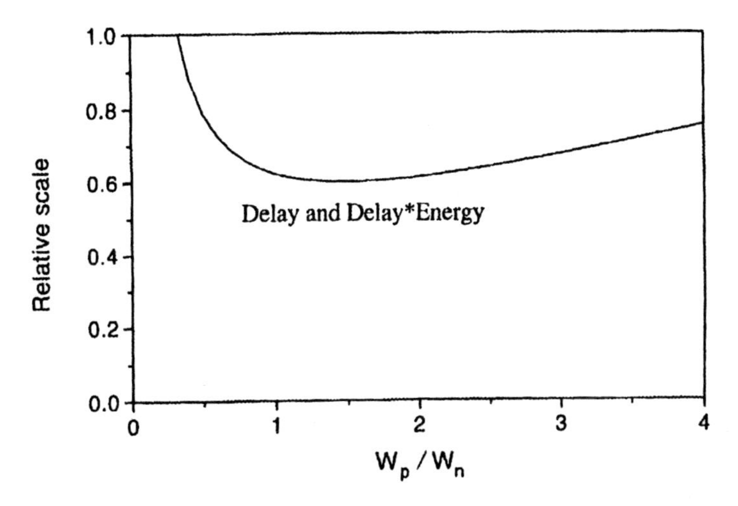

Figure 2.5 Both delay and delay-energy product have a very broad minimum around W_p/W_n=1-3, independent of load capacitance.

$$\frac{W_p}{W_n} = \sqrt{2.2} = 1.5 \tag{2.15}$$

Figure 2.5 shows a very broad minimum. $W_p/W_n = 2$ is also a good choice. C in Eq. (2.5) may be divided into a factor contributed by the driver devices and the rest of the load.

$$C = a(W_n + W_p) + C_{other} \tag{2.16}$$

What value of $W_n + W_p$ should one choose? Eqs. (2.12 and 2.13) together with Eq. (2.16) indicate that τ decreases monotonically with increasing $W_n + W_p$ as shown in Figure 2.6. It also shows that, regardless of W_p/W_n, E is minimized when the drive devices contribute the same amount of capacitance as the load devices and the interconnect.

Eq. (2.12) and Eq.(2.5b) indicate that τ is minimized when b/T_{ox}, i.e. the total oxide capacitances, contributes half the loading. Eq. (2.13) and Eq. (2.5c) indicate that τE is minimized when the total oxide capacitance is $^1/_4$ of the load. These facts are illustrated in Figure 2.7.

All the optimization analyses are summarized in Table 2.1.

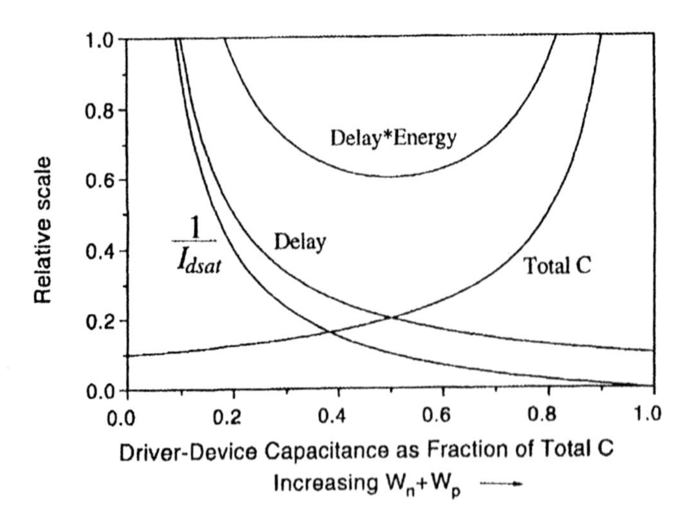

Figure 2.6 Delay decreases monotonically with increasing W_n+W_p but the delay-energy product is minimized when the drive devices contribute half of the total load capacitance.

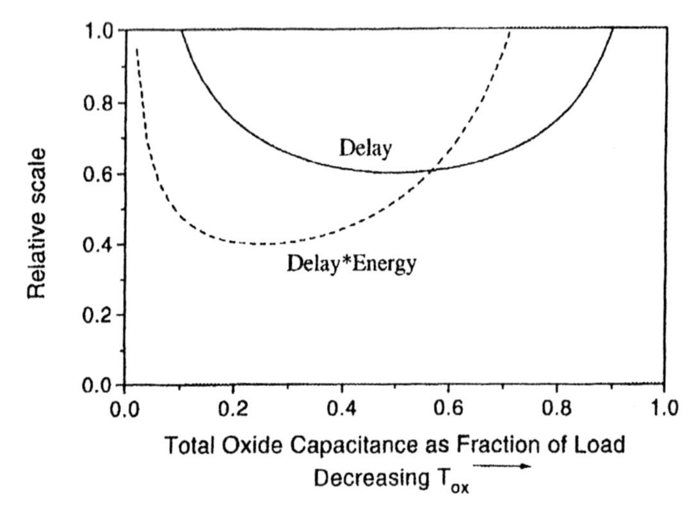

Figure 2.7 Minimum delay is obtained when T_{ox} is chosen such that oxide capacitance accounts for half the total load. The minimum delay-energy product requires thicker T_{ox} such that $^1/_4$ of the load is attributable to oxide capacitance.

2.6. Impact of Technology Scaling

Defect-free gate oxide field should be limited to 7-8MV/cm for a 20 year lifetime at 125°C [7]. Allowing margins for oxide defects, V_{dd}/T_{ox} will likely be

Delay: $\tau = \frac{CV_{dd}}{4}\left(\frac{1}{I_{dsatn}} + \frac{1}{I_{dsatp}}\right)$, Energy: $E = CV_{dd}^2$					
	L	V_{dd}	T_{ox}	W_p/W_n	W_p+W_n
to minimize τ	min	max, $>4V_t$	$C_{tox}=\frac{C}{2}$	1-3	max
to minimize τE	min	$2V_t$	$C_{tox}=\frac{C}{4}$	1-3	$C_d=\frac{C}{2}$
C: total load capacitance, C_{tox}: all load capacitances attributable to gate oxide, C_d: load capacitance attributable to driver devices.					

Table 2.1 Optimization for Delay and Delay-Energy Product

limited to 4MV/cm. This sets an important and usually dominant factor in oxide thickness selection in addition to the T_{ox} optimization summarized in Table 2.1.

A historical review and projection for future device technology is shown in Table 2.2. Note that the PFET current remains to be about half the NFET current. Figure 2.8 shows the projected trend of I_{dsat} with technology scaling [8].Even in the high-speed scenario where E_{ox} is allowed to rise with time towards a very aggressive 5 MV/cm, I_{dsat}/W basically ceases to rise with scaling beyond the 0.5 µm generation because of velocity saturation, i.e. Eq. (2.9). Figure 2.9 shows the projected inverter speed. The historical trend of speed doubling every two generations will slow down only slightly. Figure 2.9 agrees well with experimental data

Gate Length (µm)	**3**	**2**	**1.5**	**1**	**0.7**	**0.5**	**0.35**	**0.25**	**0.18**	**0.12**
V_{dd} **(V)**	5	5	5	5	5	5/3.3	3.3/2.5	2.5/2.0	1.5	1.5
V_T **(V)**	0.7	0.7	0.7	0.7	0.7	0.7/0.7	0.7/0.6	0.6/0.5	0.4	0.4
T_{ox} **(nm)**	70	40	25	25	20	15/10	9/7	6.5/5.5	4.5	4
I_{dsatn} **(mA/µm)**	0.1	0.14	0.23	0.27	0.36	0.56/0.35	0.49/0.40	0.48/0.41	0.38	0.48
I_{dsatp} **(mA/µm)**		0.06	0.11	0.14	0.19	0.27/0.16	0.24/0.18	0.23/0.19	0.18	0.24
Inverter Delay (ps)	800	350	250	200	160	90/100	70/65	50/47	40	32

Table 2.2 Impact of V_{dd}, L, and T_{ox} scaling on MOSFET current and inverter speed

down to 0.1 μm [9]. Speed in the low-power scenario is only modestly lower than in the high-power scenario.

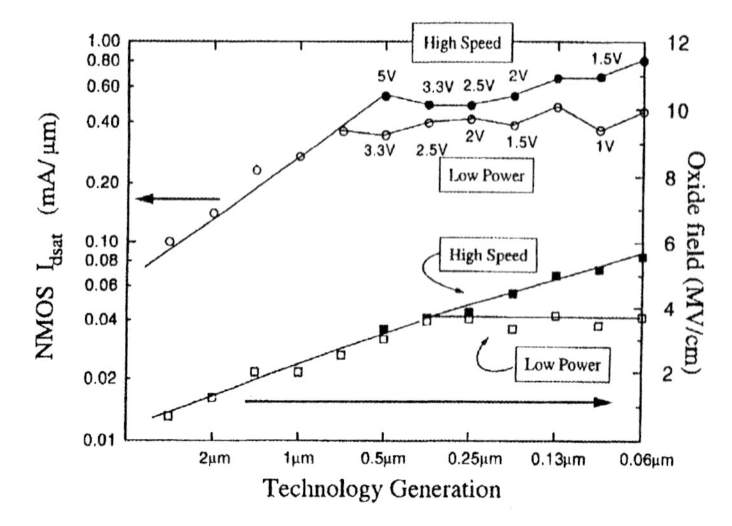

Figure 2.8 MOSFET current hardly increases beyond the 0.5 μm generation of technology due to velocity saturation even in the high-speed scenario, where oxide field rises aggressively.

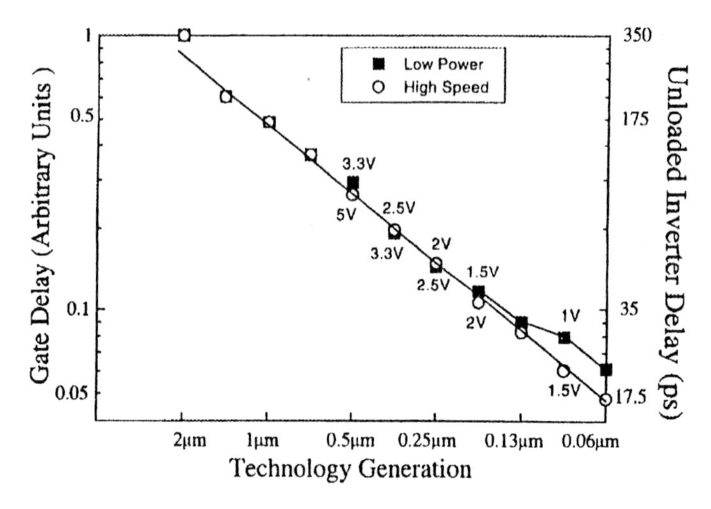

Figure 2.9 Speed in the low-power scenario lags that in the high-speed scenario, where speed doubles every 4 generations, rather than 2 generations as in the past.

2.7. Technology and Device Innovations

Many novel high-speed low-power devices have been proposed [10] based on quantum tunneling, single-electron effect, or even the motion of atoms. While these devices have excellent intrinsic switching speed and energy, they are not capable of driving the capacitance of long interconnects. In addition to the difficulty of manufacturing, there are no suitable circuit architectures that are compatible with the characteristics of these devices today. It may be more productive to look for new low power architectures first and then find new devices that match the needs of the architecture. An example of the latter model is the many device innovations spawned by the need for programmable weights and summing devices in neural networks.

The intrinsic speed-power benefit of GaAs is probably not sufficient to overcome the difference in cost and technology momentum with respect to Si except in very high speed circuits such as MMIC. In the foreseeable future, innovations in silicon CMOS technology such as those discussed below, together with the expected voltage and device scaling discussed earlier, will fuel the low power electronics industry.

Silicon-On-Insulator (SOI) technology shown in Figure 2.10 can improve delay and power through a ~25% reduction in total capacitance [11]. SOI substrates are produced by either wafer bonding or SIMOX (Separation by IMplantation of Oxygen). In wafer bonding, two bulk silicon wafers are oxidized and the two oxide surfaces are held together and bonded at a moderately high temperature. One of the starting wafers is thinned by polishing, possibly followed by

		C (SOI)	C (bulk)	C (SOI) / C (bulk)
Active Gate (F.O.=1)	C_{OX}	36.6 fF	37.6 fF	0.97
N+ Junction (1 drain)	$C_{J(N)}$	9.5 fF	18.9 fF	0.50
P+ Junction (1 drain)	$C_{J(P)}$	7.6 fF	21.6 fF	0.35
Polysilicon (10 μm^2)	C_{POLY}	0.43 fF	0.98 fF	0.44
1st Aluminum (1mm)	C_{1AL}	72.6 fF	123.2 fF	0.59
2nd Aluminum (1mm)	C_{2AL}	63.9 fF	98.4 fF	0.65

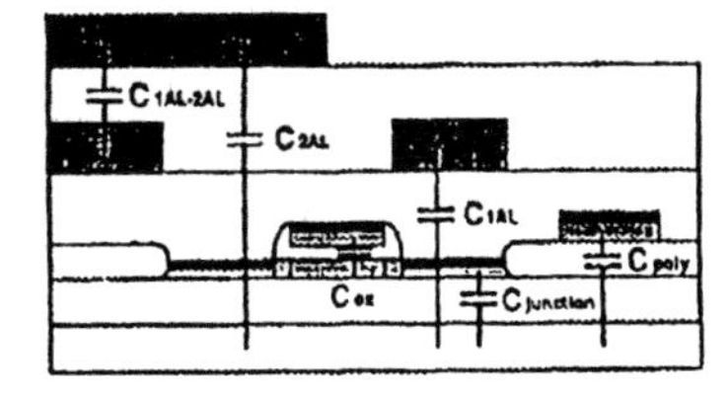

Figure 2.10 Silicon-On-Insulator technology achieves speed and power improvements through denser layout, reduced capacitances and reduced bulk charge (body) effect.

chemical or plasma etching, until a thin layer (~2000Å) of Si is left over the oxide layer. In SIMOX, oxygen is ion implanted into a Si substrate at ~150keV to a dose of ~10^{18} cm^{-2}. A 2500Å buried SiO_2 layer with flat interface is formed under a thin crystalline Si film during a ~1350° anneal.

Optimized SOI MOSFETs can have a lower capacitance and slightly higher I_{dsat} than bulk devices because of the reduced body charge and a small reduction in the minimum acceptable V_t. Together with some improvement in layout density, this potentially can result in up to 40% improvement in speed as Figure 2.11 [13]. That is two generations' worth of speed advantage. This may be traded for lower V_{dd} and obtain up to a factor 3 in power savings.

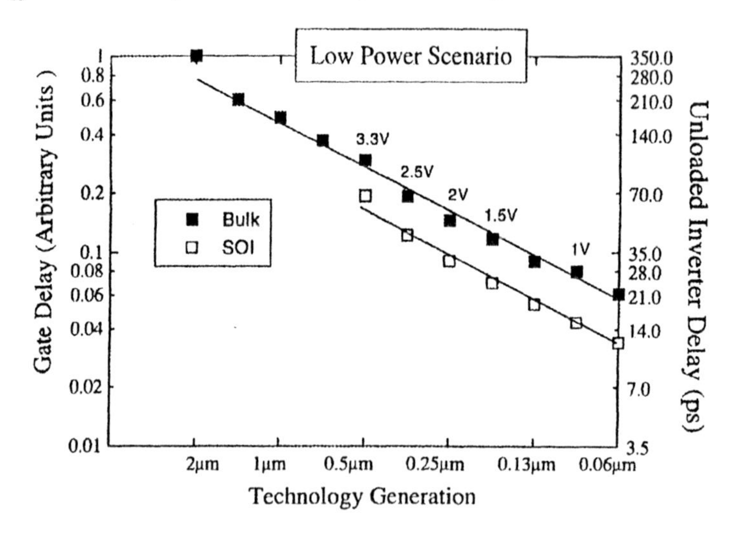

Figure 2.11 Potential speed advantage of SOI technology. Speed may be traded for lower V_{dd} to obtain a 3-fold power reduction.

Another way to reduce capacitance is to use the minimum possible width for metal interconnects that carry ac signals such as the clock and data busses. A recent breakthrough in electromigration research is the discovery that ac interconnects can carry a many times higher current density than design rules based on dc tests would allow, this due to a self-healing effect [13]. Figure 2.12 shows that electromigration lifetime is orders-of-magnitude longer under ac stress than dc stress. Similar behavior has been reported for vias and other metal systems.

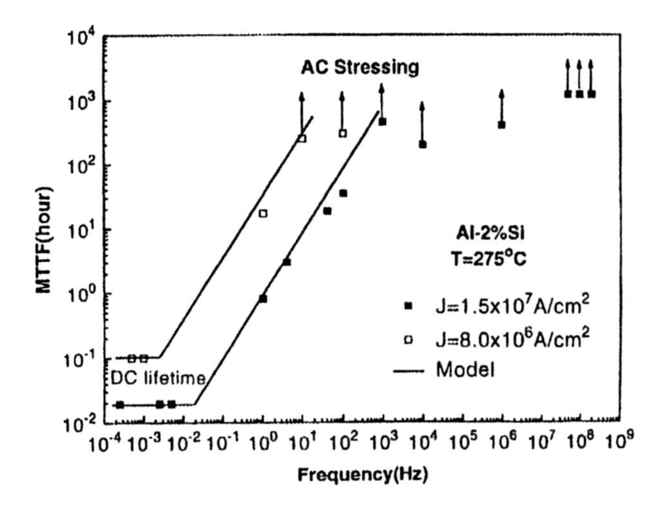

Figure 2.12 For bi-directional or pure ac current, electromigration time-to-failure (TTF) is much longer than for dc. Clock and datalines can use narrower metal lines to reduce capacitance, power and delay.

Yet another possible way to reduce metal capacitance would be to use low-permittivity insulators for inter-metal dielectrics. Relative to the CVD SiO_2 ε of ~4.2, SiOF has an $\varepsilon \sim 3.3$ and organic polymers may achieve $\varepsilon \sim 2.5$ [14].

Finally, as an example of innovative devices, consider the DTMOS (Dynamic Threshold MOS) transistor [15]. For very low power/voltage electronics at V_{dd} lower than 0.5 V, V_t should be lower than 0.15 V for speed, yet V_t should be higher than 0.3 V for low static power. This has been considered a fundamental dilemma for V_{dd} scaling. DTMOS demonstrates that a MOSFET can have a higher V_t at $V_g = 0$ for low leakage current and a lower V_t at $V_g = V_{dd}$ for high speed. In other words, V_t may be a function of V_g.

This is accomplished by simply connecting the floating body of an SOI MOSFET to the gate as shown in Figure 2.13. At $V_g = 0$, V_t is the usual threshold voltage, e.g. 0.35V. As V_g rises, V_t falls to 0.1 V due to the (reverse) body effect. Figure 2.14 shows a speed comparison between CMOS and DTMOS as obtained from simulations. A threefold speed improvement may be achieved.

2.8. Summary

V_{dd} reduction is a powerful method of reducing circuit power. Circuit speed needs not fall as V_{dd} is reduced as long as the V_{dd}/V_t ratio is kept con-

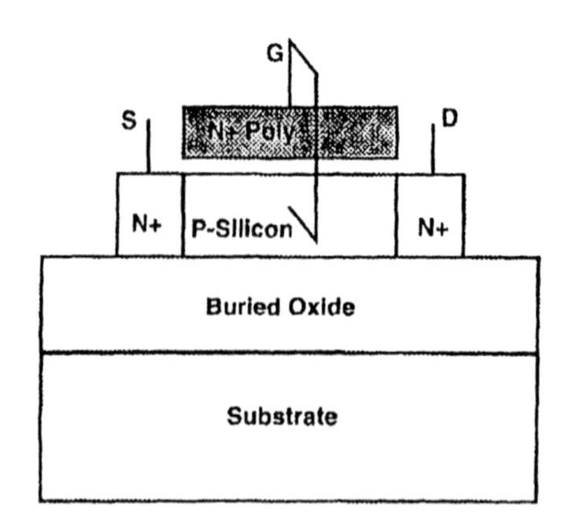

Figure 2.13 Schematic of DTMOS, an SOI MOSFET with body connected to the gate.

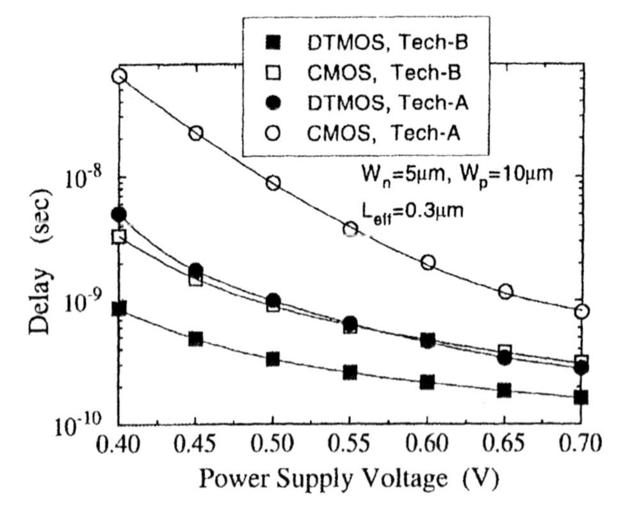

Figure 2.14 Delay of CMOS and DTMOS inverter chains.

stant. The lowest acceptable V_t is determined by the static current that can be tolerated. However, one novel device DTMOS has demonstrated that V_T can be made to vary between a high value at $V_g = 0$ and a low value at $V_g = V_{dd}$.

Although the transistor current per µm width is not expected to drop, neither is it expected to double every technology generation as in the past even if high V_{dd} is used without regard to power. Current will increase only marginally beyond the 0.5 µm generation due to velocity saturation and the oxide breakdown field limit. However, gate speed continues to improve.

Innovations for reducing circuit capacitance are being developed. SOI is an attractive evolutionary new technology. It reduces capacitance and increases current, while layout density may improve speed by 30%. This may be traded for lower V_{dd} and hence a significant reduction in power. Low dielectric-constant inter-metal layer dielectrics are being developed. Finally, the recent discovery of metal's ability to take high AC current density without electromigration should bring a significant reduction to clock and data bus widths and capacitances.

There appears to be no revolutionary low power device/technology. e.g. quantum devices, that is manufacturable or compatible with mainstream circuit architectures today. Evolutionary innovations and optimization for low voltage/power plus continued device scaling will fortunately be able to support the need of low power ULSI for a long time into the future.

References

[1] R. Brodersen et al, *ISSCC Technical Digest*, pp. 168-9, February 1193.

[2] P.K. Ko, Chapter 1 of "VLSI Electronics: Microstructure Science,", Vol. 18, Academic Press, 1989.

[3] M.S. Liang et al, *IEEE Electron Device Letters*, March 1986, pp. 409-413.

[4] Y. Mii et al, *1994 Symposium on VLSI Technology Digest of Technical Papers*, pp. 9-10, June 1994.

[5] K. Itoh, K. Sasaki, Y. Nakagome, *IEEE Symposium of Low Power Electronics*, pp. 84-85, Oct. 1994.

[6] B. Burr, J. Schott, *ISSCC*, February 1994.

[7] R. Moazzami, et al, "Projecting Gate Oxide Reliability and Optimizing Burn-in," IEEE Trans. Electron Devices, p 1643, July 1990.

[8] C. Hu, "Future CMOS Scaling and Reliability," *Proc. of the IEEE*, p. 682, May 1993.

[9] G. Shahidi, et al, "0.1μm CMOS devices," *Symposium on VLSI Technology*, p. 67, May 1993.

[10] Y. Wada, T. Uda, M. Lutwyche, S. Kondo, *1993 International Conference on Solid State Devices and Materials*, pp. 347-349, August 1993.

[11] Y. Yamaguchi, et al, *IEEE Tran. Electron Devices*, p. 179, 1993.

[12] C. Hu, "Silicon-on-Insulator for High Speed ULSI," *International Conference on Solid State Devices and Materials*, p. 137, August 1993.

[13] B.K. Liew, et al, "Electromigration Interconnect Lifetime under AC and Pulse DC Stress," *International Reliability Physics Symp.*, p. 215, 1989.

[14] J. Ida, et al, *Symposium on VLSI Technology Digest*, pp. 59-60, June 1994.

[15] F. Assaderaghi, et al, *Technical Digest IEDM*, pp. 809-812, December 1994.

3

Low Power Circuit Techniques

Christer Svensson and Dake Liu

3.1. Introduction

When CMOS (Complementary Metal Oxide Semiconductor) technology was originally introduced, low power was one of the main motivations [32]. CMOS circuits was the first (and only) digital circuit technique which did not consume any static power. Power was only consumed when the circuit was switched. By using CMOS it was believed that the power consumption problem was solved. Since then, integrated circuit complexity and speed have been continuously increased. One result of this is that also CMOS now approaches the limits of acceptable power consumption [10]. We will therefore investigate the power consumption of CMOS circuits in this chapter, and make some brief comparisons with other circuit techniques.

The scope of this chapter is thus to investigate the power consumption on the circuit level, and to compare different circuit techniques from power consumption point of view. We will limit ourselves to CMOS circuit techniques, with some references to other techniques (as pseudo-n-MOS). We will further restrict ourselves to synchronous systems.

We will start with a general model of the power consumption in CMOS circuits, taking also signal statistics into account. We will then discuss different circuit techniques, first timing circuits, as flip-flops and latches, and then logic circuits or gates. Then we will discuss high capacitance nodes, as for example I/O's separately. Finally we will give some conclusions.

3.2. Power Consumption in Circuits

Power consumption in CMOS circuits are of three main kinds [32]. Dynamic power consumption is the result of the charging and discharging of the circuit capacitances and is the dominating one. Short-circuit power consumption stems from the direct current from supply to ground during the switching of a gate. Static power consumption, finally, is normally caused by leakage currents. However, in all non-CMOS techniques (n-MOS, ECL etc.) the static power consumption is the dominating one, including the so called pseudo-n-MOS technique sometimes used in CMOS designs. Also some other special CMOS techniques, for example used in low voltage swing circuitry, may have a large static power consumption.

3.2.1. Dynamic power consumption

We will start to discuss the dynamic power consumption in the simplest CMOS circuit, the inverter, see Figure 3.1a. The inverter will serve as a general model of any CMOS gate. The picture is further simplified by assuming that all capacitances at the output node, O, can be modelled as one capacitance to ground, C_l and one to V_{dd}, C_2. Starting with a high output and a low input, let us assume that the input voltage is made high. This leads to a low output voltage and to a change in the capacitor charges. The charge of C_l is changed from $C_l V_{dd}$ to 0, as the n-transistor is shorting C_l. The charge of C_2 is changed from 0 to $C_2 V_{dd}$, a charge which is taken from V_{dd}. During the next phase, when the input voltage returns to low again, C_2 is shorted by the p-transistor and C_l is charged back to $C_l V_{dd}$, by a charge taken from V_{dd}. As a result, we take a total charge of $(C_l+C_2)V_{dd}$ from the power supply for a full switch cycle, or a total energy of $(C_l+C_2)V_{dd}^2$. By doing this f times per second, we will consume a power of $P = f(C_l+C_2)V_{dd}^2$. In order to simplify the discussion in the following, we will assume that all energy is consumed for rising outputs (equivalent to the assumption that all capacitance is concentrated in C_l). It is interesting to note that this power is consumed in the transistors as resistive losses, half is lost when a capacitor is charged and half when it is shorted by the n-transistor. It is less important where the capacitor is located, the important thing is the sum of the capacitances at the output node $C_L = C_1 + C_2$. Furthermore, when discussing synchronous circuits, it is convenient to relate the power to the clock frequency, f_c. As each node do not necessarily change at each clock cycle, we need to introduce the probability for the node to rise for each clock cycle, α. We may now express the dynamic power consumption of the inverter as:

$$P_d = \alpha f_c C_L V_{dd}^2 \tag{3.1}$$

which is the general formula for the basic power consumption in CMOS circuits [32].

3.2.2. Switching Probabilities

In order to discuss the effect of different switching probabilities at different nodes in a circuit, we will introduce the following probability factors:

α_x, probability that the node has the value x.

α_{xy}, probability that the node change from x to y during one clock cycle.

As an example, a clocked node has $\alpha_{01} = \alpha_{10} = 1$, telling that the node changes both from 0 to 1 and from 1 to 0 during each clock period. The corresponding power consumption expression will then include $\alpha_{01} f_c = f_c$. As another example, a data node must have equal probabilities to change from 0 to 1 and from 1 to 0. We further assume that the probability for any change in data during one clock period is α_a, called the activity of the node. We then have $\alpha_{01} = \alpha_{10} = \alpha_a/2$. The corresponding power expression will include $\alpha_{01} f_c = (\alpha_a/2) f_c$.

It is obvious that the value of the activity in different situations is crucial for the power consumption. Unfortunately, our knowledge about activity is quite limited. However, it has been shown, that the maximum activity of a random data signal is $\alpha_a = 0.5$ [30]. Simple logic circuits, driven by random signals, tends to have activities of 0.4-0.5 [30]. For more complex circuits, for example finite state machines, the activity tends to be lower. A set of examples indicates average activities in the range 0.08-0.18 [30]. For physical signals, like speech or video signals, the situation may be quite different. For video, for example, some examples indicate activities down to below 0.1 for the most significant bits, up to 0.5 for the least significant bits [33] (Note that this is valid for a signal which is always > 0. For bipolar signals MSB's may change very frequently if the signal is coded by 2's complement.). In conclusion, the average activity may vary between

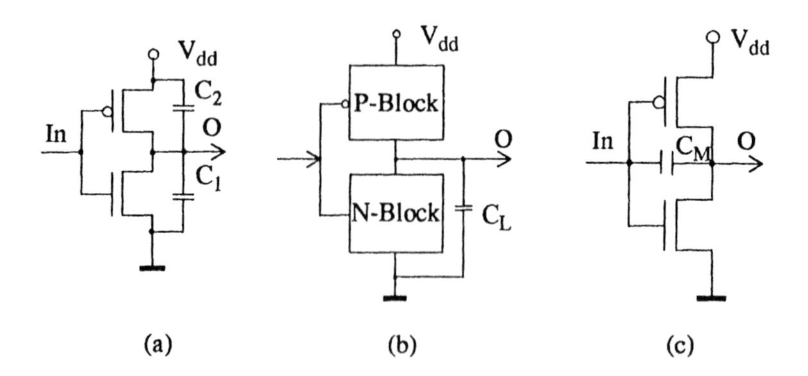

Figure 3.1 (a) Inverter Capacitances, (b) The Load Capacitance and (c) the Miller Capacitance

0.05 and 0.5 in different situations. This will affect the optimal choice of circuits.

3.2.3. Effective Capacitances

Circuit capacitances are obviously crucial for the power consumption. In the model above we assumed that all capacitances in a gate could be described as capacitances at the inverter (gate) output, C_L. We now need to generalize this. We will then assume that all capacitance in "one stage", meaning a node with a p-transistor network connected to V_{dd} and an n-transistor network connected to ground (Figure 3.1b), can be described as an effective capacitance C_L in this node. Let us first discuss the capacitance connected between output and input of an inverter, the Miller capacitance, C_M (Figure 3.1c). When the input of the inverter is made high the output becomes low. This means that the voltage over C_M changes from V_{dd} to $-V_{dd}$, its charge changed by the amount of $2C_MV_{dd}$. The charging current is taken from the input (or from the supply of the previous stage) and dumped into ground. When the input is made low instead, the same charging current flows from the supply of the present stage and is dumped via the input into the ground of the previous stage. As a result, C_M appears as an effective capacitance of $2C_M$ at both the input and the output of the stage. In Figure 3.2a C_M appears as C_{GD}.

Let us then discuss the rest of the capacitances of the inverter. These are C_G of the transistors, and the parasitic capacitances C_D, C_{GS} and C_{GD}. For the inverter case C_G+C_{GS} appears at the input of the stage and C_D at the output. Furthermore we may have wire capacitances, C_w on inputs and outputs. These capac-

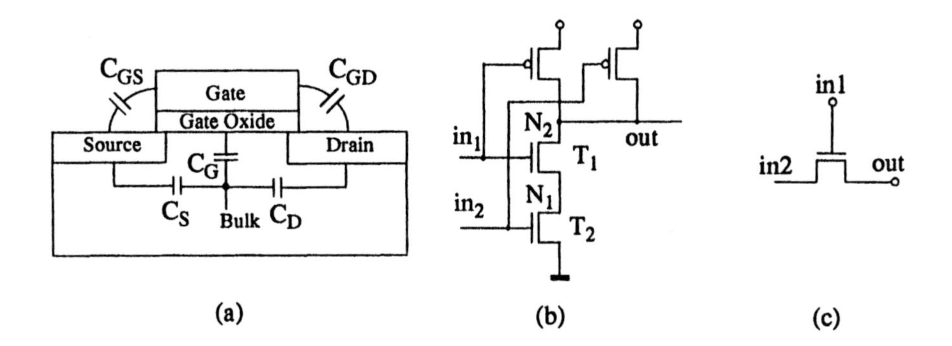

Figure 3.2 (a). Capacitances in a MOSFET (b) A CMOS Gate, (c). A Transmission Gate.

itances should include all capacitance between the wire and its environment (including the capacitance to neighboring nodes [2]).

In Figure 3.2b we show a simple static CMOS gate. In this case it is obvious that we can sum all capacitances which are related to the parallel transistor combination (the p-transistors in Figure 3.2b), in the same way as the inverter transistors. For the series combination the situation is somewhat more complicated. Here we have two nodes with capacitance, C_{N1} and C_{N2}. C_{N1} includes C_D of T_2 and C_S of T_1 and a Miller capacitance of $C_{GS}+C_G$ of T_1 and C_{GD} of T_2. It is either charged to V_{dd}-V_T (if T_1 is conducting and T_2 nonconducting when the p-net is conducting) or not charged at all (if T_1 is open when the p-net is conducting). Instead of dealing with this rather complicated model in the following, we will simplify our model by estimating the extra charge introduced by each series transistor to be the same as introduced by an extra parallel transistor. We thus add C_D+2C_{GD} from T_2 to C_{N2} and let this be the effective capacitance C_L.

As a result then, we will estimate the effective output capacitance of a gate (or more correctly of one stage according to Figure 3.2b) by:

$$C_L = C_w + nC_o + mC_i \tag{3.2}$$

$$\text{with } C_o = (C_D+2C_{GD}) \text{ and } C_i = (C_G+C_{GS}+2C_{GD})$$

Here n is the total number of transistors in the stage and m is the total number of load transistors. If transistors are different in sizes or have different capacitances of other reasons, the formula will be changed into a sum over all transistors involved. C_o and C_i represents effective transistor output and input capacitances respectively. Sometimes it may be convenient to make another partition of the capacitances between different stages, we may for example include the input capacitance of the stage in its effective capacitance and then exclude the transistors in the following stage (they are instead included there).

For a transmission gate, Figure 3.2c [32], the situation is somewhat different. This stage does not amplify the signal, therefore it do not have a Miller effect. For this gate the "input" capacitance on the transistor gates are $C_{Ti} = C_G + C_{GS} + C_{GD}$ each and the "output" capacitance at each source/drains of the transmission gate is $C_{To} = C_D + C_{DS}$. In the following we will however use $C_{Ti} = C_i$ and $C_{To} = C_o$ for simplicity.

3.2.4. Short-circuit Power Consumption

When a static CMOS gate is switched by an input signal with a nonzero rise or fall time, both n- and p-transistors will conduct simultaneously for a short time. During this time, there will be a "short-circuit current", flowing directly between V_{dd} and ground and not participating in charging or discharging C_L [32]. Using a simple static modelling. this current gives rise to a power consumption of [31]:

$$P_{sc} = \frac{\beta}{12}(V_{dd} - V_{thn} - V_{thp})^3 \frac{\tau}{T} \quad (3.3)$$

where we have assumed a simple inverter with n- and p-transistors with the same β. τ is the rise or fall time of the input signal and T is the period of the input signal (average time between transitions). Taking dynamic effects into account, gives a more accurate formula [14], which indicates that Eq. (3.3) is an overestimation with maybe 3 times of the short-circuit power consumption. Using the probabilities defined above, we may write $1/T = (\alpha_{01}+\alpha_{10})f_c$, which for a clocked node is $2f_c$ and for a logic signal is $\alpha_a f_c$. P_{sc} thus depends on the node activity in the same way as P_d and on input rise or fall time. For well designed circuits P_{sc} is normally less than 10% of P_d, so we will in most cases neglect P_{sc} in the following. Still, it can be large under special circumstances, that is for slow input signals (large τ) [31].

3.2.5. Static Power Consumption

In normal CMOS circuits the static power consumption is controlled by the leakage currents of transistors and pn-junctions. For the inverter model in Figure

3.1a, for example, the power consumption for a low input is given by the leakage current of the n-transistor, I_{d0n} and for a high input by the p-transistor, I_{d0p}. Assuming $\alpha_0 = \alpha_1$ gives a static power consumption of

$$P_s = V_{dd}(I_{d0n}+I_{d0p})/2. \tag{3.4}$$

This power consumption is normally very low and can therefore be neglected. For future processes with possibly very low threshold voltages, the leakage currents may increase, however, and become important [20].

For other circuit techniques than CMOS, the static power consumption is normally dominating. Let us take the pseudo-n-MOS inverter in Figure 3.3a [32] as an example. Here the p-transistor behaves as a current generator with current I_0, and it is sized in such a way that its current is much lower than the maximum current of a conducting n-transistor. For a low input, only the p-transistor is conducting, so the output is high. For a high input the n-transistor is conducting, making the output low.

The low output voltage is here larger than zero, as the p-transistor current, I_0, passes through the n-transistor, giving rise to a voltage drop of about $I_0/\beta_n(V_{dd}-V_T)$. The static power consumption is here given by:

$$P_s = \alpha_0 I_0 V_{dd} \tag{3.5}$$

where α_0 refers to the output signal. In this type of circuit one often relate the power consumption to the speed. Thus, assuming I_0 to be much less than the current capability of the n-transistor means that the speed is given by the rise time, which is CV_{dd}/I_0. The so called power-delay product, Pt_d, can then be estimated to:

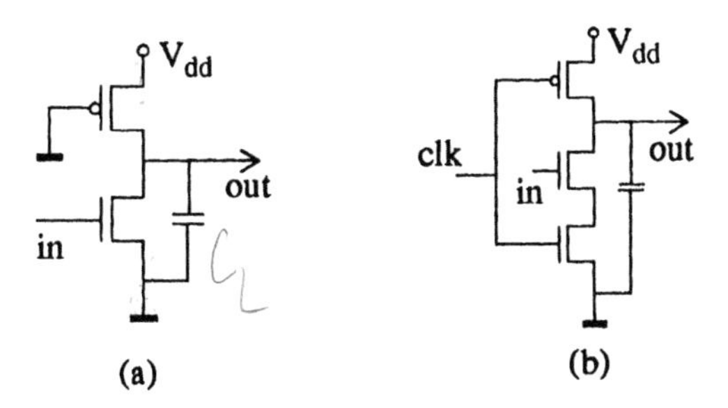

Figure 3.3 (a). A pseudo-n-MOS inverter, (b). A precharged inverter

$$Pt_d = \alpha_0 I_0 V_{dd} * CV_{dd}/I_0 = CV_{dd}^2/2 \tag{3.6}$$

where we used $\alpha_0 = 0.5$. This expression resembles the expression for dynamic power Eq. (3.1), indicating that by driving a CMOS gate with its maximum speed ($f_c = 1/t_d$) it will reach the same power consumption as its pseudo-n-MOS or n-MOS counterpart.

3.2.6. Precharged Circuits

Precharging is sometimes used in CMOS circuits to save transistors and increase speed. In Figure 3.3(b) we show a simple precharged inverter [32]. When the clock is low, the output is always high (precharge phase) and when the clock is high the output is controlled by the input (evaluation phase). The single n-transistor connected to the input can be replaced by an n-net, creating a complex logic function. The dynamic power consumption of this stage can be expressed as:

$$P_d = (\alpha_{01}+\alpha_{11}) f_c C_L V_{dd}^2 \tag{3.7}$$

where α is related to the input. By relating α_{01} and α_{11} to α_a, we have $\alpha_{01} = \alpha_a/2$ and $\alpha_{11} = (1-\alpha_a)/2$, which gives:

$$P_d = (1/2) f_c C_L V_{dd}^2 \tag{3.8}$$

The precharged stage therefore consumes power also if the input is constant (α_a = 0) [19].

3.2.7. Glitches

With glitches we mean uncontrolled appearances of signal transitions. We will here discuss two cases of glitch occurrences, one in connection with reconverting paths in logic and one in connection with precharged latches. Both are related to logic hazard, that is to spurious signals caused by different arrival times of signals to a gate.

If several parallel paths with different delays are merged into one gate, glitches may occur, see Figure 3.4a. Here the two signals, A and B, have different arrival times to the XOR gate, thus making its output change twice during the clock cycle, in spite of the fact that the final result do not change. This glitch gives rise to an unnecessary power consumption, which we may express as:

$$P_g = \gamma P_d \tag{3.9}$$

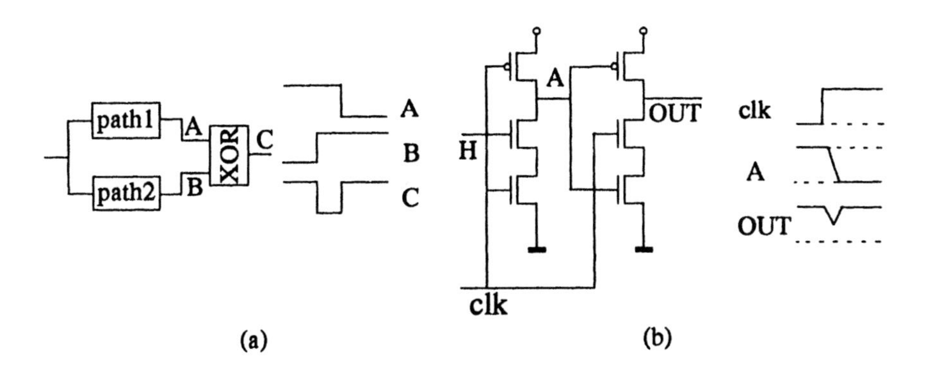

Figure 3.4 (a) A CMOS XOR gate,
(b) A glitch observed on the out node during clock low

where we have expressed the glitch power in terms of the dynamic power. Glitches may also be incomplete, that is the node voltage change is less than V_{dd} see Figure 3.4b. This gives rise to a smaller power consumption than the above expression. We let this effect be absorbed in γ in the following.

If the duration of a glitch is long enough, it may propagate through following stages. This means that large logical depths will give rise to increased glitch occurrences of two reasons, first because large logical depths creates delay variations, second because glitches in one node may propagate to following nodes (until we reach a clocked element). For short glitches, on the other hand, propagation may be terminated by low-pass filtering in the following logical stage. It is not easy to predict the glitch probability, but some investigations of typical values has been performed [30], [12]. For a series of examples Vanoostende find a glitch power of 8 - 25% of the total dynamic power, i. e. $\gamma = 0.08 - 0.25$. For more complex gates, Figueras finds γ of up to 0.67 (for a 16x16 bit multiplier, logic depth about 30).

In order to avoid glitches, one could work with smaller logical depths (by pipelining). This will instead increase the clocking power. Another principle is to perform a careful logic design with balanced delays to avoid hazards. However, glitches may still occur as a result of statistical delay variations. Finally, glitches are prevented by using precharged (domino) logic. This is because after the precharge of a domino chain no node can be changed more than once. In this case however, precharge power will dominate instead, so we do not save power.

The second case of glitch may occur as a result of internal hazard in, for example, a precharged latch, see Figure 3.4b. This kind of glitch may be avoided by making the output n-transistors slower or by connecting the output n-transistors to the bottom n-transistor in the previous stage instead of to ground [35].

3.2.8. Reduced Voltage Swing

Sometimes power can be saved by reducing the voltage swing. This method is particularly useful for high capacitance nodes as long buses, I/O's etc. We may have two low voltage swing cases, one CMOS-like with no static power consumption, and one n-MOS-like case with static power consumption.

A CMOS-like driver may be designed as in Figure 3.5a. Here T_1 and T_2 are used as voltage limiters, they will limit the output voltage to $V_2 - V_T$ to $V_1 + V_T$, or the swing to $\Delta V = V_2 - V_1 - 2V_T$. The charge needed to recharge the load capacitor is thus $C_L \Delta V$. As this charge is taken from the power supply at V_{dd}, it will consume an energy of $C_L \Delta V V_{dd}$. The power consumption is therefore:

$$P_d = \alpha_{01} f_c C_L V_{dd} \Delta V \tag{3.10}$$

where α_{01} relates to the output.

An n-MOS-like driver may be designed as in Figure 3.5b. Here the bottom transistor acts as a current generator with the current I_0, chosen in such a way that $I_0 R = \Delta V$, where ΔV is the wanted output swing. In this case we will have a static power consumption of:

$$P_s = \alpha_0 I_0 V_{dd} \tag{3.11}$$

where again α_0 is related to the output. As in the previous discussion on pseudo-n-MOS gates above (section 3.2.5.), we may estimate the power-delay

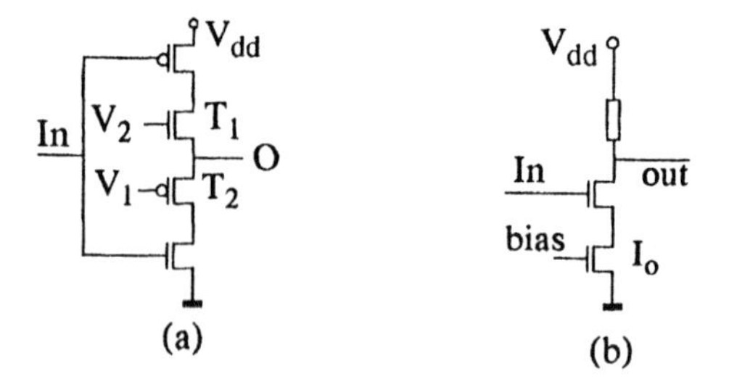

Figure 3.5 Basic circuits for reduced voltage swing.

product by first estimating the delay, t_d:

$$t_d = RC_L \tag{3.12}$$

Using this value of t_d we may estimate the power delay product:

$$P_s t_d = \alpha_0 C_L V_{dd} \Delta V \tag{3.13}$$

which expression is very similar to Eq. (3.10) above. Here we may chose R, and therefore I_0 in such a way that the performance constraints are fulfilled (for example $t_d = 1/2f_c$). In this case P_s includes also the dynamic power consumption as the charge taken from V_{dd} to charge C_L is equal to the charge not taken from V_{dd} during the time when C is discharged.

3.3. Flip-flops and Latches

Flip-flops and latches are used for controlling the timing of computations in synchronous systems. A flip-flop is here defined as a circuit which makes the output equal to the input at a clock edge (normally only one of the two clock edges). At any other time, the output is kept constant. A latch is here defined as a circuit which is open during one clock phase (clock low or high) and closed during the other phase. When the latch is open, its output is equal to its input, otherwise its output is kept constant.

CMOS flip-flops can be static or dynamic. Static flip-flops are of two main types, made from gates or transmission gates, see Figure 3.6a and b. They may also contain asynchronous set and/or reset. Dynamic flip-flops may be of many types; we will discuss three types here, C^2MOS, precharged TSPC (true single phase clock) and non-precharged TSPC, Figure 3.7.

Latches are normally dynamic. We will discuss three types, C^2MOS, precharged TSPC, and non-precharged TSPC.

3.3.1. Static flip-flops

We will discuss two types of static flip-flops, gate-based and transmission-gate based flip-flops [32]. In order to be able to make a reasonable comparison between the different circuits, we will include clock buffering in each flip-flop and assume that only one clock signal is available from outside. In Figure 3.6a we show a positive edge-triggered static flip-flop. It is designed from 4 3-input gates and two inverters, and therefore includes 28 transistors. By assuming a data activity of α_a, we may estimate the dynamic power consumption of this

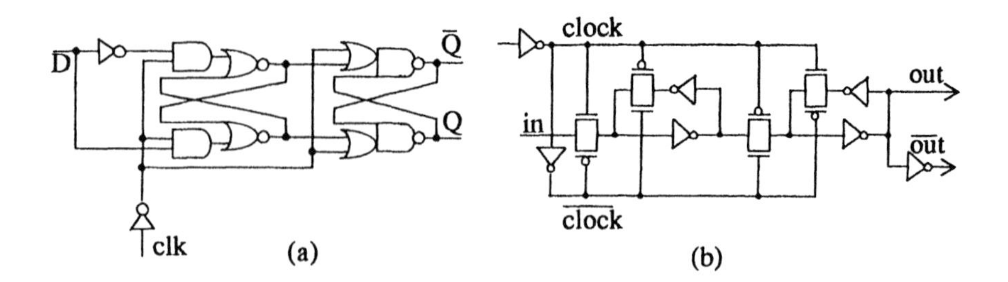

Figure 3.6 (a) A static D flip-flop made from gates,
(b) A static D flip-flop made from transmission gates.

flip-flop according to the following. We have 4 clocked inputs and 1 inverter, with a total of 10 transistor input capacitances and 2 output capacitances. Furthermore, we have two And-Or-Inv, and two Or-And-Inv gates, each gate including 6 transistors, and an inverter. There are 10 inputs driven by data signals, each with 2 transistors and also one inverter output (with 2 connected transistors) and 4 gate outputs, each with 6 transistors connected. Thus totally:

$$P_d = 10f_cC_iV_{dd}^2 + 2f_cC_oV_{dd}^2 + (10+2)(\alpha_a/2)f_cC_iV_{dd}^2 + 2(\alpha_a/2)f_cC_oV_{dd}^2 + 24(\alpha_a/2)f_cC_oV_{dd}^2$$

or

$$P_d = [10C_i + 2C_o + 12(\alpha_a/2)C_i + 26(\alpha_a/2)C_o]\, f_cV_{dd}^2 \qquad (3.14)$$

In Figure 3.6b we have a positive edge-triggered static flip-flop based on transmission gates. This circuit is made from 7 inverters and 4 transmission gates, that is a total of 22 transistors. Following the same procedure as above, we may estimate the power consumption by noting that 12 transistor gates are clocked, 2 inverter outputs are clocked, and 5 inverters and 4 transmission gates have a data signal on their inputs and outputs. This gives a total power consumption of:

$$P_d = [12C_i + 4C_o + 10(\alpha_a/2)C_i + 26(\alpha_a/2)C_o]f_cV_{dd}^2 \qquad (3.15)$$

where we have treated transmission gate inputs and outputs as inverter outputs from a capacitance of view (this may be a slight overestimation as the transmission gate do not have a Miller effect, see section 3.2.3.).

3.3.2. Dynamic flip-flops

Dynamic flip-flops may be of many types. Again, we will assume that only one clock signal is available from outside and that clock buffering is included in

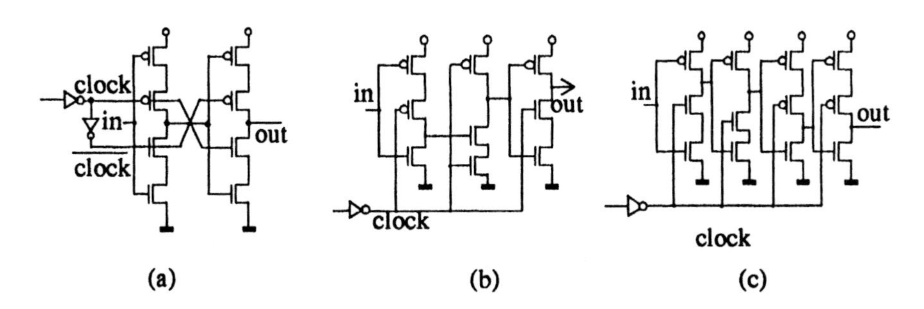

Figure 3.7 Dynamic MS D flip-flops, (a) C^2MOS, (b) TSPC, (c) Non-precharged TSPC

each flip-flop. We will treat 3 types below. We start with an edge-triggered flip-flop made from two C^2MOS latches, Figure 3.7a. This circuit contains 12 transistors, of which 8 are clocked at their gates and 4 at their outputs. Furthermore there are 4 transistors with data on their inputs and 8 transistors with data on their outputs. From this we may estimate the power consumption as:

$$P_d = (8C_i + 4C_o + 4(\alpha_a/2)C_i + 8(\alpha_a/2)C_o)f_c V_{dd}^2 \quad (3.16)$$

In Figure 3.7b we have a precharged TSPC (true single phase clock) flip-flop containing 11 transistors, of which 6 are clocked on their inputs, 2 are clocked on their outputs, there are 3 transistors with simple data on their inputs, 2 transistor gates connected to a precharged node, 2 data nodes with 3 transistors and 1 precharged node with 3 transistors. Totally we estimate the power consumption to:

$$P_d = (6C_i + 2C_o + 3(\alpha_a/2)C_i + 6(\alpha_a/2)C_o + 2(1/2)C_i + 3(1/2)C_o)f_c V_{dd}^2 \quad (3.17)$$

We may also create an edge-triggered flip-flop from two non-precharged TSPC latches, see Figure 3.7c. In this case we have 14 transistors, of which 6 are clocked on their inputs, 2 are clocked on their outputs and the others have data signals. From this we get:

$$P_d = (6C_i + 2C_o + 8(\alpha_a/2)C_i + 12(\alpha_a/2)C_o)f_c V_{dd}^2 \quad (3.18)$$

3.3.3. Latches

Common latches are normally dynamic. In our case we can derive these from the dynamic flip-flops discussed above. A C^2MOS latch thus consists of one of the stages from Figure 3.7a, a precharged TSPC latch is made from the two last stages in Figure 3.7b and a non-precharged TSPC latch from the two first (or the two last) stages of Figure 3.7c. The power consumption of these circuits can then be estimated as above, and a comparison between the latches will approximately follow a comparison of the corresponding flip-flops.

For the special case of precharged CVSL logic, an RS flip-flop is a suitable static latch, see Figure 3.8a [25],[32]. The particular flip-flop in Figure 3.8a is set or reset by low inputs. In this case the two outputs from the logic stage are precharged high so that the flip-flop holds its previous data. When the clock is made high, one of the outputs of the logic stage will go low, thus setting or resetting the flip-flop. The power consumption of this latch can be written:

$$P_d = [4(1/2)C_i + 4(\alpha_a/2)C_i + 8(\alpha_a/2)C_o]f_c V_{dd}^2 \tag{3.19}$$

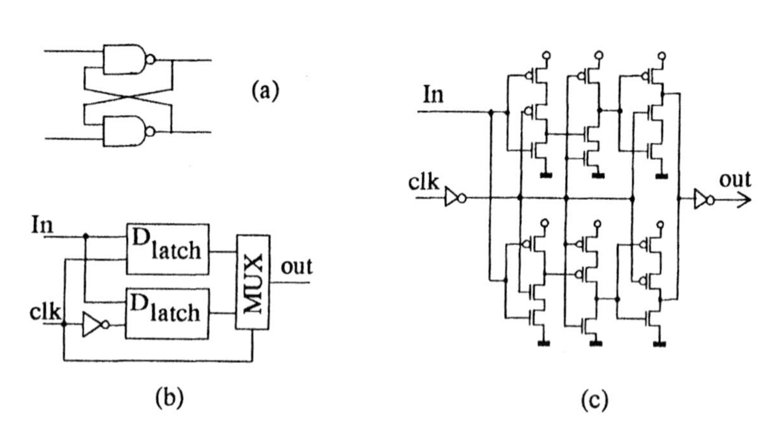

Figure 3.8 An RS flip-flop, (b) Double edge triggered flip-flop
(c) TSPC Double edge triggered flip-flop

3.3.4. Double-edge triggered flip-flops

By combining two flip-flops (or latches) and one multiplexer, it is possible to form a double-edge-triggered device (DET), which utilizes both clock phases for timing, see Fig. Figure 3.8b [15]. In Figure 3.8c we show an example of a DET flip-flop [1]. Using the same principles as above, we estimate the power consumption of this circuit to:

$$P_d = [10C_i + 2C_o + 4(1/2)C_i + 6(1/2)C_o + 8(\alpha_a/2)C_i + 14(\alpha_a/2)C_o]f_c V_{dd}^2 \tag{3.20}$$

or approximately the power consumption of a precharged flip-flop doubled (Eq. (3.17)).

3.3.5. Discussion

Section 3.3. treats the circuits which are used for timing control in synchronous logic. By using the formulas given above, we may make some comparisons between the various techniques used. We will choose to do this under simplified assumptions, that is assuming $C_o = C_i$. We may then calculate the power consumption of each flip-flop above in terms of $f_c C_i V_{dd}^2$, using α_a as parameter, see Figure 3.9.

From Figure 3.9 we can conclude that the gate-based static flip-flop (GFF) has lower power consumption than the transmission gate flip-flop (TGFF), in spite of the fact that it uses considerably more transistors. The reason is that it has fewer clocked transistors. Of the dynamic flip-flops, the non-precharged TSPC (NPTSPC) consumes least power, about the same power as the GFF at very low α_a and about half of the static one at high α_a. The RS-flip-flop latch (RSFF) appears very efficient, but must be considered in connection with the precharged CVSL logical gate (in this special case the latch is indirectly clocked through the logic).

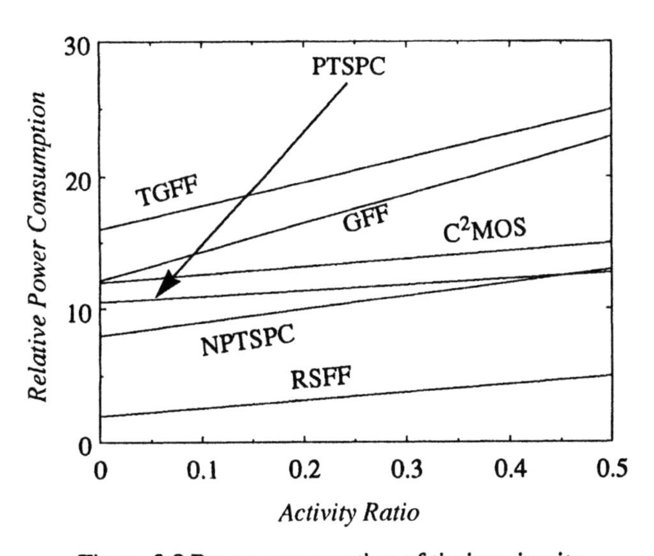

Figure 3.9 Power consumption of timing circuits

Considering the double-edge triggered flop-flop (DETFF), the example above has a power consumption which is double compared to the corresponding simple flip-flop (PTSPC in this case). However, for a fixed data throughput it needs only half the clock frequency, making its power consumption for a given data throughput equal to the corresponding single edge triggered flip-flop.

3.4. Logic

Logic circuits or gates are performing boolean functions. We will here consider single-valued functions of n variables (the gate have a fan-in of n). The classical CMOS gate consists of two transistor networks connected to the output, one n-net connected to ground and one p-net connected to the supply voltage (V_{dd}), see Figure 3.2a [32]. Both networks performs the boolean function, in a complementary way, so that the output is always connected to ground or V_{dd}. For simple functions (nor and nand), we need 2n transistors for performing the function (Figure 3.2b). For more complicated functions we may need up to $2^n n$ transistors, depending on the function. Note also, that one static gate stage can only perform an inverting boolean function, for a general function we need also inverted inputs, Figure 3.10a.

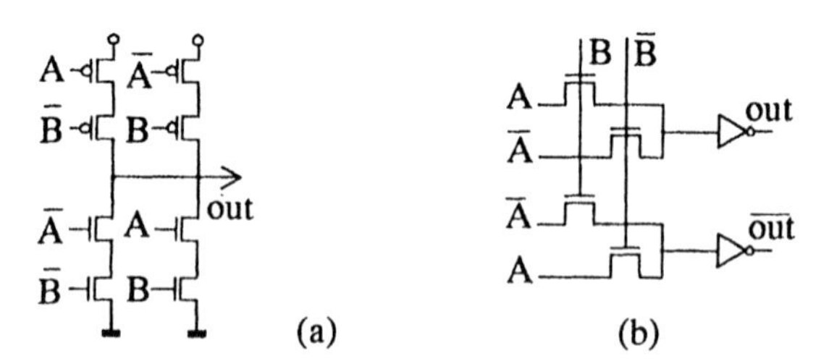

Figure 3.10 XOR gate of (a) Static CMOS (b) CPL logic

In the comparisons between different types of logic circuit techniques below, we will use simple nand or nor-gates as examples, that is we assume that the number of transistors in each net is equal to the fan-in, n. We will also, as in section 3 above, estimate the power consumption for each gate by including both input and output capacitance of the gate. In this section we will assume that the activity, α_a, is the same of all nodes (output and inputs).

For simple static logic we may estimate the dynamic power consumption to:

$$P_d = (\alpha_a/2)(2nC_i + 2nC_o)f_c V_{dd}^2 \quad (3.21)$$

Where we have combined the results from sections 3.2.1., 3.2.2., and 3.2.3..

The static gate is characterized by a direct relation between the power consumption and the activity, and also a proportionality to the number of transistors in the circuit (which in general is between 2n and $2^n n$, where n is the number of inputs).

3.4.1. Alternative CMOS Logic Circuit Techniques

Many different CMOS circuit techniques has been proposed [32]. We will here discuss a limited number of techniques, which hopefully are representative for most proposed techniques. The techniques treated are Standard static logic (Figure 3.1b), CPL (Complementary Pass-transistor logic, using n-type pass transistors and inverter buffers) [34], Precharged logic (representing NORA [13], TSPC [35] etc.), Domino logic (chain of precharged stages, and CVSL logic (static and dynamic) [32][7].

Complementary pass transistor logic creates the logic function by pass transistors, see Figure 3.10b. As the pass transistor network can not restore the logic levels, we need to add an inverter for level restoring. Also, normally we need both data and inverted data on the inputs. In the same time some gates, as for example XOR, are easier to implement in pass transistor logic than in other forms. We will, just for illustration, choose a gate of the type in Figure 3.10b, as our example. Its power consumption can be written:

$$P_d = [(\alpha_a/2)2(n-1)(2C_i + 2C_o + 2\nu C_o) + 2(\alpha_a/2)(2\nu C_i + 2C_o)]f_c V_{dd}^2 \quad (3.22)$$

where the first term represents the pass transistor inputs and output (the term with ν) and the second term represents the inverter. $\nu = (V_{dd}-V_T)/V_{dd}$ represents the lower voltage swing of $V_{dd}-V_T$ at the inverter input. n in the above expression is the number of inputs (n=2 in Figure 3.10b). For n>2 we assume a simple NAND or NOR type of gate (as for the static logic above), see [34]. For a more general gate, the number of transistors will be larger. n inputs may need up to $2^{(n-1)}(n-1)$ transistors, corresponding to a binary tree with depth (n-1) restructured into a matrix (see [32]).

In Figure 3.3b, we show an example of a precharged gate (section 3.2.6.). Note, that for a gate the input n-transistor is replaced by an n-net performing the appropriate logical function. For this case we can estimate the dynamic power

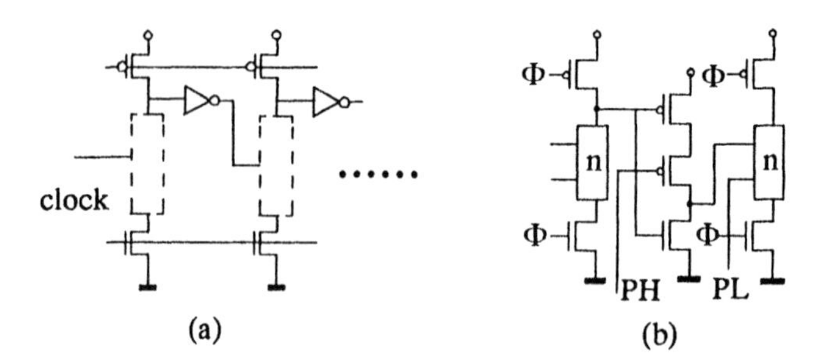

Figure 3.11 Logic styles, (a) Domino logic (b) DCPL

consumption as follows. The inputs is assumed to have an activity of α_a, thus consuming the power $(\alpha_a/2)nf_cC_iV_{dd}^2$. The output consumes the power $(1/2)(n+2)f_cC_oV_{dd}^2$. Finally the clocked transistors consumes a clock power (on node clk) of $2f_cC_iV_{dd}^2$. The total dynamic power consumption can therefore be written:

$$P_d = ((\alpha_a/2)nC_i + ((n+2)/2)C_o + 2C_i)f_cV_{dd}^2 \quad (3.23)$$

Domino logic is just a chain of precharged stages, Figure 3.11a, where each stage can be treated separately from power consumption point of view. Still it differs from a chain of static logic stages in that the domino chain is free from race-induced glitches (see section 3.2.7.). An interesting variation of domino logic is the DCPL (Data-Clock Precharged Logic) domino chain, Figure 3.11b [36]. Here the first stage is a normal precharged stage but the following stages are precharged by the data rather than by the clock. Through the first stage we create a data in the form of a "return to 1" signal. This means that in the whole following chain each second stage will be precharged to 1 and the rest to 0. When real data arrives, each node can only change once, so no race glitches occur. The stages can be static stages or may have somewhat fewer transistors, see [36]. Compared to the precharged stages, the DCPL stages do not consume clock power and do not need the clocked transistors, instead the number of logic transistors may be somewhat increased. We may estimate the power consumption by just removing the clocking power from the expression for the precharged gate.

CVSL logic is based on differential signals, that is we use both data and data inverse on both inputs and outputs. In Figure 3.12a we show a static CVSL gate. The logic function is created twice (as in static logic), using two comple-

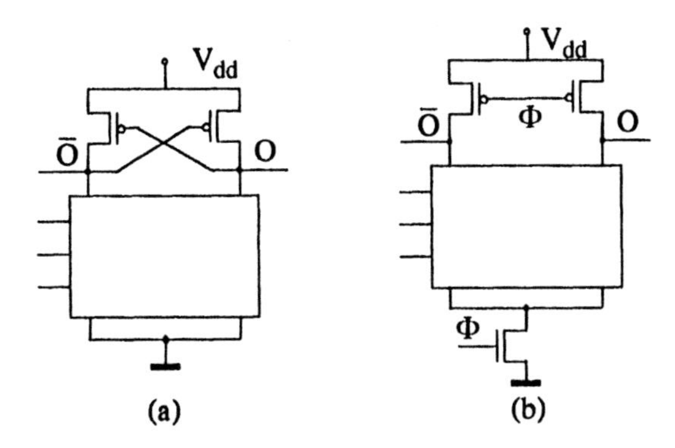

Figure 3.12 CVSL gates

mentary n-nets. For each input vector one of the two n-nets is conducting, this making its output low. Through the cross-coupled p-transistors the other output is then made high. In this case we may estimate the dynamic power consumption as:

$$P_d = (\alpha_a/2)2(n+1)(C_i+C_o)f_cV_{dd}^2 \tag{3.24}$$

In order to have the static CVSL gate to work properly, the p-transistors must not have a too strong driving capability (we will however not consider transistor scaling in the power estimation). This makes the CVSL gate relatively slow. The speed can be improved by precharging the output nodes. This can be done by keeping or by removing the cross-coupled pair, we will remove the cross-coupled pair below (meaning that the circuit becomes dynamic), Figure 3.12b. The precharged CVSL may then be treated in a similar way as the precharged gates above:

$$P_d = [(\alpha_a/2)2nC_i + (2n+3)(1/2)C_o + 3C_i]f_cV_{dd}^2 \tag{3.25}$$

3.4.2. Discussion

The view of the power consumption of logic given in this section is oversimplified and must therefore be used with great care. There are four problems in comparing different circuit techniques. First, transistor counts depends strongly on the actual logical function to be implemented. A striking example can be found in [34], where it is found that a full adder implemented in SL (static logic) needs 40 transistors, whereas it needs only 28 transistors if implemented in CPL (complementary pass logic). This differs from our simplified view (Eq. (3.21) and Eq. (3.22)). However, also [34] is contradictory, it turns out that the sum part of the adder needs 28 and 12 transistors and the carry part 12 and 16 transistors in CL and CPL respectively. Second, characterization by using only a simple activity value (α_a) is not realistic. In practice α_a of different nodes depends on the actual logical function and on how it is implemented [4]. Also, glitch probability depends on which logical technique is used. Third, we have not taken transistor scaling into account. Normally, for example, p-transistors are designed wider than n-transistors (by a factor of 2-2.5). This will change the weighting factors of different transistors in our formulas and make n-MOS rich designs more advantageous (as for example CPL and CVSL). Fourth, finally, we have not taken speed into account. The different circuit techniques discussed have different speed capabilities.

Let us anyway try to make some comparisons among the different circuit techniques discussed. In Figure 3.13 we depict the power consumption versus activity, using the simplified formulas derived above. We can directly see that precharge logic is very disadvantageous from power point of view, especially for low activities. It also appears that static logic is the most advantageous technique. We believe that these are correct general conclusions, see also [18]. Static logic, SL, is a simple, robust and power-lean technique, well suited for all applications, including cell system, gate array and full custom design styles [24]. It is also

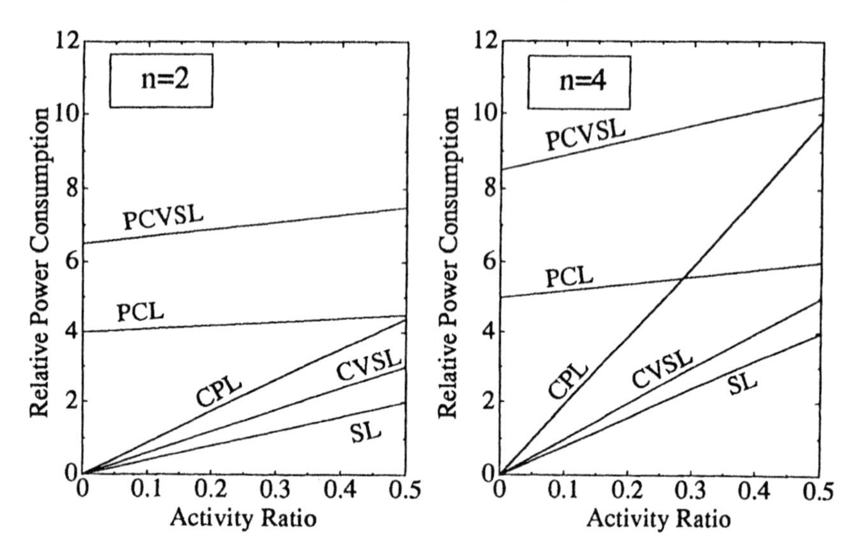

Figure 3.13 Combinational logic power consumption versus activity ratio

quite clear that precharge logic (PCL or PCVSL) is very power-hungry. Still, PCL is probably the fastest technique available, so it may be motivated in special cases [35].

Concerning CPL and CVSL, both are differential techniques. This means that some logical functions are more easily (and more efficiently) implemented in these techniques than in single-ended logic. The full adder (or the sum part of it) discussed above is a good example. Because of the lower transistor count of the CPL full adder it will consume less power than its SL counterpart [34]. An important drawback of these techniques is that the differential outputs calls for 2 wires per output. For general cases, like cell system or gate array design styles, this will be devastating for power consumption, as the total interconnection length may be more than doubled. Even in the non-differential case, interconnections tends to consume more power than the logic [21]. The differential technique can however be advantageous in compact, self-contained blocks, like adders, multipliers etc. Concerning a comparison between CPL and CVSL, CPL normally have a larger transistor count, but is also faster than CVSL [34]. Also CPL has an important drawback, that is it will perform worse at low supply voltages. This is serious from power point of view as low supply voltage is a very important route to low power consumption.

3.5. High Capacitance Nodes

High capacitance nodes occur in a number of situations, for example in clock networks, in clock drivers, at long buses or interconnections and at the outputs (I/O's). Normally, we need extra strong drivers to drive high capacitance nodes, in order to keep pace with the required speed. It may also be dangerous to allow too long rise-times in the system, as it will lead to large short-circuit power consumption in the following stages (section 3.2.4.). We will discuss the driver problem below.

In order to reduce the power consumption in connection to large capacitance nodes, there are two possibilities. One is to reduce the activity of these nodes, the other is to reduce the voltage swing. Reducing activity is an algorithm/architecture problem and will not be treated here [4]. Voltage swing reduction is discussed below.

3.5.1. Drivers for Large Loads.

Quite often we need to drive large capacitive loads with reasonable speed and/or with short rise and fall times. A typical example is a clock network, where we may have quite strong demands on slopes [17]. Clock driving is also very important from the power consumption point of view, as the clock power may be 20-40% of the total chip power [21], [10].

The standard method to drive large loads is to use a tapered inverter chain, Figure 3.14a [32], [14]. Let us study such a chain loaded by C_L and using N stages and a tapering factor of f. f is given by $f^N = Y = C_L/2C_i$. For such a chain it can be shown that there exist an optimum value of f for which the total delay of the chain is minimum [14]. For $C_o = C_i$ (as before), this optimum will occur at f = 3.5. The delay is however a relatively flat function of f, so also larger values of f may be acceptable.

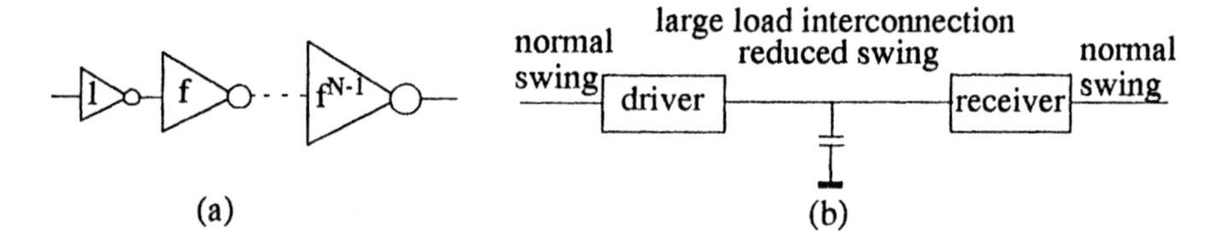

Figure 3.14 Driving the large load (a) A driver chain (b) Off chip driving

Let us study the power consumption of the inverter chain. The total switched capacitance of the chain can be written [14],[6]:

$$C_{driver} = 2(C_i+C_o)(Y-1)/(f-1) \tag{3.26}$$

or, if we instead calculate the total driver capacitance divided by the load capacitance:

$$C_{driver}/C_L = (1+C_o/C_i)(1-1/Y)/(f-1) \tag{3.27}$$

This expresses the power consumption of the driver chain in terms of the power consumption in the load. For large values of C_L, we may neglect 1/Y. Let us furthermore again let $C_o = C_i$. Using the above optimum value of f=3.5 leads to a power consumption of the driver which is 80% of the power consumption in the load, or in other words, the driver causes an excess power consumption of 80% over the load. This is thus quite a large overhead.

By instead minimizing the power-delay product of the driver chain, we will arrive to a larger value of f and therefore a lower value of the excess power [6]. However, there is no reason why this particular choice should be more interesting than another one. Instead, we should minimize the power consumption, by choosing as large f as possible, taking delay and slope constraints into account. If we, for example, increase f from 3.5 to 9, the power consumption overhead will be reduced from 80% to 25% of the load power, at the cost of a delay increase of 20%.

It is also possible to work with non-constant tapering factors, but the improvements is very limited [6]. Furthermore, if the load consists of gates and inverters (as in the case of a clock network), the short-circuit power of these circuits may increase with a slower driver (large f). We judge this effect as quite small.

3.5.2. Low Voltage Swing

An efficient way to save power is to reduce the voltage swing on high capacitance nodes, see Figure 3.14b. Here we depict an interconnection, at which we first convert the signal to a low swing and then restore it to full swing again. In this scheme we save power in two ways. First, the charge needed for charge/discharge C_L is lowered. Second, as the current to be delivered by the driver to charge/discharge C_L in a certain time is lowered, the driver size can be reduced, so the driver itself will consume less power. The problems are to produce a small swing with small static power consumption and to amplify the signal with a small power consumption. We will discuss a few examples of circuit techniques for low voltage swing below.

One approach for swing limitation is to let the driver limit the voltage swing. A straight-forward method is to use an ordinary inverter with special supply voltages, Figure 3.15a [23]. The special supply voltages can be generated on chip with limited power overhead [23]. As the effective drain voltages of the driver transistors are reduced, to the actual swing in this case, we still need the same transistor sizes as for the full swing driver. Therefore the power consumption of the driver stage input and on the pre-drivers will be quite large. Another approach with similar properties is the one proposed in Figure 3.5a. In this case we may obtain some swing reduction also without extra voltages, by letting

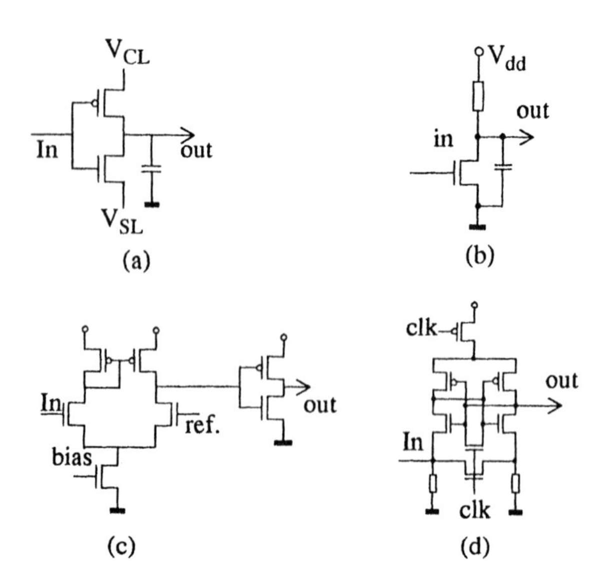

Figure 3.15 Low swing circuit techniques

V_2=V_{dd} and V_1=0 (giving a swing of V_{dd}-2V_T). Also, as the drain voltages is larger than the output swing, we may use smaller transistor sizes than in the full swing case.

Another straight-forward approach is to use a simple n-MOS driver with a resistive load (as in Figure 3.5b). Even simpler, we may size the n-transistor to give the correct current, I_0, as in Figure 3.15b. The drawback with this circuit is that it consumes static power (it can however be put in a standby mode just by setting the output to high). The benefits are many. The n-transistor runs at full drain voltage, so its size can be reduced considerably compared to the corresponding full swing CMOS driver. For the same speed, we need only $\Delta V/V_{dd}$ of the current needed in the full swing case. Furthermore, compared to an inverter, we need no p-transistor. As the p-transistor normally is scaled larger than the n-transistor, this gives very large savings on the drivers input capacitance For example, assuming a voltage swing of $\Delta V = 0.2V_{dd}$ and a normal p-transistor size of twice the n-transistor size, will make the input capacitance of the driver in Figure 3.15b 15 times smaller than that of a full swing inverter driver. Besides the power savings on the load itself (Eq. (3.13)), we will save considerably on the driver input and on the pre-drivers. The n-MOS driver with a resistive load can also be implemented as an open-drain output driver (with the load external to the chip), it can be used as a terminated driver for driving transmission lines [28], and similar circuits can be used to produce ECL-compatible outputs [27].

On the receiving side we need to amplify the low voltage swing back to normal again. This can be accomplished by a normal amplifier, for example as in Figure 3.15c [28]. By using a differential amplifier, it is easier to set the switch voltage to an appropriate value. It may also be used for detecting differential signals directly [28]. A drawback is that the amplifier consumes static power. In [23] another amplifier is proposed, which do not consume static power (if the swing is not too small). A third possibility is to use a clocked sense amplifier, as is used at the bit lines in memories, see for example Figure 3.15d [3]. Such a solution do not consume static power either.

A second approach for swing limitation is to let the receiver limit the voltage swing. To some extent, we have already discussed this case. If we just move the resistor in Figure 3.15b to the receiver side, we can say that the receiver is limiting the voltage swing. The idea is thus to have a current-mode driver and a low-impedance receiver. We may for example use an inverter driver (still with reduced size), combined with a low input impedance receiver. By using a current conveyor input stage, it is possible to obtain an extremely low input impedance and therefore a very low voltage swing [29]. Tan demonstrated that this principle can reduce power consumption more than 40 times. Again, all current-mode approaches consumes static power. An important application of this approach is the memory bit-lines. Here, the driver is the memory cell, where we can not accept any overhead. Therefore, even if the bit-line swing normally is small, it can not be controlled by the memory cell, so the swing may become large if the memory is run slowly or if the process is faster than typical. The only way to control the power consumption of the bit-lines is then to limit their swing by the receiver (sense amplifier in this case).

In conclusion, there are many possibilities to save power by utilizing low swing on large capacitive loads. Important examples are chip I/Os, which normally have much higher capacitance than internal nodes, bit-lines in memories, which have large capacitance and no opportunity to use large drivers (the memory cell is driver) and long wires or buses. Less obvious is to use low voltage swing on the clock distribution, even if such schemes has been proposed [8]. The reason is of course that we want to use the clock directly on switch transistors and also that we need to minimize clock delay. For I/Os, the need for power savings is one important motivation for creating a new I/O standard, IEEE Std 1596.3 [11], which is based on a differential current mode transmitter and a low impedance (100 Ω) receiver and which uses a voltage swing between 250 and 400 mV.

3.6. Summary

The aim of this chapter was to describe the main parameters controlling circuit power consumption and to discuss and compare different low power circuit techniques.

We conclude that combinational static logic with simple, non-precharged, dynamic flip-flops appears to be best suited for low power applications. This is also in agreement with the branch-based approach for low power [24]. Also static flip-flops based on gates are quite good. Other techniques, like for example combinational complementary pass transistor logic, may be advantageous in complex, self-contained blocks of logic. By using double-edge-triggered flip-flops, the clock frequency can be halved for a given data rate. This can lead to large power savings in the clock distribution network.

Driving large loads always consumes much power. When full swing is necessary, special care should be taken not to consume too much power in the drivers. Otherwise, reducing the voltage swing is an efficient method to save power. Several techniques for low voltage swing interconnects were discussed.

References

[1] M. Afghahi, J. Yuan, Double edge-triggered D-flip-flops for High Speed CMOS Circuits, IEEE J. of Solid-State Circuits, Vol. 26, No. 8 pp 1168-1170, 1991.

[2] H. B. Bakoglu, Circuits, Interconnections, and Packaging for VLSI, Reading, Addison-Wesley Publishing Company, 1990.

[3] T Blalock and J Jaeger, A High-Speed Clamped Bitline Current-Mode Sense Amplifier, IEEE J. of Solid State Circuits, Vol. 26, pp 542-548, 1991.

[4] A. Chandrakasan, S. Sheng and R. W. Brodersen, Low-Power CMOS Digital Design, IEEE Journal of Solid-State Circuits Vol. 27, No. 4, 1992.

[5] A. Chandrakasan et al.,, Design of Portable Systems, Proc. of IEEE 1994 Custom Integrated Circuits Conference, pp 12.1.1-12.1.8, 1994.

[6] J.-S. Choi and K. Lee, Design of CMOS tapered Buffer for Minimum Power-Delay Product, IEEE Journal of Solid-State Circuits Vol. 29, No. 9, pp 1142-1145, 1988.

[7] K. M. Chu and D. L. Pulfrey, A Comparison of CMOS Circuits Techniques: Differential Cascade Voltage Switch Logic Versus Conventional Logic, IEEE J. of Solid-State Circuits, Vol. 22, No. 4, pp 528-532, 1992.

[8] E. De Man and M. Schöbinger, Power Dissipation in the Clock System of Highly Pipelined CMOS Circuits, Proc. 1994 Int. workshop on Low Power Design, pp133-138.

[9] D. Deschacht, M. Robert, and D. Auvergne, Explicit Formulation of Delay in CMOS Data Paths, IEEE Journal of Solid-State Circuits Vol. 23, No. 5, pp 1257-1264, 1988.

[10] D. W. Dobberpuhl et al, A 200-MHz 64-b Dual-Issue CMOS Microprocessor, IEEE J. of Solid-State Circuits, Vol. 27, No. 11, pp 1555-1566, 1992.

[11] Draft 1.00 IEEE Std 1596.3-1994, IEEE Standard for Low-Voltage Differential Signal for SCI (LVDS).

[12] M. A. Ortega and J. Figueras, Bounds on the Harzard Power Consumption in Modular Static CMOS Circuits, a talk on PATMOS'94, Barcelona, Spain, unpublished.

[13] N. F. Goncalves, H. J. DeMan, NORA: A Race-free Dynamic CMOS Technique for Pipelined Logic Structures, IEEE J. of Solid-State Circuits, Vol. 18, June 1983.

[14] N. Hedenstierna and K. Jeppson, CMOS Circuit Speed and Buffer Optimization, IEEE Tr. of Computer Aided Design, Vol. 6, March 1987, pp. 270 - 281.

[15] R. Hossain, L. D. Wronski, and A. Albicki, Low Power Design Using Double Edge Triggered Flip-Flops, IEEE Tr. on VLSI Systems, Vol.2, No. 2, 1994, pp 261-265.

[16] M. Ishibe et al. High-Speed CMOS I/O Buffer Circuits, IEEE J. of Solid-State Circuits, Vol. 27, No. 4, pp 671-673, 1992.

[17] P. Larsson, C. Svensson, Noise in Digital Dynamic CMOS Circuits, IEEE Journal of Solid-State Circuits, Vol. 29. No. 6, pp 655-662, 1994.

[18] W. Lee et al, A Comparative Study on CMOS Digital Circuit Families for Low Power Applications, Proc. 1994 Int. workshop on Low Power Design, pp129-132.

[19] D. Liu, C. Svensson, Comparison of power consumption in CMOS synchronous logic circuits, Proc. of European workshop on power and timing modelling pp 31-37, 1992

[20] D. Liu, C. Svensson, Trading Speed for Low Power by Choice of Supply and Threshold Voltages, IEEE Journal of Solid-State Circuits Vol. 28, No. 1, pp 10-17, 1993.

[21] D. Liu, C. Svensson, Power Consumption Estimation in CMOS VLSI Chips, IEEE Journal of Solid-State Circuits Vol. 29, No. 6, pp 663-670, 1994.

[22] C. Metra, Minimal Power-Delay Product CMOS Buffer, 4th International Workshop on PATMOS, Oct. 1994, Barcelona, pp 150-157.

[23] Y. Nakagome et al. Sub-1-V Swing Internal Bus Architecture for Future Low-Power ULSI's, IEEE Journal of Solid-State Circuits Vol. 28, No. 4, pp 414-419, 1993.

[24] C. Piguet, J-M. Masgonty, S. Cserveny, E. Dijkstra, Low-Power Low-Voltage Digital CMOS Cell Design, 4th Int. Workshop on PATMOS, Oct. 1994, Barcelona, pp 132-139.

[25] D. Renshaw and C. H. Lau, Race-Free Clocking of CMOS Pipelines Using a Single Global Clock, IEEE J. of Solid-State Circuits, Vol. 25, No. 3, pp 766-769, 1990.

[26] W. Roethig, E. Melcher, and M. Dana, Probabilistic Power Consumption Estimations in Digital Circuits, Proc. European Workshop on power and timing modelling pp 7-15, 1992.

[27] M. S. J. Steyaert et al., ECL-CMOS and CMOS-ECL Interface for 150-MHz Digital ECL Data Transmission Systems, IEEE JSSC, Vol. 26, No. 1, pp 18-23, 1991.

[28] C Svensson and J Yuan, High Speed CMOS Chip to Chip Communication Circuit, Proc. 1991 Int. Symp. on Circuits and Systems, pp. 2221-2231.

[29] N. Tan and S. Eriksson, Low Power Chip-to-Chip Communication Circuits, Electronics Letters, Vol. 30, No. 21, 1994, pp 1732-1733.

[30] P. Vanoostende et al, Evaluation of the limitations of the simple CMOS power estimation formula: comparison with accurate estimation, Proc. European workshop on power and timing modelling pp 16-25, 1992.

[31] H. J. M. Veendrick, Short-Circuit Dissipation of Static CMOS Circuitry and its Impact on the Design of buffer circuits, IEEE JSSC Vol. 19, No.4, 1984.

[32] N. H. E. Weste, K. Eshraghian, Principles of CMOS VLSI Design, (Second edition), Reading, Addison-Wesley Publishing Company, 1993.

[33] M Winzker, Influence of Statistical Properties of Video Signals on the Power Dissipation of CMOS Circuits, Proc. of PATMOS'94, Barcelona, 17-19 Oct. 1994, pp 106-113.

[34] K. Yano, et al, A 3.8-ns CMOS 16X16-b Multiplier Using Complementary Pass-Transistor Logic, IEEE J. of Solid-State Circuits, Vol. 25, No. 2, pp 388-395, 1990.

[35] J. Yuan C. Svensson, High Speed CMOS Circuit Techniques, IEEE J. of Solid-State Circuits, Vol. 24, No. 1, 1989, pp 62-70.

[36] J. Yuan, C. Svensson and P. Larsson, New domino logic precharged by clock and data, Electronics Letters, Vol. 29, No. 25, pp 2188-2189, 1993.

4

Energy-Recovery CMOS

William C. Athas

With energy-recovery CMOS, circuit energy that would otherwise be dissipated as heat is instead conserved for later reuse. From such an approach to low power, there are no *a priori* limits to the degree to which energy and power dissipation can be reduced inside a circuit. By externally controlling the length and shape of the signal transitions, the energy expended to transition a signal between logic levels can be reduced to an arbitrarily small degree. In contrast, unconventional digital CMOS circuits, the energy dissipated to "flip a bit" from one to zero or vice versa is a fixed and non-negotiable quantity on the order of $(1/2)CV^2$. Methods and techniques to reduce dissipation are thus limited to minimizing the number of transitions required to carry out a computation, or reducing the magnitude of the individual capacitances (C) or the voltage swing (V) between logic levels zero and one. These methods offer significant energy and power reduction but they are ultimately limited by $(1/2)CV^2$.

For logic circuits built from field-effect transistors (FETs), there is nothing intrinsic about the devices which mandates that at least this much energy be dissipated. Electrical charge on the gate of the FET controls the current that flows between source and drain. Since the channel charge and gate charge are always kept separate, there are no unavoidable losses due to the recombination between charges of opposite polarity. This property of FETs can be exploited to reduce the energy dissipated per transition by operating the FETs according to two principles which have been stated in a number of different ways. The definition of Frank and Solomon is a succinct and clear statement of the principle [11]:

1. When the FET is off, the source and drain are brought to the same potential.
2. When the FET is on, the source potential can vary while the drain floats. The variations must be sufficiently slow so that the drain follows fairly closely, with only a small potential difference between source and drain.

The term adiabatic charging has been coined to describe FET circuits that operate in this way. The term "adiabatic" is taken from thermodynamics both literally and figuratively in that literally, the asymptotic dissipation is zero, and figuratively, because the FET's drain "adiabatically" follows the source.

This approach to low-power operation is applicable to any system where the circuit load, which is attached to the drain of the FET, can be characterized as capacitive. For CMOS circuits, conceptually the most important capacitance is the gate-to-channel capacitance because by charging or discharging that capacitance, the channel current is controlled. Practically, the parasitic capacitances can play a far more important role to the circuit's power dissipation than the gate-to-channel capacitances.

Figure 4.1 defines the fundamental relationships between the basic elements of an energy-recovery computing system. In this system, a computation consists of charging and discharging the capacitive circuit nodes, adiabatically or otherwise. The role of the middle component, labelled the *regenerator*, is all important since it determines the ultimate efficiency of the energy-recovery process. The regenerator is also the *governor*, in that it determines the length and the shape of the transitions for the purpose of charging and discharging the logic circuits' signal capacitances. It serves the dual purpose of power source for energy and clock source for timing.

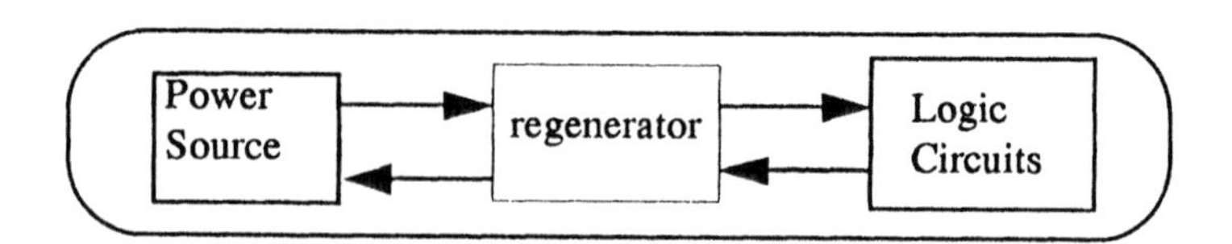

Figure 4.1 Basic energy-recovery computing system

It is possible to build regenerators through a variety of mechanisms such as commuting the energy between electrical and magnetic forms[1], or using banks of switched capacitors to incrementally step energy on to and off of the circuit load. The theory and design of circuits that work in this way is an important topic but its theory is not explored in this chapter. For the sake of completeness, a simple, though inefficient design is presented in Appendix 4.A. There is emerging research on how to design highly efficient regenerators [26][4].

To adiabatically cycle the energy between the logic circuit loads and the regenerator requires that there always be an adiabatic path connecting the two. A simple way to meet this requirement is to *reverse* each transition by backing out the energy along the same set of wires and transistors that were used to steer it in. This has been called temporal [26] or trivial [3] reversibility and requires that the logic circuit's state be otherwise held constant throughout the entire charge/discharge cycle. As will be explored later, separate circuits for injecting and ejecting charge are possible and often desirable.

Readers familiar with charge-steering and precharge design techniques can immediately begin to think up schemes for building logic circuits that could work in this manner. Indeed, there are myriad ways to invent FET-based circuits that operate adiabatically or semi-adiabatically as well as circuits that are either fully or partially reversible. Although the principles are simple, analyzing energy and performance is typically subtle and tedious. Therefore, the first part of this chapter investigates a simple logic style and some rudimentary circuit building blocks based on CMOS transmission gates (T-gates). These circuits are simpler to analyze and have been successfully demonstrated in some small chip-building experiments. However, they have some serious practical performance limitations. Later in this chapter, the T-gate logic is compared to some emerging logic styles which offer significantly better performance but are more difficult to model.

4.1. A Simple Example

To illustrate a simple example of a practical energy-recovery circuit, consider the design of a clock-AND gate, which is an AND gate with one input tied to the clock [19]. Figure 4.2 depicts a "conventional," non-adiabatic CMOS design. The input x is assumed to be well-behaved and only changes when the clock signal, clk, is low. With the clock low, the output is held low by the n-channel FET

---

1. See Appendix 4.A

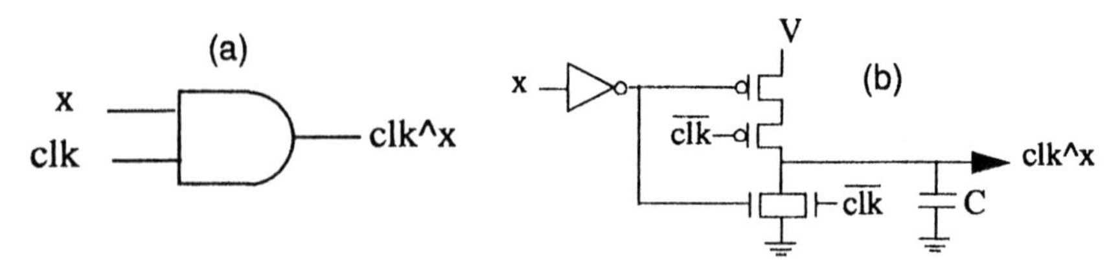

Figure 4.2 Clock-AND gate (a) Logic symbol (b) CMOS schematic

(nFET) attached to $\overline{\text{clk}}$. Changes on the x input do not affect the output. When the clock swings from low to high and the x input is high, the two p-channel FETs (pFETs) connected in series between the output and the supply rail at voltage V, will conduct. The capacitance C, which models the input capacitance of the logic driven by the output of this circuit, will then charge up to V.

The energetics of this circuit (excluding the input inverter) will be analyzed through two separate, but equal approaches. First, energy dissipation will be determined from the law of energy conservation and by doing all the bookkeeping as charge flows through the various potential drops. Second, the FETs will be modelled as single-valued resistances and the dissipation will be derived for when those resistors connect the output capacitance to either V or ground. Issues such as parasitic capacitances and dissipation due to short-circuit currents will be neglected because they do not change the basic result and greatly detract from the clarity of the presentation. Suffice it to say that the results derived are a lower bound to the real dissipation since including these other effects only serves to increase the actual dissipation.

From first principles, it is known that the charge quantity to be delivered from the power supply to the capacitance C is $Q=C \cdot V$ The power supply must deliver at least this much charge to the source of the top pFET at voltage V. The work done by the power supply to do this is simply:

$$E_{inj} = Q \cdot V = CV^2 \tag{4.1}$$

Also from first principles, of this energy,

$$E_{bit} = (1/2)CV^2 \tag{4.2}$$

will be stored on C once it has been fully charged to V.

From the conservation of energy law, the difference between E_{bit} and E_{inj} must be accounted for between the source of the top pFET and the drain of the bottom pFET. Since there are no places to *store* energy inside the channel of the pFET, it must therefore be *dissipated* in the channels of the two transistors. This leads to the conclusion:

$$E_{diss} = E_{inj} - E_{bit} = (1/2)CV^2 . \tag{4.3}$$

An important observation at this point is that from this explanation, it was unnecessary to mention how fast the input signals x or $\overline{\text{clk}}$ switched nor the size of the pFETs. The energy dissipation is independent of the transistor design or the input switching characteristics.

For the falling clock edge, the pFET path will stop conducting and the output capacitance will be discharged through the nFET connected to $\overline{\text{clk}}$. For this transition, the power supply does no work. The energy stored on the capacitor changes from E_{bit} to *0*. Similar to the argument for the pFET, this energy must be dissipated in the nFET. The energy returned to the power supply is $E_{rtn} = Q \cdot 0 = 0$. The energy dissipated per cycle is:

$$E_{cycle} = E_{inj} - E_{rtn} = CV^2 - 0 = 2 \cdot E_{bit} . \tag{4.4}$$

This derivation determines where the energy was dissipated. To delve into the details as to why there is this much dissipation, the individual FETs are modelled as single-valued resistance when they are conducting. When the upper two pFETs are on, they form a resistance, R, between the power-supply rail at voltage V and the output capacitance. With this configuration, current as a function of time is:

$$i(t) = \frac{dq}{dt} = \frac{V}{R}e^{-\frac{t}{RC}} . \tag{4.5}$$

The magnitude of the current is determined by the voltage drop from the power-supply rail to the capacitor. From resistance and the current, the instantaneous power dissipation in the resistance is:

$$P(t) = i(t)^2 R = \left(\frac{V}{R}e^{-\frac{t}{RC}}\right)^2 R = \frac{V^2}{R}e^{-\frac{2t}{RC}} . \tag{4.6}$$

The energy dissipation is the integral of power from 0 to T:

$$E_{diss} = \int_0^T P(t)dt = \int_0^T \frac{V^2}{R} e^{-\frac{2t}{RC}} dt = \frac{1}{2}CV^2\left(1 - e^{-\frac{2T}{RC}}\right) \quad (4.7)$$

The asymptotic energy dissipation, i.e., the limit as T approaches infinity, equals that of Eq. 4.3, which was derived by tracking the potential differences that the charge was forced to traverse through. The analysis for the discharging of C is similar and the details are not presented.

One way to summarize these results is to notice that at the beginning of each cycle, the energy injected by the power supply is CV^2, which is twice the bit energy. To deliver a "bit" to the circuit, the power supply must inject the energy of two bits through the pFET pullup network. To discard a bit, its energy is dissipated through the nFET pulldown network. Viewed in this way, conventional circuits represent a maximum in wastefulness. Note that this wastefulness stems from how the FETs are connected and operated.

To investigate the energy dissipation for an energy-recovery version of this circuit, the starting point is a "discrete" solution which will lead to a continuous, adiabatic solution. In Eqs. 4.5–4.7, current, power, and energy are exponential in time relative to the circuit's RC time constant. For example, current asymptotically approaches zero and energy dissipation asymptotically approaches $(1/2)CV^2$. After $3 \cdot RC$, the output voltage is within 95% of its final value, current is less than 5% of its initial value, and energy dissipation is within 97.5% of its final value.The assumption is made then, with only a negligible amount of error, that after $3 \cdot RC$, the circuit has reached its final output value.

Assume that the allowed transition time is twice this, or $6 \cdot RC$. With a conventional circuit, the extra charging time would bring the output ever closer to the supply rail. Another idea is to apply a variable voltage in the form of a "step" voltage to the circuit, by adding a second power supply that outputs a voltage of $V/2$. The $V/2$ supply powers the circuit for the first $3 \cdot RC$, and then the V supply powers the circuit from $3 \cdot RC$ to $6 \cdot RC$. The energy dissipation of Eq. 4.7 is then split into two parts:

$$E_{diss} = \int_0^{3RC} P(t)dt + \int_{3RC}^{6RC} P(t)dt \cong \frac{1}{2}C\left(\frac{V}{2}\right)^2 + \frac{1}{2}C\left(\frac{V}{2}\right)^2 = \frac{1}{4}CV^2 . \quad (4.8)$$

Doubling the transition time T and halving the maximum voltage drop reduces the net dissipation by a factor of two. Through the same line of reasoning, quadrupling the transition time and quartering the maximum voltage drop into four steps yields a net energy reduction of a factor of four. In general, by stretching the transition time by a factor of n and keeping the maximum voltage drop to V/n, the energy savings scales by a factor of n. This approach to energy recovery is called *step-wise* charging [22].

The next case to be considered is where the regenerator continuously varies the input voltage V so that current remains constant throughout the entire charge/discharge process, i.e.,

$$i(t) = \frac{Q}{T} = \frac{CV}{T}. \tag{4.9}$$

The energy-dissipation equation is [19]:

$$E_{diss} = \int_0^T i(t)^2 R dt = R\int_0^T \left(\frac{CV}{T}\right)^2 dt = \frac{RC}{T}CV^2. \tag{4.10}$$

Energy dissipation is directly inversely proportional to the transition time. Starting with this new equation to reducing dissipation, the design problem is to devise a circuit that has the same functionality as that of Figure 4.2 but exploits adiabatic charging. The first sub-problem to solve is how to exploit the regenerator as a constant-current, variable-voltage source. One solution is to introduce a new power source, but this is not necessary because the clock signal can be employed for this purpose. A suitable circuit is shown in Figure 4.3. In this circuit, the clock signal, which is powered by the regenerator, directly drives the output. Following the example of Denker, the clock signal is renamed to be the *power-clock* [10], because of its twofold role. As before, whenever x is high the output will be clamped low. If x is low, then the T-gate of M1 and M2 will pass

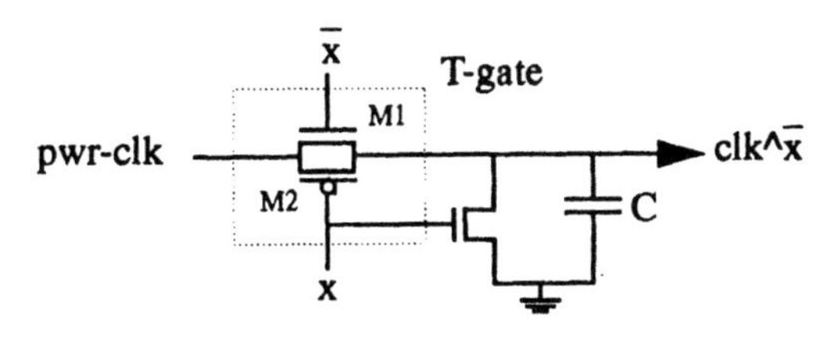

Figure 4.3 Adiabatic, Clocked-CMOS Inverter

the power-clock to the output. The nFET (M1) connected to $\bar{x}$ serves two purposes. The first is for when x is high and the power-clock is low. In this case, M1 pulls the output down to the low-voltage level of the power-clock. The second reason is more subtle. Without M1, the pFET (M2) would not turn on until the power-clock's voltage was above M2's threshold voltage. This would incur a non-adiabatic bump in the dissipation between the input and output.

Dynamically adjusting the power-clock voltage to comply with constant-current charging of Eq. 4.10 achieves an adiabatic-charging effect. The power supply and the clock generator are now one and the same, and the circuit has exactly the same input/output functionality as the circuit in Figure 4.2.

4.2. A look at some practical details

In the preceding section, the FET channels were modelled as single-valued resistances and the gates as lumped capacitances. The ideal situation would be to be able to directly apply such simple energy equations to CMOS circuits. To this end, FETs can be modelled by an "on-resistance," but, this resistance is a highly non-linear function of the gate and channel voltages. This section investigates how much additional detail is necessary.

Even though the resistances are highly non-linear, the general results about energy dissipation hold because they are based on how charge is transported between the logic circuits and regenerator, and not about the exact nature of the switching devices. Following the analysis in Section 4.1, time-varying on-resistance could be substituted into Eqs. 4.7–4.10, and all the integrations could be redone to get the energy dissipations. However, for even the simple case of RC charging, the equations can become quite involved. It is easy to construct circuit topologies which are difficult to analyze because of this complication. Aware of this pitfall, a simple circuit style is presented based on the CMOS T-gate, which can be reasonably modelled by a single resistance when it is conducting [2].

As in the example of Figure 4.3, a T-gate is an nFET and a pFET connected in parallel. To turn on the T-gate with minimal on-resistance, the pFET's gate is grounded and the nFET's gate is tied to V. For a small potential difference across the channel, which is the intended region for adiabatic circuits, the small-signal model [24] for the conductances, G_p and G_n, of the FETs is applied.

$$G_p = \frac{C_p}{K_p}(V - V_{ch} - V_T) \qquad C_p = C_{ox}LW_p \qquad K_p = \frac{L^2}{\mu_p} \tag{4.11}$$

$$G_n = \frac{C_n}{K_n}(V_{ch} - V_T) \qquad C_n = C_{ox}LW_n \qquad K_n = \frac{L^2}{\mu_n} \tag{4.12}$$

Channel length L, oxide capacitance per unit area, C_{ox}, and mobilities μ_n and μ_p, are well-known simulation parameters for modelling CMOS FETs. For bulk CMOS, the threshold voltage parameter V_T will vary with channel voltage because of the body effect. For a starting point, V_T is set equal to V_{TO}, which specifies threshold voltage for a channel-to-bulk voltage of zero volts. From these parameters, thee process constants K_p and K_n are defined. Channel length is assumed to be the minimum allowable for smallest on-resistance. C_{ox} could also be included in these constants, but since the analyses are mostly about energy in terms of capacitance, resistance, and voltage, it is more convenient to separately consider the gate capacitances of the FETs by C_p and C_n, which are calculated from L, C_{ox}, and channel width W. This last parameter is under direct control of the designer who can choose the gate capacitance by varying the channel width. The equations do not take into account body effects or the difference in threshold voltage between nFETs and pFETs, which limits accuracy.

By selecting the widths of the two FETs such that

$$K_n/C_n = K_p/C_p, \tag{4.13}$$

the sum of the two conductances simplifies to:

$$G_p + G_n = \frac{C_n}{K_n}(V - V_{ch} - V_T + V_{ch} - V_T) = \frac{C_n}{K_n}(V - 2V_{th}) \tag{4.14}$$

The on-resistance of the T-gate is the reciprocal of this sum:

$$R_{TG} = \frac{K_n}{C_n(V - 2V_T)} \tag{4.15}$$

In this model, on-resistance is independent of channel voltage. The next question is then: is there enough detail in Eq. 4.15 to reasonably model energy dissipation from Eq. 4.10? To answer this question, the on-resistance from Eq. 4.15 is inserted into Eq. 4.10 to calculate the energy dissipated for the circuit of Figure 4.3.

$$E_{diss} = \frac{R_{TG}C}{T}CV^2 = \frac{C}{C_n} \cdot \frac{1}{T} \cdot \frac{K_n}{(V - 2V_{Tc})}CV^2 \tag{4.16}$$

L (µm)	C_{ox} (fF/µm^2)	V_{th} (V)	μ_n (cm^2/Vs)	μ_p (cm^2/Vs)	K_n (ps)	K_p (ps)	C_n (fF)	C_p (fF)
2µm	0.83	0.82	552	185	2.16	6.42	6.6	19.7

Table 4.1 SPICE Model Parameters and Calculated Constants

Next, values calculated from this equation are compared to measurements taken from the device-level, circuit-simulation program SPICE for a constant-current source driving a large capacitance through a T-gate.

Values for C_n, C_p, K_n, K_p, etc., are calculated from the level-two SPICE model parameters for a MOSIS 2 µm process. They are listed in Table 1. The width of the nFET is 4 m (the minimum size for this process is 3 m) and the pFET is 12 m. The calculated on-resistance is 3.2 k Ω. The load capacitance is selected to be 250 fF so that it is much larger than the combined gate capacitance, otherwise all or part of the gate capacitance would have to be included as part of the load capacitance. The constant-current source is configured so that charge delivery is spread out over intervals of 1ns to 10ns. Appendix 4.B is a complete pSPICE listing for the simulation experiment.

The graph of Figure 4.4 compares the simulation results to Eq. 4.16 for an output swing of 0 to 5 volts. The RC product is 820 ps. The values from the equation consistently show smaller energy dissipation than the measurements. Part of this discrepancy is due to the body effect and to the mismatch between the threshold voltages which was exacerbated in this experiment because it was necessary to bias the pFET body voltage to 20% more than the output voltage swing. This was necessary because to maintain constant current for the shorter transition times, the power-clock voltage went above V. Regardless, even with this simple model, the equation is within 25% of the SPICE measurements. Furthermore, for fabrication technologies such as Silicon-on-Insulator (SOI), body effects will not be present and Eq. 4.16 should be of higher accuracy.

One way to improve the T-gate's performance is to increase the gate capacitances C_p and C_n. However, the energy requirements for driving these capacitances should be included in the calculated dissipation since they also contribute to the total dissipation. There are two general cases to consider. The first is when conventional logic powers the inputs. In this case all of the input energy is dissipated. The second case is when inputs are driven adiabatically.

For the conventional case, the input energy is simply:

$$E_{in} = \kappa C_n V^2 \qquad \kappa = 1 + \frac{\mu_n}{\mu_p}. \tag{4.17}$$

The quantity κC_n models the gates of an nFET and a pFET as a single lumped capacitance. To determine how output energy dissipation scales as a function of κC_n, Eq. 4.16 is re-written in terms of V and the RC product of the input capacitance and on-resistance of the T-gate, called τ_{TG}.

$$E_{out} = \frac{C}{C_n}\frac{\tau_{TG}}{T} C V^2 \qquad \tau_{TG} = R_{TG} \cdot C_n \tag{4.18}$$

The total system dissipation is:

$$E_{sys} = E_{in} + E_{out} = \kappa C_n V^2 + \frac{C}{C_n}\frac{\tau_{TG}}{T} C V^2 = \left(\kappa C_n + \frac{C}{C_n}\frac{\tau_{TG}}{T}\right) V^2. \tag{4.19}$$

Eq. 4.19 defines an important trade-off in energy-recover CMOS circuits. When the channel width, and thereby the input capacitance, is increased, the energy dissipated in driving the load decreases, but the energy dissipated in driving the inputs increases proportionately. The optimization process is straightforward. The optimal value for C_n is found given all the other parameters:

$$C_{n_{opt}} = \sqrt{\frac{2\tau_{TG}}{\kappa T}} \cdot C. \tag{4.20}$$

Inserting Eq. 4.20 into Eq. 4.19 yields the following expression for the minimum dissipation:

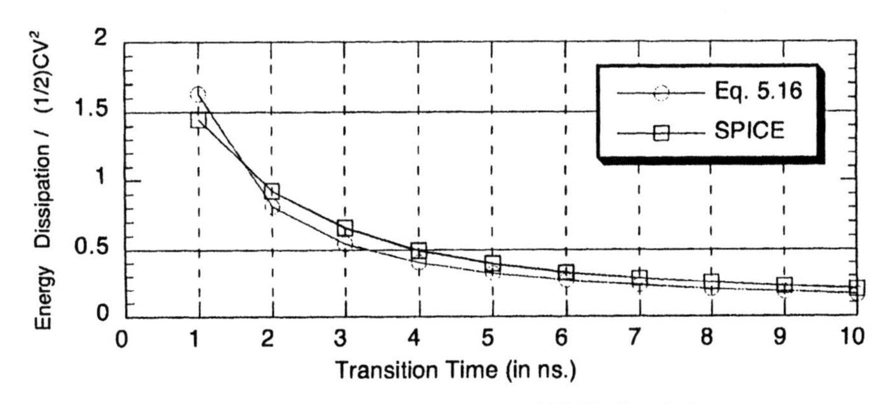

Figure 4.4 Comparison of Eq.4.16 to SPICE simulation

$$E_{sys_{min}} = 2\kappa C_{n_{opt}} V^2 = \sqrt{\frac{8\kappa\tau_{TG}}{T}} CV^2 \tag{4.21}$$

Clearly, something important has been lost at this point. Originally, by increasing the transition time, the dissipation would drop off linearly. By Eq. 4.21, the derivation for the optimal powering of an adiabatic load from a conventional source yields a far less attractive result: the dissipation only scales as $T^{-1/2}$.

An obvious improvement is to cascade two circuits, where the smaller drives the input of the larger. By doing so, the total minimum energy dissipation is [5]:

$$E_{total_{min}} = 8\left(\frac{\kappa\tau_n}{T}\right)^{3/4} CV^2 \tag{4.22}$$

From Eq. 4.21 to 4.22, the energy scaling improves from $T^{-1/2}$ to $T^{-3/4}$. It has been conjectured that for a cascading of n such stages, the energy dissipation scales as:

$$E_{total_{min}} \sim T^{2^{-n}-1} \tag{4.23}$$

This result is impractical since the parasitic capacitances can no longer be neglected when the capacitances of the final gate-drive FETs become comparable to that of the load. Also, for a constant cycle time interval, the transition time for each of the steps must decrease at least linearly in the number of cascaded driver stages.

In summary, to keep the energy-recovery scaling linear in transition time, the inputs and outputs must be powered adiabatically. The next two sections explore the implications to logic design when all of the logic including the inputs and outputs are driven adiabatically.

4.3. Retractile Logic

Figure 4.5 depicts a two-level tree decoder of two inputs and four outputs built from six T-gates. This decoder can be configured to implement the functions of NAND, NOR, XOR, and XNOR by selectively tying together different mutually-exclusive outputs. The tree decoder can be a useful energy-recovery circuit but it is often neither the smallest or fastest. It does concisely establish that any of the readily known universal logic elements can be built from T-gate logic.

Implementing logic functions when there are more than two inputs is difficult to do efficiently. One way to proceed is to build a single logic network of T-gates through which the power-clock signal is steered to one of two mutually-exclusive outputs, e.g., the tree decoder. Each series connections between power-clock and output implements an ANDing of the inputs and each parallel path an ORing of the inputs. Dual-rail encoding, which is needed otherwise to drive the complementary inputs of the T-gates, can be put to good use here in that logical NOT is trivially implemented by interchanging the two signal rails.

Such circuits become quite large and slow because signal delay through n pass transistors increases as n^2 [18]. For reasons of performance and also for complexity management, it is desirable to compose small logic cells, such as the tree decoder, as building blocks for larger functions. As it turns out, composition of adiabatic circuits is hard because signals that ripple through logic blocks can cause non-adiabatic charge flow.

One way to ensure that all charge flow is adiabatic between composed or "cascaded" logic blocks is to operate them in a *retractile* fashion [13]. Figure 4.6 diagrams how retractile cascades would be arranged for $n=4$. The generalization to larger n is straightforward. Each logic block requires a separate power-clock. The first power-clock is the first to go high and the last to go low. The second is the next to go high and the next-to-last to go low. The fourth is the shortest pulse, is the last to go high, and the first to go low. Note also that in this diagram, as in the diagrams that will follow, the multiple signal lines that flow between logic blocks are considered to be bundled together and are drawn as a single thick line.

The necessity for sequencing the powering of the logic blocks from one to n is because T-gate inputs for each block must be at stable logic levels before the outputs of the block can be adiabatically cycled. The necessity for retracting the power-clock signals in the *reverse* order is a bit more subtle. Referring back to the

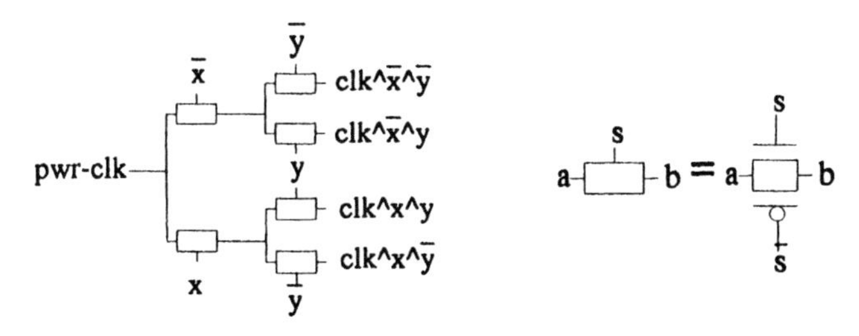

Figure 4.5 Two-level tree decoder built from T-gates

clock-AND circuit in Figure 4.3, the input x was stable during the power-clock pulse. The input x must be stable on the rising edge of the power-clock, otherwise the output would not be a full-swing signal. A requirement of the timing discipline was that x was always stable during the power-clock transition.

If the input x changes while the power-clock is high, there are no immediate problems. The energy of the output is not returned to the regenerator but neither is it dissipated. This does, however, lead to a complication which will result in an unavoidable, non-adiabatic dissipation. If x changes from high to low while the power-clock is high, the bit energy will not be returned to the regenerator, but will be marooned. One idea would be to try to recover it the next time the power-clock goes high again. This will work if somehow the circuit remembers that the charge was left behind last time. In which case there is the added requirement that x changes from a low to high only after the power-clock has gone high again. If x changes from low to high when the power-clock is low, then there will be a large potential drop across the conducting T-gate and a non-adiabatic dissipation will occur. If x changes from low to high only after the power-clock signal has gone high again and the circuit remembered wrong, that is the output was low and not high, then again there will be a non-adiabatic dissipation.

The reader may be a bit uneasy, and rightly so, about the need for an extra memory bit to keep track of the state of the signal to be retracted since in essence, the isolated charge on the wrong side of the T-gate is in fact a dynamic memory bit. This conundrum is pursued in the next section on reversible pipelines.

The retractile cascades avoid this problem by holding the inputs for each stage constant until the output energy has been returned. This approach is sufficient, in that any task which can be done with combinational logic can be done in this manner. However, the approach has significant undesirable properties. Foremost is the low duty factor on each of the logic stages, which are only cycled

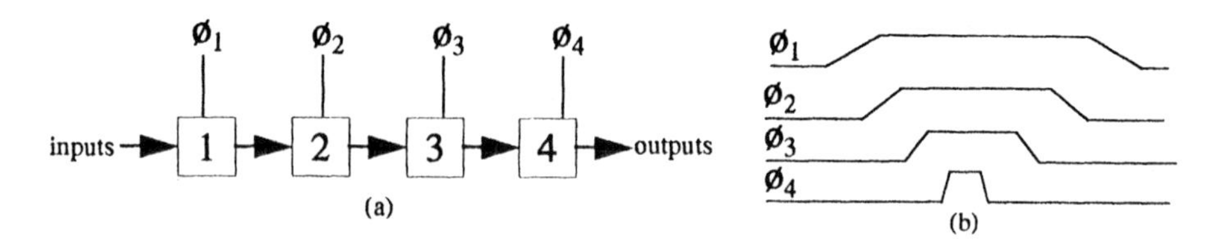

Figure 4.6 (a) Retractile cascade of 4-stages (b) and its timing

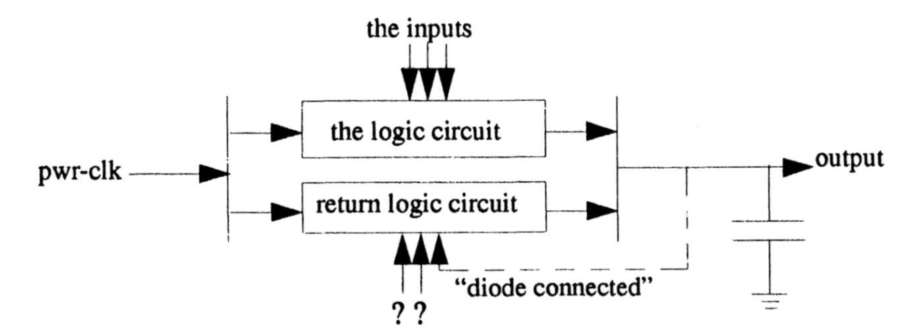

Figure 4.7 General scheme for retractile-free, adiabatic operation

1/*n*th of the time. Secondly, the recursive nature of the retractile clock signals requires that the number of power-clock signals and their timing depends on the number of logic stages employed. Ideally, the number of power-clocks and their timing would be independent of the logic depth.

4.4. Reversible Pipelines

The state of the retractile cascade before and after a cycle is the same—fully de-energized with all of the power-clocks low. In the process of computing, the incremental application of the power-clocks advances the computation one stage/step towards the final answer. Once the end is reached, the computation reverses one stage at a time back towards the initial input. The retractile cascade is a reversible-computing system.

At every point in the operation of the retractile cascade, the path for steering charge to an output is also the same path along which the charge returns to the regenerator. The T-gate inputs along the path must remain stable while the associated power-clock is active. This imposes a serious restriction to throughput. By decoupling the changing of the inputs from the switching of the outputs, the stages could operate concurrently on different data. To implement this, a way is needed to return the energy of the outputs although the input has changed.

A first implication is that there must be a second path for the energy to return along, as diagrammed in Figure 4.7. The power-clock and the inputs to the logic circuit function as they did before, while the return-path circuit brings the energy back to the regenerator once the inputs have changed. The fundamental question is then: where do the return-path inputs come from?

One idea is to apply the energy of the output itself to drive the inputs for the return path, as shown by the dotted line. This approach cannot be completely successful since the controlling energy cannot do double duty as the controlled energy, at least not entirely. The proof is simple for situations where the switching device requires a minimal amount of energy on its control element, i.e. the FET's gate. As energy is transferred from the control element through the channel, a minimum energy level is reached below which the channel no longer conducts. The remaining energy is marooned on the output. For a "diode-connected" FET, this energy is $(1/2)CV_T^2$. This problem of self reference [6] will exist for any switching device that requires a minimal energy to keep the device turned on. Energy cannot be used to keep the switch on while at the same time, transport the same energy through the switch. Some portion of the energy must remain while the rest is returned. This stranded or marooned energy contributes to an inevitable and unavoidable, non-adiabatic dissipation since the next time the switch is turned on, the voltage drop across the switch will result in a dissipation of $(1/2)CV_T^2$.

It is tempting to try to engineer a way out of this problem with a clever circuit that introduces an auxiliary circuit to control the devices in the return path. As with the retractile cascades, another power-clock could be introduced to adiabatically make a copy of the output and then "reverse" the original output from the copy. This idea is essentially to introduce a memory element to latch the output. It also has a problem of a circular definition, since the outcome of this procession is the original problem except that it is the copy (i.e., latched value) that must now be reversed.

This leads to a non-intuitive conclusion about the relationship between information and energy. Bits can be adiabatically generated, and they can be adiabatically copied, but they be cannot adiabatically erased. However, through copying, the non-adiabatic dissipation can be *postponed*. The consequence of erasure causing an inevitable dissipation is a famous result from the theory of the thermodynamics for computing engines [16]. However, postponing the dissipation leads to a small loophole that minimizes the overall impact of erasure.

Perhaps the simplest, practical circuit example of the postponement principle is a shift register. The timing for the shift register and its structure can then be generalized to implement *reversible* pipelines.

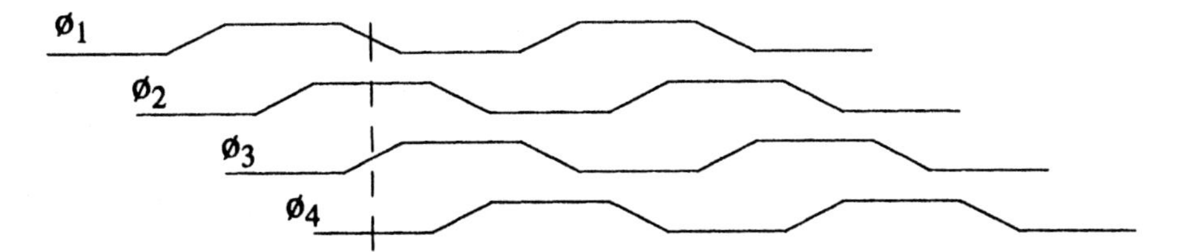

Figure 4.8 Four-phase overlapping power-clock scheme

With two power-clocks, dissipation can be deferred by copying the output and then adiabatically erasing the original using the copy. At the end of this action, the original problem remains but is relocated from the original to the copy. In principle, by recursively applying this technique the dissipation could be postponed indefinitely, providing it was possible to keep introducing additional copy circuits and power-clocks. From a circuit perspective, the net effect is to move data along a chain of circuit elements, which is, in essence, the function of a shift register. An important observation is that data can be adiabatically load new data into a circuit cell once the bit it was storing had "moved on" and the cell had been returned to its initial, "unloaded," condition. This leads to a novel shift-register construction which has the following property: data is only erased when it reaches the end of the shift register. For an n-bit shift register, when a new bit is shifted in and one bit is shifted out, the non-adiabatic dissipation is independent of n.

There are myriad ways of building and clocking such a circuit and a few have been reduced to practice [3], [25]. Described here is a T-gate approach that is in some sense minimal in circuitry and power-clocks.

The timing requirements of the retractile cascades resulted in a peculiar relationship between the power-clocks. A power-clock from one stage fully overlapped all of the power-clocks in the stages that followed it (see Figure 4.6(b)). This was necessary to ensure that the charge path was stable for both the rising and falling edge of the "steered" power-clock. With the introduction of the return path, this requirement is relaxed so that each rising edge of a power-clock must occur while the power-clock from the previous stage is high, while each falling edge must occur while the power-clock from the *successor* stage is high. From these requirements, a four-phase scheme, as diagrammed in Figure 4.8, is sufficient and has the added benefits of symmetric signals of equal duty cycle.

Figure 4.9 depicts a block diagram and T-gate circuit for a shift register. Each of the shift register's cells is composed of two T-gates: one for the logic cir-

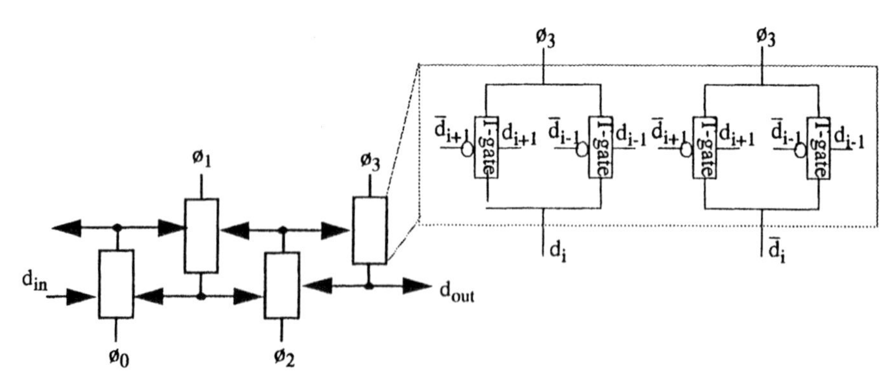

Figure 4.9 A 4-stage, shift-register cell built from T-gates

cuit and the other for the return circuit. The logic-circuits inputs are from the previous stage and the return-circuit inputs from the successor stage. The overlapping power-clocks establish three distinct intervals for each data bit: *set*, *hold*, and *reset*. The *set* interval occurs when the power-clock transitions from low to high while the previous-stage inputs are stable. The *hold* interval occurs while the power-clock signal is high. During hold, the output serves two purposes: to reset the previous stage when the power-clock signal for the previous stage goes low and to set the successor stage when the power-clock signal for the successor stage goes high. After the hold interval, the stage is reset when the power-clock goes low while the successor-stage inputs are stable.

For the shift register, the non-adiabatic dissipation at the end can be eliminated entirely by connecting the end back to the beginning. In this configuration as a circular shift register, there is no net flow of information into or out of the system. Once the register is loaded with a bit pattern, the pattern will recirculate indefinitely and adiabatically. The circular shift register is a curious contraption. In and of itself, it is a self-sustaining, adiabatic system. The laws of thermodynamics have not been broken. By reversing the sequencing of the power-clocks, it can as easily shift in one direction as it can in the other. It is a fully reversible system.

Systems of this type are not limited to shift registers. Younis and Knight have invented and demonstrated general-purpose pipeline stages that work in a similar fashion as the shift register [25]. One way to visualize these general-purpose systems is diagrammed in Figure 4.10. The lines between the adjacent stages are in fact bundles of data wires. As long as combinational logic is included along

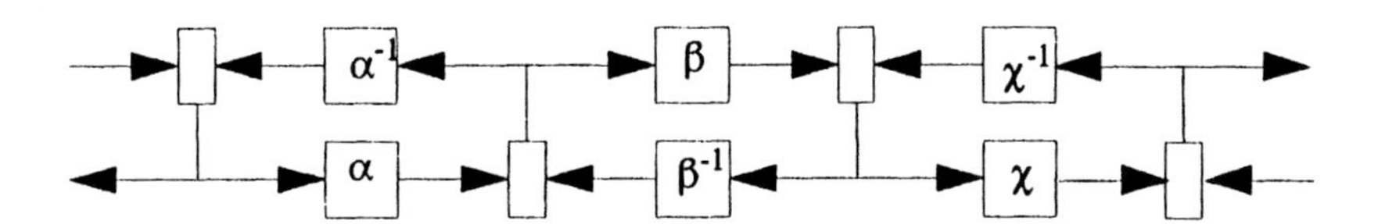

Figure 4.10 Reversible pipeline using the shift register approach

the return path that can undo what was done along the forward path, e.g., computing the *inverse* function of the forward logic, the situation is the same as that of the shift register.

For example, using the nomenclature of Figure 4.10, let the functions α, β, and χ be $x+1$, $\sqrt{x}$, and $Tan(x)$. Then providing α^{-1}, β^{-1}, and χ^{-1} are $x-1$, x^2, and $ArcTan(x)$, i.e., the inverses of the original functions, then the flow of data through the pipeline will be the same as it was for the shift register.

There is an important complication though, which will increase the number of power-clocks and possibly the number of signal rails per bit [25] for pipelines more sophisticated than the shift register. Each signal in the shift register goes through the events of set, hold, and reset. Before set and after reset, an output signal is either undefined or held at a valid logic level called *idle*. In the shift register example, the positive-rail output was set low and the negative-rail output was set high during idle. This configuration naturally turned off any T-gates connected to the output. This will work as long as the rails are not interchanged, i.e., a logic NOT is performed between the stages. If this does occur, then the interchange will naturally turn-on idle T-gates connected to the output, possibly resulting in non-adiabatic charge flow.

The problem can be solved by adding "guard" T-gates [3] (Figure 4.11) and more power-clocks to control the guard T-gates. These additional power-clocks

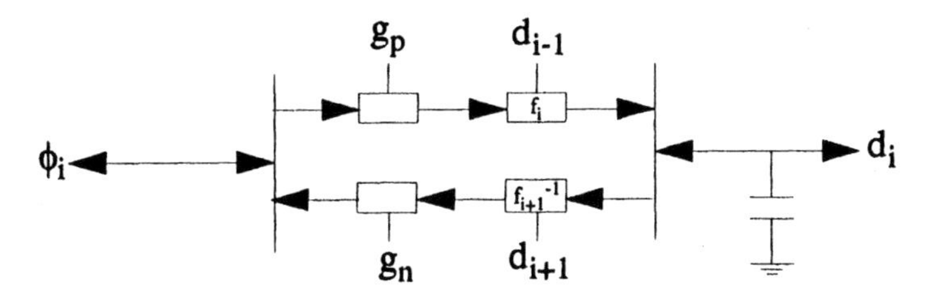

Figure 4.11 The reversible logic cell with guard T-gates

can be exploited to drive other pipeline stages. In practice, eight power-clocks is sufficient for a fully-static, pipeline structure [21]. Two bit-level-pipelined chips, a reversible parity generator and a reversible three-bit adder [4], were built and demonstrated with this pipeline structure. A higher performance approach using a single-rail, ternary encoding but more power-clocks was invented and demonstrated by Younis and Knight[26].

4.5. High-Performance Approaches

From Eq. 4.10, transition time and energy dissipation are inversely proportional. For example, for a fixed voltage swing and capacitive load, doubling the transition time decreases the energy dissipated by a factor of two. For the T-gate logic, the cost for this inverse relationship was dual-rail logic and signalling. For the reversible pipelines, there was the logic overhead to implement the inverse functions for the return paths. Since for conventional circuits, these requirements would be either optional or superfluous, it is important to investigate their corresponding dependence between energy and time.

By far the most direct and well-known method to reduce energy and power conventionally is to decrease the voltage swing that in turn is equal to the potential difference between the power supply rails. If the voltage swing cannot be decreased, for example, because of application requirements, then energy recovery is the only means to decrease energy and power for a fixed capacitive load. In this case, efficiency depends on the magnitude of the voltage swing, the transition time, and how the energy-recovery circuitry inputs are powered. Assuming equal input and output voltage swing, then as the swing is increased, the magnitude of the signal energy increases as the voltage squared while the from Eq. 4.15, T-gate on-resistance will decrease linearly. Hence, the *relative* energy savings increases approximately linearly with voltage swing.

When the voltage swing can be decreased, doing so will have desirable and undesirable ramifications for energy-recovery and conventional circuits. For both, the magnitude of the signal decreases as the square of the voltage swing, and in both cases, energy dissipation is proportional to signal energy. For energy recovery, however, decreasing voltage swing *increases* the on-resistance and thus *decreases* the efficiency of the energy-recovery process. For conventional, reducing voltage swing increases the transition time or delay.

One consequence of reducing the signal energy is the impact upon noise margins. Fortunately, internal noise, i.e., inductive, capacitive, and resistive, becomes proportionately less severe as voltage swing is reduced [8]. External noise, of course, depends on the application environment. The power supply is another noise source and may become the limiting factor for many portable applications because of supply-voltage ripple from dc-dc conversion [8].

To proceed then, noise margin issues are neglected and the focus is upon the energy-versus-delay properties of FETs driving FETs. This reduced problem is still difficult to analyze concisely because of the many remaining factors involved such as, choice of circuit topology, process technology, transition time, and how the circuit is to be used. Conclusions from one example or comparison do not generalize to other, larger examples.

Therefore, the nature of the FETs will be characterized as it pertains to the relationship between energy and delay. The starting point is a charge-model viewpoint of the FET with the important assumption that the FET behavior can be modelled independent of the instantaneous channel voltage. Between energy recovery and conventional then, the FET, as a mechanism, is used in two significantly different ways:

1. Conventional logic: FETs operate in the saturation region; signal energy is dissipated.
2. Energy-recovery logic: FETs operate in the linear region; signal energy is recovered.

For conventional logic, energy and voltage can be related to transition time by modelling saturation current via the well-known Shichman and Hodges equation [20]:

$$I_{sat} = \frac{1}{2}\mu_n\left(C_n/L^2\right)(V - V_T)^2 \qquad C_n = W \cdot L \cdot C_{ox} \tag{4.24}$$

Delay or transition time is then the time to discharge the capacitor at voltage V from the current of Eq. 4.23.

$$T_S = \frac{Q}{I_{sat}} = \frac{CV}{(1/2)\,\mu_n\left(C_n/L^2\right)/(V - V_T)^2} = \frac{C}{C_n}\frac{2K_nV}{(V - V_T)^2} \tag{4.25}$$

This expression does not take into account the linear region or sub-threshold conduction near V_T and thus is inapplicable when V is near V_T. However, excluding the region near V_T, it has been used elsewhere with reported good results [9]. It will be used here because of this positive experimental experience, and because it meets the requirement for channel-voltage independence.

To further simplify the expression of Eq. 4.25 for our purposes, the supply voltage V is expressed as a multiple m_S of the threshold voltage.

$$V = m_S \cdot V_T \tag{4.26}$$

By introducing m_S, the expression for transition time can be neatly partitioned into separate factors for parameters that are process related versus those that are circuit-load related.

$$T_S = 2 \cdot \left(\frac{K_n}{V_T}\right) \cdot \left(\frac{C}{C_n}\right) \cdot \frac{m_S}{(m_S - 1)^2} \tag{4.27}$$

K_n/V_T has units of time and contains all of the process-related parameters. C/C_n is the ratio of the input to output load capacitance. Delay varies approximately linearly with m_S, while the energy varies as the square of m_S.

The energy-recovery case could be stated in terms of the T-gate formula of Eq. 4.15. It will be advantageous, however, to generalize the equation by considering the two contributing factors of gate-to-source voltage, V_{gs}, and the *effective* threshold voltage, V_{TE}. The small-signal model, conductance formula for the nFET is then:

$$G_n = \frac{C_n}{K_n}(V_{gs} - V_{TE}) . \tag{4.28}$$

V_{gs} and V_{TE} will depend on the energy-recovery circuit topology. To simplify and parameterize these dependencies, V_{gs} and V_{TE} are approximated by the linear relationships:

$$V_{gs} = \alpha m V_{th} \qquad V_{TE} = \beta V_T . \tag{4.29}$$

Applying these definitions to Eq. 4.28, on-resistance can be expressed as:

$$R_{on} = \frac{1}{G_n} = \left(\frac{K_n}{V_T}\right)\left(\frac{1}{C_n}\right)\frac{1}{(\alpha m_R - \beta)} . \tag{4.30}$$

For the T-gate on-resistance of Eq. 4.15, α equals one and β equals two.

Energy dissipation as a function of m_R and T is expressed by inserting Eq. 4.30 into Eq. 4.10.

$$E_R(T, m) = \frac{R_{on}C}{T}CV^2 = \left(\frac{1}{T}\right)\left(\frac{K_n}{V_T}\right)\left(\frac{C}{C_n}\right)\frac{m_R^2}{(\alpha m_R - \beta)}CV_T^2 \tag{4.31}$$

Eq. 4.31 is factored similarly to that of the conventional delay formula. A useful observation about this formula is that there is an optimal m_R for which the dissipation in the energy-recovery circuit is minimal:

$$m_R = 2\frac{\beta}{\alpha} \qquad E_R(T) = E_R(T, m_R) = \left(\frac{1}{T}\right)\left(\frac{K_n}{V_T}\right)\left(\frac{C}{C_n}\right)\left(\frac{4\beta}{\alpha^2}\right)CV_T^2 \quad . \tag{4.32}$$

The time scale for Eq. 4.32 is in units of seconds. To relate how the energy-recovery case compares to the conventional case, the delay is set equal to T_S. This time-scale normalization to the conventional eliminates the process and circuit-load constants.

$$E_R(T_S) = 2\frac{\beta}{\alpha^2}\frac{(m_S - 1)^2}{m_S}CV_T^2 \tag{4.33}$$

The final step is to normalize the energy dissipation to the conventional.

$$\frac{E_R(T_S)}{E_s} = \frac{2\frac{\beta}{\alpha^2}\frac{(m_S - 1)^2}{m_S}CV_T^2}{(1/2)\,Cm_S^2V_T^2} = 4\frac{\beta}{\alpha^2}\frac{(m_S - 1)^2}{m_S^3} \tag{4.34}$$

The graph of Figure 4.12 depicts the behavior of this normalized energy dissipation as a function of m_S for T-gate logic. That is, m_S determines the delay of the conventional circuit by fixing the supply voltage at $m_S \cdot V_T$. This delay is then used for the transition time of the T-gate circuit with voltage swing $m_R \cdot V_T$.

In this graph, y-axis values less than one indicate where the energy-recovery dissipation is less than the conventional dissipation for the *same* transition

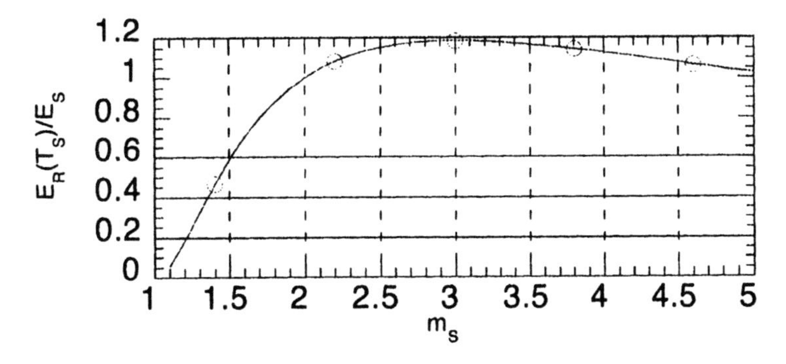

Figure 4.12 Energy dissipated normalized to conventional

time. The graph should be interpreted for its shape rather than exact numeric value. Different values for α and β will shift the graph up or down. Similarly, the larger capacitance of the energy-recovery circuit will also shift the graph up or down either linearly or as the square of the capacitance, depending upon where the capacitance exists in the circuit. In the comparisons that follow, the quantity $4(\alpha/\beta^2)$ will be used as the *figure of merit* for evaluating the efficiency of the energy-recovery mechanisms.

What may be concluded from this analysis is that energy recovery can offer better energy-versus-delay when m is small and also possibly for when m is large. The relative flatness of the curve for large m is misleading because in reality, the FETs will become velocity saturated at the higher voltages. Hence higher voltage will not contribute to increased speed, only higher dissipation.

In the critical region of a few multiples of V_T, if energy-recovery is to be superior, then the efficiency of the energy-recovery process, capacitive overhead, and other contributing factors have to made significantly smaller than that of the T-gate logic. To this end, some high-performance approaches are now examined.

4.5.1. Split-Rail Charge-Recovery Logic

The T-gate logic required dual-rail versions of every signal for the complementary inputs to each T-gate, which in turn required that each logic function be implemented twice. Furthermore, there was the performance degradation due to the body effect which is largely absent for conventional logic because FET source nodes are often tied to the same potential as the body. A fully-adiabatic logic that either solves or diminishes these problems is Split-Rail Charge-Recovery Logic (SCRL). It is based on conventional, static, logic structures [26], thus dual-rail logic and signalling are not needed and because of how the signals swing, body effects are less severe. Figure 4.13 depicts how the venerable clock-AND gate would be cast in SCRL. A pivotal idea to SCRL is to encode the idle interval as $V/2$. In contrast, for the T-gate logic, this interval was encoded by keeping both rails at zero, or as in the shift register, at their "off" values. An SCRL multiplier organized as a reversible pipeline has been designed and implemented as a CMOS chip [26].

Following the clock-AND example, the two power-clocks $\Phi 1$ and $/\Phi 1$ are initially set to $V/2$ while the output is also at $V/2$. Excluding the first inverter, changes on the input x will not cause dissipation in the FETs because there are no potential drops in the circuit. With the input x stable at 0 or V, the power-clocks

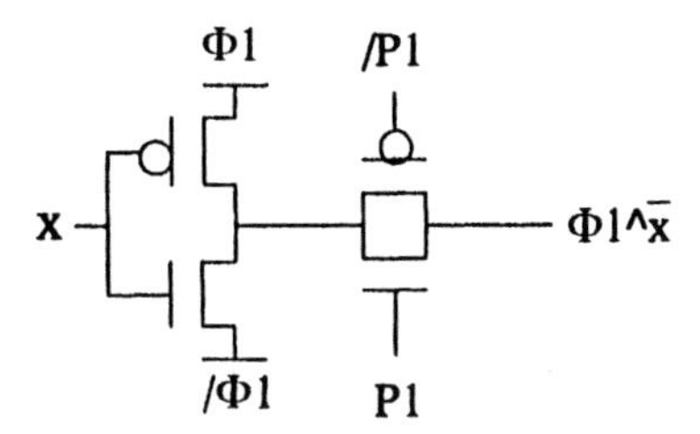

Figure 4.13 SRCL version of the clocked-CMOS inverter [25].

split apart, with Φ1 swinging from $V/2$ to V and /Φ1 swinging from $V/2$ to 0. The output adiabatically tracks the power-clock that it is connected to. The signals /P1 and P1 are full-swing signals that latch the output before the power-clocks merge back to V. These signals and their T-gate are not necessary for the clocked-inverter example.

There are advantages and disadvantages to SCRL. Of the advantages, the logic swing is reduced to a maximum of $V/2$; thus the magnitude of the dissipation is proportional to $V^2/4$. This will shift the graph of Figure 4.12 down and increase the range over which energy-recovery outperforms conventional logic. Furthermore, the effective electric field between gate and channel is larger than that for the T-gates because of the opposite direction of the voltage swing on the output relative to the input signal level. This contributes to an effective higher conductance, i.e., lower on-resistance. Also, because of the reduced swing, body effects are reduced. To compare SCRL to the T-gate logic, SCRL's α can be estimated as 3/4 by using the midway point for the output-voltage swing. β equals one. The figure of merit ($4\beta/\alpha^2$) equals 7.1, which is slightly better than that of T-gate logic. However, since voltage swing is reduced by a factor of two, the total dissipation will be reduced overall by a factor of four, which will shift the curve in Figure 4.12 down by a factor of four.

There are, however, some non-trivial disadvantages. Probably the most significant is the number of power-clocks needed to sustain a reversible SCRL pipeline. Power-clock requirements range from sixteen signals for two-phase dynamic logic to twenty for four-phase static logic [26]. A second drawback is that SCRL signals typically cannot be directly interfaced to conventional logic because of the voltage level for the idle state.

4.5.2. Semi-Adiabatic Approaches

SCRL goes a long way towards providing both high performance and adiabatic operation. However, the low throughput of the retractile cascades and the logic overhead of the reversible pipelines still loom as practical barriers. In simple situations such as the clocked-CMOS inverter, there is a simple rationale for exploiting energy recovery. In contrast, for more complex situations involving sequential operation, there is the extra logic and intricate timing. The goal is to find schemes where adiabatic and non-adiabatic circuits are judiciously interposed to get the simplicity of the former and the functionality of the latter.

This new operating regime, where the balance between the adiabatic and non-adiabatic dissipation comes into play, leads to an "Amdahl's law" [1] for semi-adiabatic circuits. Assume that a fraction p of the energy of a chip can be recovered adiabatically while the rest must be dissipated. The dissipation for this fraction p will decrease, say linearly, with increasing switching time. Energy dissipation can be modelled as:

$$E_{SA} = p \cdot \frac{E_{chip}}{T} + (1-p)E_{chip} \qquad (4.35)$$

The dissipation, if the chip were operated conventionally, would be E_{chip}. The potentially energy-savings factor is the ratio of the non-adiabatic to the semi-adiabatic:

$$\frac{E_{NA}}{E_{SA}} = \frac{E_{chip}}{\frac{p}{T}E_{chip} + (1-p)E_{chip}} = \frac{1}{\frac{p}{T} + (1-p)} \qquad (4.36)$$

The asymptotic savings is:

$$\lim_{T \to \infty} \frac{1}{\frac{p}{T} + (1-p)} = \frac{1}{1-p} . \qquad (4.37)$$

The asymptotic savings is inversely proportional to $1-p$. If 90% of the chip's capacitance can be driven adiabatically, then the reduction is limited to a factor of ten. The primary interest is therefore in chip designs for which better than 90% of the chip can be driven adiabatically. In practice, this may not be difficult to achieve since depending on the design, a large portion of the switched capacitance may be in driving large loads such as data buses and decoder lines

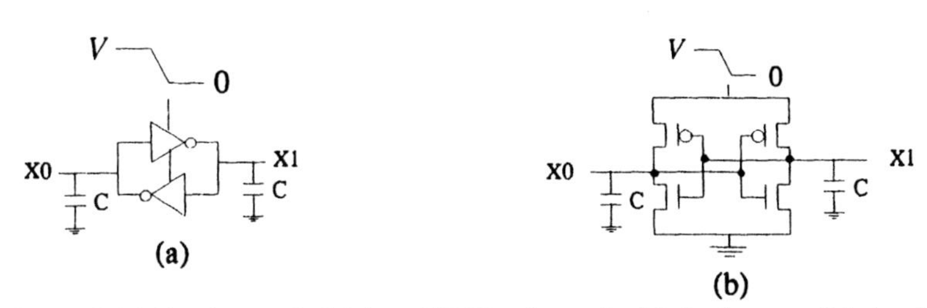

Figure 4.14 Logic symbol (a) and FET schematic (b) for irreversible latch operated semi-adiabatically.

[7]. These loads can be trivially reversed and are typically one to two orders of magnitude larger than the minimum gate capacitances.

Partially-Erasable Latches

A self-contained latch that could be adiabatically erased would be an extremely useful component, since it would reduce the complexity of the reversible pipeline to that of a conventional one. However, it is impossible to build a latch that would be perfectly adiabatic when its content is erased. An alternative is to build a latch that can be partially erased through adiabatic discharging and to compare its dissipation to the ideal, fully-adiabatic discharge. An approach that is representative of the general idea is to start with a conventional latch composed of two cross-coupled inverters and replace the dc supply voltage with a power-clock that swings between 0 and V. Assume initially that X1 and X0 of Figure 4.14 are set to V and 0, respectively, and the power-clock is at V. The power-clock then ramps from V to 0 as before. The pFET attached to X0 will conduct until the input voltage reaches V_T. At this point the voltage drop from gate to source will be equal to V_T and the pFET will turn off. Consequently the latch "forgets what it is trying to forget" and the output X1 will remain at V_T. The stranded energy left on the output capacitance C is $(1/2)CV_T^2$ as opposed to the conventional $(1/2)CV^2$ dissipation. This idea of starting with an irreversible latch and using the pFETs to recover the major portion of the total energy was proposed and SPICE simulated for a small shift-register chip [15].

A significant advance over this approach was made with the 2N-2N2P logic family [10]. Figure 4.15 depicts a 2N-2N2P NAND/AND gate. The earlier shift-register experiment assumed that the residual $(1/2)CV_T^2$ energy would be dissipated by clocked nFETs. With 2N-2N2P, the output is not discharged. Referring again to Figure 4.15, the two cases to consider are when the output does or does not change after the power-clock has returned to zero. For the case when it does not, then when the power-clock is reapplied, the output that ramped down to V_T will charge up from V_T to V adiabatically. If the bit did flip, i.e., the output was discharged by the pulldown FETs, then when the power-clock is applied, both outputs will try to simultaneously charge up once the power-clock voltage exceeds V_T. This will be a source of non-adiabatic dissipation. The output that is not held low by the pulldown nFETs will charge up while the other will stay clamped low. There will be some short-circuit current through the pulldown nFETs of the clamped-low output which will decrease as the other output charges up. Neglecting the short circuit current, the dissipation for charge up will be at least CV_T^2.

Simplicity in clocking is an important advantage to 2N-2N2P. The logic can be powered and clocked by two-phase, possibly partially overlapping, clock signals that can be compatible with conventional logic and constitute a major improvement from the eight-to-twenty overlapping power-clocks required for the reversible, fully adiabatic approaches. Chip area to implement many-input gates is reportedly smaller than that of conventional gates and noise immunity is reportedly excellent. There are a few disadvantages, such as the need for dual-rail logic and signalling. A more serious limitation is the modest scalability of the approach. Using the previously introduced definition for m, the dissipation per transition for the conventional case can be expressed as $(1/2)m^2CV_T^2$. In contrast, approaches bases on partially-erasable latches such as 2N-2N2P have a dissipation per transition of roughly CV_{th}^2. The asymptotic benefit depends therefore on m, which in the critical region of the graph in Figure 4.12 is between two and five.

Bootstrapped Gates

A key advantage to the 2N-2N2P approach is that a small signal energy of $(1/2)CV_T^2$ could be adiabatically increased to $(1/2)CV^2$. Before bits are

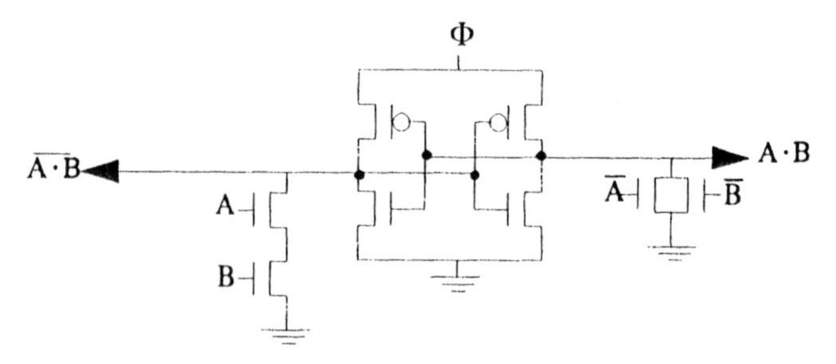

Figure 4.15 NAND/AND gate in the 2N-2N2P logic family

flipped their energies are first reduced to the small levels, then flipped, and then restored to the full levels so that they can drive the inputs of other 2N-2N2P circuits. The drawback is that the energy dissipation can only be reduced to $(1/2)CV_T^2$.

With bootstrapped circuits [14], [19], [12], the output energy dissipation can be fully adiabatic while the input is semi-adiabatic. The uplifting idea is to exploit the voltage-dependent capacitances that form between the gate and source or drain of a FET. Below threshold, the capacitance is small because charge carriers of polarity opposite to those on the gate are located deep in the substrate. Increasing the gate voltage causes charge carriers to draw near the surface region under the gate. In addition to enhancing this region to form a channel, this also greatly increases the capacitance between gate and channel which is inversely proportional to the distance between the opposite charges.

It is possible to design logic gates and dynamic latches that exploit this effect. Figure 4.16 depicts once again the clock-AND gate. With x low, the output is clamped low. If x is high, the isolation transistor M1 will charge-up the boot node to $V - V_{TE}$, and then turn off. This charging can be carried out adiabatically. The power-clock can then swing the boot transistor's drain (M2) from 0 to V. In the style of Tzartzanis [23], the direction of the bootstrap action is indicated by drawing an arrow at one end of the FET. The voltage that is capacitively-coupled from drain to gate will be less than V because of the capacitive divider formed between the gate-to-channel capacitance and the other (e.g., parasitic) capacitances present on the node. Nevertheless, if sufficient charge is deposited on the gate, the capacitively-coupled voltage will *add to* the original voltage and raises the gate voltage above V while M1 remains off. This is the bootstrapping action.

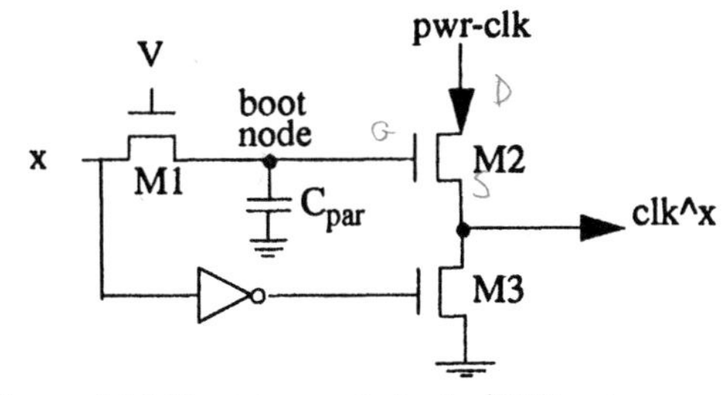

Figure 4.16 Bootstrapped clock-AND gate.

Relating the performance of this approach to the performance model, α represents the effectiveness of the capacitance divider which increases with the gate capacitance of M2, i.e.,

$$\alpha = \frac{C_{boot}}{C_{boot} + C_{par}} \tag{4.38}$$

Typical values for α range from 0.6 to 0.9. β equals two because of an additional threshold drop through M1.

For α=0.8, the quantity $4\beta/\alpha^2$ equals ten which is somewhat worse than the value of eight for the T-gate logic. However, there are some important mitigating factors. Foremost is the absence of a ratioed pFET, which means that for the same amount of input capacitance, the on-resistance will be a factor of two to three smaller. Also, dual-rail logic is not necessary, though is often desirable for better noise immunity and higher speed [23].

Bootstrapped gates are potentially appealing because of the possibility of high performance and adiabatic operation, but as always, there are some disadvantages. For example, the presence of pFET devices can be a source of latch-up, the logic is dynamic, and the circuits are sensitive to circuit noise. These problems are topics for future research.

4.6. Summary

This chapter has explored the prospect of reducing power dissipation in digital CMOS circuits by employing techniques to recover energy that would otherwise be dissipated as heat. At the device level, adiabatic charging is the key to exploiting useful amounts of energy recovery.

At the circuit level, power-clock signals directly drive the large capacitive loads of a chip. Starting with T-gates to implement charge-steering logic and the power-clock signals, it was shown how to implement arbitrarily combinational logic functions in CMOS that exploited the adiabatic-charging principles. The situation for sequential logic was more involved and introduced some intriguing theoretical issues about the fundamental nature of computation as a physical process. Regardless, energy recovery can be used for circuits that exhibit sequential behavior, as was shown with the reversible, adiabatic pipelines.

It can be done. The question at hand then is: When, if ever, is it worth doing? In addressing this question, the issues of input and output capacitance, voltage swing, process performance, and transition time are all important. When input and output capacitance, and voltage swing are fixed, energy recovery is the only way to reduce dissipation. For situations where the voltage swing can be reduced, then for conventional circuits, doing so will reduce the energy dissipation but increase the transition time. In principle, the voltage swing can be reduced down to the threshold voltage without giving up conventionality. For energy-recovery circuits to work adiabatically, however, they must operate at voltages significantly above the threshold voltage. The choice of which is more energy efficient depends on many factors. The necessity that energy-recovery circuits operate at voltages well above threshold is also a virtue, in that there is wide range of circuit-design styles and approaches to choose from. One such approach, that of bootstrapping, appears to offer both high-performance and efficient energy recovery.

Ultimately, the practicality of energy-recovery CMOS depends on the efficiency of the energy-recovery process in both the logic circuits and the regenerator that produces the power-clocks. In this chapter, the regenerator has been treated as a black box, since its circuitry deals principally in a different branch of electronics and because, if energy-recovery CMOS is to be successfully used for designing VLSI chips, then there must be a separation of concerns between the power-supply design and the logic-circuit design. Appendix A outlines some of the issues in building an efficient regenerator.

Acknowledgments

The author wishes to express his sincere thanks to his colleagues Lars "Johnny" Svensson Jeff Koller, Nestoras Tzartzanis, Tom Knight, Saed Younis, Paul Solomon, and Dave Frank. Thanks to Chuck Seitz and Sven Mattison for many of the original good ideas. The work described in this paper was supported by the Advanced Research Projects Agency, contract DABT63-92-C0052.

References

[1] G. M. Amdahl, "Validity of Single-processor Approach to Achieving Large-scale Computing Capability", *Proc. AFIPS Conf*erence, pp. 483-485, Reston VA., 1967.

[2] W.C. Athas, J.G. Koller, L. Svensson, "An Energy-Efficient CMOS Line Driver Using Adiabatic Switching," Fourth Great Lakes Symposium on VLSI, Mar. 1994.

[3] W.C. Athas, L. Svensson, J.G. Koller, N. Tzartzanis, E. Chou, "A Framework for Practical Low-Power Digital CMOS Systems Using Adiabatic-Switching Principles," 1994 International Workshop on Low Power Design, Apr. 1994.

[4] W.C. Athas, L. Svensson, "Reversible Logic Issues in Adiabatic CMOS," Proc. of the 1994 Workshop on Physics and Computation, Nov. 1994.

[5] W.C. Athas, L. Svensson, J.G. Koller, N. Tzartzanis, E.Y.-C. Chou, "Low-Power Digital Systems Based on Adiabatic-Switching Principles," *IEEE Trans. on VLSI Systems*, Vol. 2, No. 4, Dec. 1994.

[6] W.C. Athas, "Low Power Design Methodologies," Kluwer Academic Press, 1995.

[7] J.D. Bunda, "Instruction-Processing Optimizations Techniques for VLSI Microprocessors," Ph.D. thesis, Univ. of Texas at Austin, May, 1993.

[8] J.B Burr, A.M. Petersen, "Energy Considerations in Multichip-Module based Multiprocessors," *IEEE International Conference on Computer Design*, 1991.

[9] A.P. Chandrakasan, S. Sheng, R.W. Brodersen. "Low-Power CMOS Digital Design," *IEEE JSSC*, vol. 27, no. 4, pp. 473-484, Apr. 1992.

[10] J.S. Denker, "A Review of Adiabatic Computing," *1994 IEEE Symposium on Low Power Electronics*, Oct. 1994.

[11] D.J. Frank, P. Solomon, "Electroid-Oriented Adiabatic Switching Circuits," *1995 Int'l Symposium on Low-Power Design*, Apr. 1995.

[12] L.A. Glasser, D.W. Dobberpuhl, "The Design and Analysis of VLSI Circuits," Addison-Wesley, Reading, 1985.

[13] J.S. Hall, "An Electroid Switching Model of Reversible Computer Architectures," Proc. ICCI '92, Fourth International Conf. on Computing and Information, 1992.

[14] R.E. Joynson, J.L. Mundy, J.F. Burgess, C. Neugebauer, "Eliminating Threshold Losses in MOS Circuits by Bootstrapping Using Varactor Coupling," *IEEE Jnl. of Solid-State Circuits*, SC-7,No. 3, June 1972.

[15] J.G. Koller, W.C. Athas, "Adiabatic Switching, Low Energy Computing, and the Physics of Storing and Erasing Information," Proc. of the 1991 Workshop on Physics and Computation, Nov. 1991.

[16] R. Landauer, "Irreversibility and Heat Generation in the Computing Process," *IBM Journal*, July 1961.

[17] D. Maksimovic. "A MOS Gate Drive with Resonant Transitions," Proc. IEEE Power Electronics Specialists Conference, IEEE Press, 1991.

[18] C.A. Mead, L. Conway, "Introduction to VLSI Systems," Addison-Wesley, Reading, 1980.

[19] C.L. Seitz, A.H. Frey, S. Mattisson, S.D. Rabin, D.A. Speck, J.L.A. van de Snepscheut, "Hot-Clock nMOS," Proc.of the 1985 Chapel-Hill Conf. on VLSI, Apr. 1985.

[20] M. Shichman, D.A. Hodges, "Modeling and simulation of insulated-gate field-effect transistor switching circuits," *IEEE Jnl. of Solid-State Circuits*, SC-3, 285, Sept. 1968.

[21] L. Svensson, "Logic Style for Adiabatic Circuits," ISI Internal Report, Mar. 1994.

[22] L. Svensson, J.G. Koller "Driving a Capacitive Load without Dissipating fCV^2," 1994 IEEE Symposium on Low Power Electronics, Oct. 1994.

[23] N. Tzartzanis, W.C. Athas, "Design and Analysis of a Low Power Energy-Recovery Adder," Fifth Great Lakes Symposium on VLSI, Mar. 1995.

[24] N.H.E. Weste, K. Eshraghian, "Principles of CMOS VLSI Design: A Systems Perspective," Addison-Wesley Publishing Company, 1994.

[25] S.G. Younis, T.F. Knight, "Practical Implementation of Charge Recovery Asymptotically Zero Power CMOS," Proc. of the 1993 Sym. on Integrated Systems, MIT Press, Apr. 1993.

[26] S.G. Younis, "Asymptotically Zero Energy Computing Using Split-Level Charge Recovery Logic," Ph.D. thesis, Massachusetts Institute of Technology, June 1994.

[27] S.G. Younis, T.F. Knight, "Switchless Non-Dissipative Rail Drivers for Adiabatic Circuits," Proc. of the 1995 Chapel-Hill Conference on VLSI, Apr. 1995.

Appendix 4.A

A simple regenerator for the power-clock is the resonant-transition-gate drive [17] that was originally invented for improving the efficiency of controlling the power MOSFET in switching power supplies. The overall idea is to build up a large current in an inductor without the capacitive load in the current loop. Energy stored in the inductor is then transferred to the capacitor at quasi-constant current. Dissipation is minimized by turning the switches on and off when the voltage drop across them is close to zero.

Figure 4.17 is a simplified schematic. To generate a quasi-constant-current pulse of magnitude CV/T, the switch S1 is closed. With one end of the inductor at $V/2$ and the other grounded, a current will *linearly* build up. When the current reaches CV/T, switch S1 is opened and the circuit enters "RLC" mode. Current will flow from the inductor into the capacitor C through resistance R at a quasi-constant rate of CV/T. Once the capacitor has fully charged to V, switch S2 is then closed, which will clamp the output to V. The current in the inductor will begin to decrease and then reverse direction. When it reaches $-CV/T$, the switch S2 opens and charge is drawn from the capacitor back to the voltage source. If the $V/2$ voltage source is in practice implemented as a large capacitor then the energy can be recovered.

The principal source of energy dissipation for the power supply is I^2R losses through the switches during current build up and build down. The interval for current build-up and build-down must be significantly larger than T to model the charging approximately constant-current. To take this into account, auxiliary variable b models the time for current build-up and build-down as $b \cdot T$. With this assumption, the dissipation is:

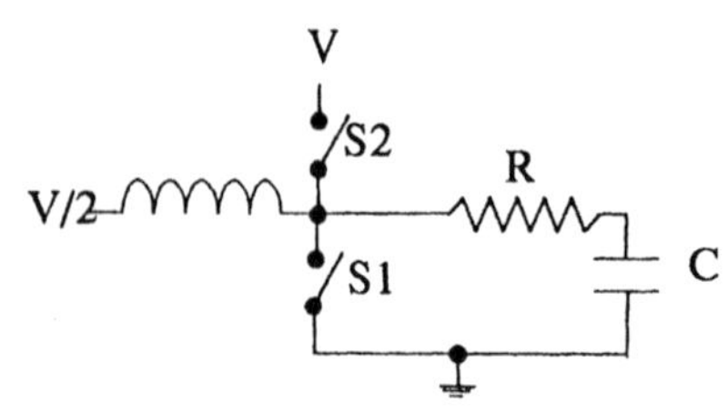

Figure 4.17 Resonant-Transition-Gate Drive Schematic

$$E_{RT} = \int_0^{(b \cdot T)} P dt = R_{on} \int_0^{(b \cdot T)} \left(\frac{CV}{bT} \cdot t\right)^2 dt = \left(\frac{b}{3}\right) \frac{R_{on}C}{T} CV^2 \tag{4.39}$$

For comparison, for a highly under-damped rail-driver circuit operating at a frequency well below the critical-damping point [2][3]:

$$E_{RD} = \frac{\pi^2}{8} \frac{R_{on}C}{T} CV^2 \tag{4.40}$$

The resonant-transition-gate drive compares poorly since a practical value for m is ten or larger, but there are some important mitigating factors:

1.The critical-damping point is much higher for the resonant transition gate drive than for the rail driver: $T > (RC)/(b \cdot \pi)$ versus $T > (\pi/2) \cdot RC$.

2.There are no body effects if the switches are implemented at bulk FETs.

3.The on-resistance of the switches is not in series with the load resistance R. If the switches are implemented in the same technology as the VLSI circuits then, as has previously been described [2][3], the combined energy dissipation will scale only as the square root of the switching time. For asymptotically-zero energy dissipation behavior, this is a severe limitation. However, for the critical region in the graph of Figure 4.12, and the first factor of ten below the critical region, it is possible [26] to appeal to a better device technology that has an RC product smaller than the VLSI chip's CMOS transistors. For this scenario, the energy dissipation will initially decrease linearly with switching time.

A second, and perhaps more important concern, is that for circuit approaches such as bootstrapped logic, the charge transfer is not equal in both directions because of the charge-trapping effect. The implications of this asymmetry need to be investigated. There is also the problem that for transitions of length T, the cycle time is $2(b+1)T$ and the output waveform is of 50% duty cycle, which may not be not directly compatible with system clocking requirements. However, it may be possible to employ a multiplicity of resonant-transition-gate drives to generate the required waveforms.

Appendix 4.B: SPICE file for the T-Gate Experiment

```
T-Gate Experiment for Section 4.2

.probe
.param Vdd=5
.param Cload=250fF
.param Q={(Cload+25fF)*Vdd}
.param Lambda=1um
.param Trise=1ns
.step param Trise LIST 1ns 2ns 3ns 4ns 5ns 6ns 7ns 8ns 9ns 10ns

Vsup 1 0 DC {Vdd}
Vbb  9 0 DC {1.4*Vdd}

* The constant current pulse
I1 0 3 pulse  (0A {Q/(Trise+0.1ns)} 1ns 0.1ns 0.1ns {Trise} 100ns)
Vm 2 3 DC 0V

* The T-gate
M1  2 1 4 0 cmosn L={2*Lambda} W={4*Lambda} NRD=0.0 NRS=0.0
M2  2 0 4 9 cmosp L={2*Lambda} W={12*Lambda} NRD=0.0 NRS=0.0

C1 4 0 {Cload} IC=0V

* record instantaneous power
Epower 7 0 value = {i(Vm)*(v(4)-v(2))/(0.5*Q*Vdd)}

.tran 0.1ns 15ns UIC

.MODEL CMOSN NMOS LEVEL=2
+       PHI=0.600000 TOX=4.1700E-08 XJ=0.25000U TPG=1
+       VTO=0.8236 DELTA=5.1270E+00 LD=1.8070E-07 KP=4.5744E-05
+       UO=552.4 UEXP=1.5150E-01 UCRIT=9.2630E+04 RSH=2.6550E+01
+       GAMMA=0.5049 NSUB=5.2670E+15 NFS=1.9800E+11 NEFF=1.0000E+00
+       VMAX=5.8570E+04 LAMBDA=2.9840E-02 CGDO=2.2445E-10
+       CGSO=2.2445E-10 CGBO=3.9785E-10 CJ=9.1699E-05 MJ=0.8283
+       CJSW=4.1092E-10 MJSW=0.297720 PB=0.800000
.MODEL  CMOSP PMOS LEVEL=2
+       PHI=0.600000 TOX=4.1700E-08 XJ=0.25000U TPG=-1
+       VTO=-0.9584 DELTA=3.6660E+00 LD=3.1970E-07 KP=1.5386E-05
+       UO=185.8 UEXP=3.0820E-01 UCRIT=1.0260E+05 RSH=5.1080E+01
+       GAMMA=0.7510 NSUB=1.1650E+16 NFS=3.270E+11 NEFF=1.0000E+00
+       VMAX=9.9990E+05 LAMBDA=4.2020E-02 CGDO=3.9711E-10
+       CGSO=3.9711E-10 CGBO=3.9682E-10 CJ=3.2659E-04 MJ=0.5680
+       CJSW=3.1624E-10 MJSW=0.260598 PB=0.800000
.end
```

5

Low Power Clock Distribution

Joe G. Xi and Wayne W-M. Dai

5.1. Power Dissipation in Clock Distribution

The clock frequencies of CMOS digital systems are approaching gigahertz range as a result of the submicron technology. At the same time, power dissipation is becoming a limiting factor in integrating more transistors on a single chip and achieving the desired performance. With the recent advances in portable wireless computing and communication systems, we have seen the shift from a performance driven paradigm to one in which power drain becomes the most important attribute.

Layout or physical design has traditionally been a key stage in optimizing the design to achieve performance and cost goals. As the demand for minimizing power increases, low-power driven physical design has attracted more and more attention. Recently, low power considerations were given in the floorplan stage of chip layout [3] while placement and routing techniques for minimizing power consumption were also studied [10],[13]. An iterative method for transistor sizing and wire length minimization has been applied to semi-custom design [14]. In this chapter, we consider a physical design problem which has more global impact on the total system power consumption —the physical design of clock distribution.

The synchronization of a digital system requires one or more reference signals to coordinate and insure correct sequences of operations. Fully synchronous operation with a common clock has been the dominant design approach for digital systems. A clock tree is usually constructed to globally distribute the clock signal to all modules throughout the system. The transitions of clock signal provide the reference time for each module to latch in data, trigger operations and transmit outputs. Clock has been recognized as the most important signal in determining system performance and total power consumption.

The amount of power dissipated by a CMOS circuit consists mainly of two parts:

- **Dynamic power** — due to charging and discharging capacitive loads at every clock cycle.
- **Short-circuit power** — due to the short-circuit current through PMOS and NMOS transistors during the switching interval.

Carrying large loads and switching at high frequency, the clock is a major source of dynamic power dissipation in a digital system. For example, the DEC Alpha chip uses a clock driver which drives 3250 pF capacitive load [7]. When operating at 200 MHz, with a 3.3 V supply voltage, the dynamic power dissipation alone on the clock is 7.08 W, which amounts to 30% of the chip's total power dissipation. In fact, the dynamic power dissipated by switching the clock can be given by:

$$P_{clk} = f V_{dd}^{2} (C_L + C_d) \qquad (5.1)$$

where C_L is the total load on the clock and C_d is the clock driver capacitance. Given the total number of clock terminals N, the nominal input capacitance at each terminal, c_g, the unit length wire capacitance, c_w and the chip dimension, D, assuming an H-tree based global clock routing of h levels [2], C_L can be given by:

$$C_L = N c_g + 1.5 (2^h - 1) D c_w + \alpha \sqrt{N 4^h} c_w \qquad (5.2)$$

where the second and third terms are the global and local wiring capacitance respectively, α is an estimation factor depending on the algorithm used for local clock routing [5].

Eq. (5.1) and Eq. (5.2) suggest that the dynamic power dissipated by clock increases as the number of clocked devices and the chip dimensions increase. A global clock may account for up to 40% of the total system power dissipation [7]. For low power clock distribution, measures have to be taken to reduce the clock terminal load, the routing capacitance and the driver capacitance.

Clock skews are the variations of delays from clock source to clock terminals. To achieve the desired performance, clock skews have to be controlled within very small or tolerable values [11]. Clock phase delay, the longest delay from source to sinks, also has to be controlled in order to maximize system throughput [6],[1].

Traditionally, the amount of capacitive load carried by the clock was largely due to the loading capacitance of clock terminals (or sinks). However, as technology advances into deep-submicron, device sizes are shrinking rapidly which reduces the clock terminal capacitances. The increases of chip dimensions also make the interconnect capacitance increasingly important [2]. Moreover, because of the high frequency requirement, performance driven clock tree construction methods such as adjusting wire lengths or widths to reduce clock skew [19],[23] increase the interconnect capacitance to a more dominant part of the total load capacitance on clock. Reducing the interconnect capacitance may significantly reduce the overall system power consumption. On the other hand, low power systems with reduced supply voltages require increasing the device sizes to maintain the necessary speed, i.e. the sizes of clock drivers are increased substantially to ensure fast clock transitions. This exacerbates both the dynamic and short-circuit power dissipated by clock. Therefore, clock distribution is a multi-objective design problem. Minimizing clock power consumption has to be considered together with meeting the constraints of clock skew and phase delay. In the following sections, we address some of the issues and propose some solutions in low power clock distribution.

5.2. Single Driver vs. Distributed Buffers

5.2.1. Clock Driving Schemes

To ensure fast clock transitions, buffers have to be used to drive the large load capacitance on a clock [2],[7]. There are two common clock driving schemes: In the *single driver scheme* as shown in Figure 5.1a, a chain of cascaded buffers with a very large buffer at the end is used at the clock source, no buffers are used elsewhere; in the *distributed buffers scheme* as shown in Figure 5.1b, intermediate buffers are inserted in various parts of the clock tree. At each inser-

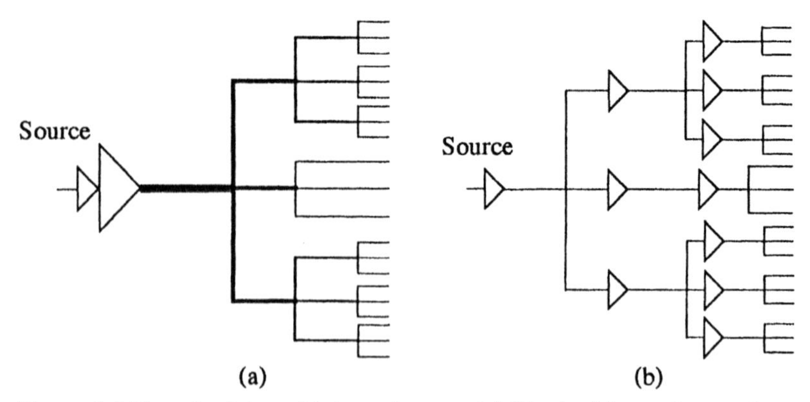

Figure 5.1 Two clock tree driving schemes: (a) Single driver scheme where drivers are at clock source and no buffers else where; (b) Distributed buffers scheme where intermediate buffers are distributed in the clock tree.

tion location, one or more buffers may be cascaded.

The *single driver scheme* has the advantage of avoiding the adjustment of intermediate buffer delays as in the *distributed buffers scheme*. Often in conjunction with this scheme, wire sizing is used to reduce the clock phase delay. Widening the branches that are close to the clock source can also reduce skew caused by asymmetric clock tree loads and wire width deviations [23],[16].

The *distributed buffers scheme* has been recognized to have the advantage that relatively small buffers are used and they can be flexibly placed across the chip to save layout area. Also, for a large clock tree with long path-length, the intermediate buffers (or repeaters) can be used to reduce clock phase delay [6],[2].

Figure 5.2 illustrates the effects of wire widening and intermediate buffer insertion on delay reduction. In high speed design, a long clock path as shown in Figure 5.2a can be treated as a distributed RC delay line. Widening the line as shown in Figure 5.2b will make it a capacitive line with smaller line resistance. By adjusting the sizes of buffers at the source, the line delay will be reduced. The other way to reduce line delay is to insert intermediate buffers along the line as shown in Figure 5.2c. The intermediate buffers partition the line into short segments each of which has small line resistance. This makes the delay of the line more linear with the length. While widening wires requires increasing the sizes of buffers at the source and hence the short-circuit power dissipation, the small

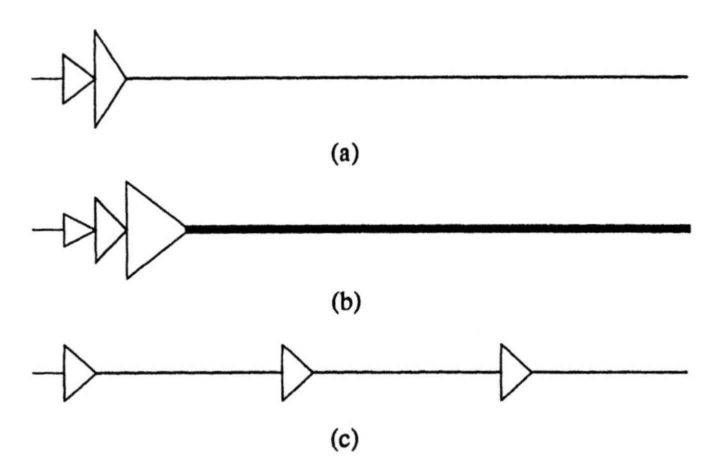

Figure 5.2 (a) A long clock line can be considered as a distributed RC line. Two ways to reduce delay of the long clock line: (b) by wire widening; (c) by intermediate buffers insertion.

intermediate buffers used to drive the short wire segments impose little penalty on power dissipation.

For power minimization in a clock tree, the *distributed buffers scheme* is preferred over the *single driver scheme*. We consider a simple example of an equal path length tree and its delay model as shown in Figure 5.3, where $l_1 = l_2$. The lengths and widths of the tree branches are l_0, l_1, l_2 and w_0, w_1, w_2. The load capacitances at sinks, s_1 and s_2 are C_{L1} and C_{L2} respectively. The skew between s_1 and s_2 can be derived as:

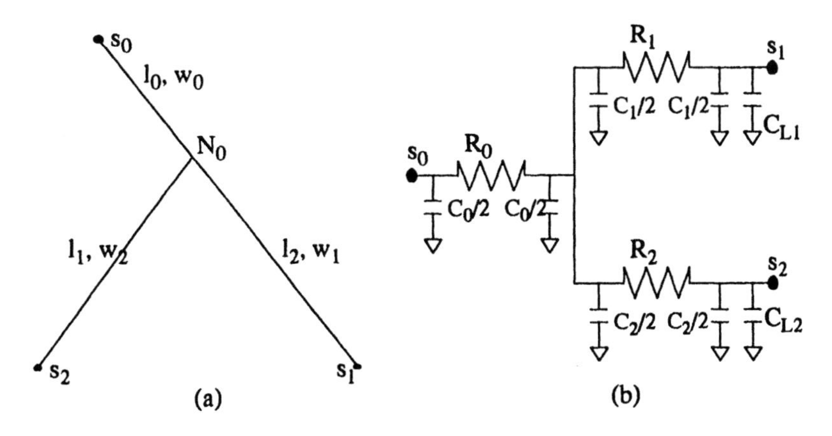

Figure 5.3 (a) An equal path-length clock tree; (b) The delay model.

$$t_s = \frac{rl_1}{w_1}C_{L1} - \frac{rl_2}{w_2}C_{L2} \tag{5.3}$$

where r is the sheet resistance of the wire. The skew variation in terms of wire width variations can be stated as:

$$\Delta t_s = \frac{\partial t_1}{\partial w_1}\Delta w_1 + \frac{\partial t_2}{\partial w_2}\Delta w_2 = -\frac{rl_1 C_{L1}}{w_1^2}\Delta w_1 + \frac{rl_2 C_{L2}}{w_2^2}\Delta w_2 \tag{5.4}$$

Assuming the maximum width variations $\Delta w = \pm 0.15w$, the worst case additional skew is:

$$\Delta t_s = 0.15\left(\frac{rl_1 C_{L1}}{w_1} + \frac{rl_2 C_{L2}}{w_2}\right) \tag{5.5}$$

Eq. (5.3) and Eq. (5.5) indicate that without wire width variations, skew is a linear function of path length. However, with wire width variations, the additional skew is a function of the product of path length and total load capacitance. Increasing the wire widths will reduce skew but result in larger capacitance and power dissipation. Reducing both the path length and load capacitance also reduces skew and minimum wire width can be used such that wiring capacitance is kept at minimum. Therefore, if buffers are inserted to partition a large clock tree into sub-trees with sufficiently short path-length and small loads, the skew caused by asymmetric loads and wire width variations becomes very small.

Additionally, if the chip is partitioned in such a way that clock distributed to subsystems can be disabled (or powered down) according to the subsystem functions, significant amount of power can be saved. This can be done by replacing the buffers with logic gates.

Clock buffers dissipate short-circuit power during clock transitions. The intermediate buffers also add gate and drain capacitance to the clock tree. But compared to the *single driver scheme* which uses large buffers at the clock source, this scheme does not introduce more short-circuit power dissipation if buffers are not excessively inserted and overly sized. In the mean time, both dynamic and short-circuit power dissipation of buffers have to be minimized.

5.2.2. Buffer Insertion in Clock Tree

Different buffer delays cause phase delay variations on different source-to-sink paths. For simplicity, we budget the given tolerable skew of a buffered clock tree, t_s into two components, the tolerable skew for buffer delays, t_s^b and the skew allowed for asymmetric loads and wire width deviations after buffer insertion, t_s^w,

$$t_s = t_s^b + t_s^w \quad (5.6)$$

To balance the source-to-sink delays in a clock tree, we start with an equal path-length clock tree [4],[22]. An equal path-length tree, T consists of a source s_0 and a set of sinks $S = \{s_1, s_2, ..., s_m\}$. The buffer insertion problem is to find the locations on the clock tree to insert intermediate buffers. We call these locations *buffer insertion points(BIPs).*

Formulation 1: *Given an equal path length clock tree T with minimum width for all branches and the tolerable skew constraint,* t_s^w *, the problem of buffer insertion is to determine the minimum number of BIPs in T, such that the skew due to asymmetric loads and wire width variations is less than* t_s^b *.*

To meet the skew constraint, the buffer insertion scheme should try to balance the buffer delays on source-to-sink paths independent of the clock tree topology. Our balanced buffer insertion scheme partitions the clock tree into a number of levels and BIPs are determined at the cut-lines. The resulting clock tree has the following properties: Each source-to-sink path has the same number of buffers(levels); all sub-trees rooted at a given level are equal path-length trees. We select the cut-lines so as to form *iso-radius levels.* An *iso-radius level* of the clock tree is a circle centered at the clock source such that all the paths from the source to the BIPs are of the same length. For a clock tree with path length of L, we choose the radius θ of the first level cut-line (nearest to the clock source) as $\theta = L/(\phi+1)$ where ϕ is the designated number of levels of buffers. The radius is θ for the first iso-radius level, 2θ for the second iso-radius level, ..., $\phi\,\theta$ for the ϕth iso-radius levels. We determine the minimum number of buffer levels ϕ^* that satisfies the skew constraint by iteratively evaluating the worst case skew of the clock tree with $\phi = 1, 2, ...,$ until we find some number of levels, ϕ^* for which the worst case skew due to asymmetric loads and wire width variations is less than d_s^w. The paradigm is depicted in Figure 5.4.

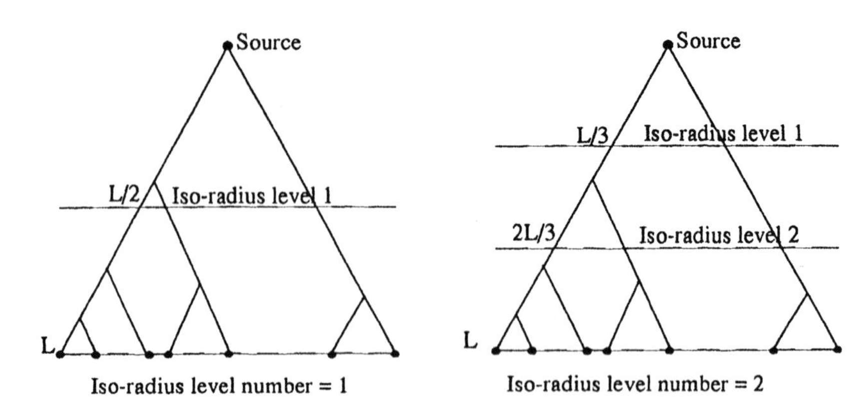

Figure 5.4 Buffer levels are increased until the tolerable skew bound is satisfied.

An example of the buffer insertion scheme is shown in Figure 5.5(a). Previous methods [17] insert buffers level-by-level at the branch split points of the clock tree as shown in Figure 5.5(b). This works well in a full binary tree where all sinks have the same number of levels. In the case of a general equal path-length tree, such as the case in Figure 5.6, different numbers of buffers are inserted on different source to sink paths. Depending on the clock tree topology, some large sub-trees may still require wire widening to reduce skew.

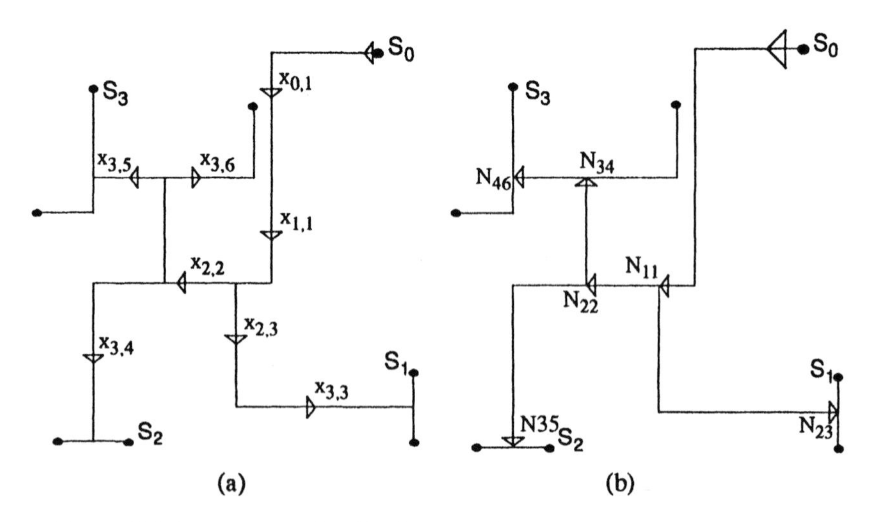

Figure 5.5 An example of buffer insertion in an equal path-length clock tree: (a) using the balanced buffer insertion method; (b) using the level-by-level method.

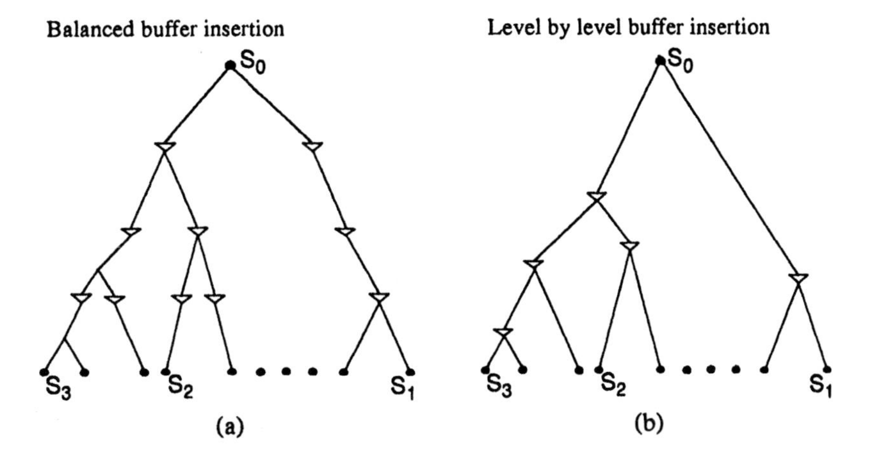

Figure 5.6 Comparison of buffer insertion in a general equal path-length clock tree using: (a) the balanced buffer insertion method; (b) the level-by-level method.

5.3. Buffer and Device Sizing under Process Variations

The sizes of buffers can be further adjusted to reduce power dissipation. We formulate buffer sizing as a constrained optimization problem with power as the minimization objective and skew and phase delay as the constraints.

Formulation 2: *Given a clock tree T with intermediate buffers inserted by the balanced buffer insertion algorithm, the problem of power minimization by buffer sizing(PMBS) is to determine the size of each buffer in T to minimize total power, P_{tot}, subject to the phase delay constraint, d_p and skew constraint, t_s^b:*

$$\max (d_i) \le t_p \tag{5.7}$$

$$\max (d_i - d_j) \le t_s^b \tag{5.8}$$

where d_i and d_j are the phase delays from s_0 to any sinks s_i and s_j respectively.

The above formulation assumes typical process parameters and a fixed PMOS/NMOS transistor ratio for CMOS buffers, i.e. $wp/wn = 2.0$. In reality, MOS device parameters such as carrier mobilities and threshold voltages may vary in a remarkably wide range from die to die for the same process. Moreover, process spread causes PMOS and NMOS device parameters to vary independently from die to die in a chip production environment [18],[2]. This type of process variations are becoming more significant as feature sizes and supply voltages are being scaled down. Additional skew will arise from the buffer delay variations even when delay is balanced with typical process parameters.

The processing spread for a given CMOS technology is usually characterized by extracting three sets of process parameters for PMOS and NMOS devices: fast (or high current) parameters, typical (or medium current) parameters and slow (or low current) parameters. The rise time of a buffer with slow parameter of PMOS can be as many as 2 to 4 times the rise time of a buffer with fast parameter of PMOS. In a chip fabrication environment, a die can have fast PMOS and slow NMOS, or slow PMOS and fast NMOS or other combinations of the three sets of parameters[1]. This type of process variation is different from the device or wire geometry variations which can be overcome by increasing the device sizes or wire widths [17].

The variations of device driving capability can usually be characterized by f_P for PMOS and f_N for NMOS, where $f_N, f_P \geq 1$. If the rise time of a buffer with typical process parameters is T_r, then the rise time with fast and slow parameters, T_r^f and T_r^s respectively, are:

$$T_r^f = \frac{T_r}{f_P}, T_r^s = T_r f_P \tag{5.9}$$

Similarly, if the fall time of a buffer with typical process parameters is T_f, then the fall time with fast and slow parameters, T_f^f and T_f^s respectively, are:

$$T_f^f = \frac{T_f}{f_N}, T_f^s = T_f f_N \tag{5.10}$$

Figure 5.7 demonstrates the process variation effects on phase delays. In Figure 5.7(a), the paths from the source to two different sinks x_1 and x_2 have the same number of buffers but with different sizes. Let d_1 be the delay with typical process parameters from source to x_1, which is the sum of the three buffer delays d_{11}, d_{12} and d_{13} on that path, d_2 is the delay with typical process parameters from source to x_2. We have $d_1 = d_{11} + d_{12} + d_{13}$ and $d_2 = d_{21} + d_{22} + d_{23}$. If the delays of two paths are balanced with typical process parameters,

$$d_1 = d_{11} + d_{12} + d_{13} = d_{21} + d_{22} + d_{23} = d_2 \tag{5.11}$$

[1]. Here we assume that parameter variations of PMOS transistors in the same die are negligible, same for NMOS.

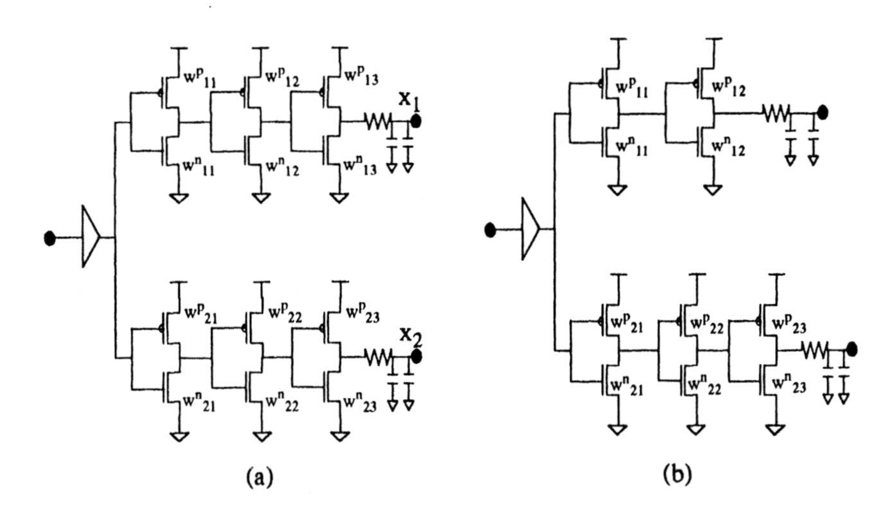

Figure 5.7 Delay variations due to process parameter variations.
(a) Two paths with equal number of buffers but buffers are of different sizes;
(b) Two paths with unequal number of buffers.

In case PMOS devices have fast parameters and NMOS devices have slow parameters and the input is switching from logic 0 to logic 1, the delays become:

$$d_1' = f_N d_{11} + d_{12}/f_P + f_N d_{13} \tag{5.12}$$

$$d_2' = f_N d_{21} + d_{22}/f_P + f_N d_{23} \tag{5.13}$$

Even when $d_{11} \neq d_{21}$ or $d_{12} + d_{13} \neq d_{22} + d_{23}$, we could have $d_1 = d_2$. But $d_1' \neq d_2'$.

This problem gets more serious if two paths have different number of buffers as in Figure 5.7(b) [18].

One solution to this problem is to separately balance the delays through PMOS and NMOS devices in the two paths [18]. If $d_{11} + d_{13} = d_{21} + d_{23}$ and $d_{12} = d_{22}$, then,

$$d_1' = f_N(d_{11} + d_{13}) + d_{12}/f_P = f_N(d_{21} + d_{23}) + d_{23}/f_P = d_2' \tag{5.14}$$

which means $d_1' = d_2'$ regardless of the variations of either PMOS or NMOS parameters. Therefore, by separately equalizing the delays through PMOS devices (or the pull-up path) and the delays through NMOS devices (or the pull-down path) on different paths, this type of process variation effects can be eliminated.

Let d_i^p, d_j^p be the pull-up path delays and d_i^n, d_j^n be the pull-down path delays from clock source to any two sinks s_i, s_j. Then the phase delays with typical process parameters from source to s_i and s_j are $d_i = d_i^p + d_i^n$ and $d_j = d_j^p + d_j^n$.

If the pull-up and pull-down path delays with typical process parameters satisfy:

$$d_i^p - d_j^p \le \varepsilon^p,\ d_i^n - d_j^n \le \varepsilon^n \tag{5.15}$$

where the pull-up path skew and pull-down path skew,

$$\varepsilon^p = \frac{t_s}{2f_P},\ \varepsilon^n = \frac{t_s}{2f_N} \tag{5.16}$$

then in the worst case when both PMOS and NMOS have slow parameters, $d_i = f_P d_i^p + f_N d_i^n$, $d_j = f_P d_j^p + f_N d_j^n$, the skew constraint, t_s, can still be satisfied:

$$d_i - d_j = f_P (d_i^p - d_j^p) + f_N (d_i^n - d_j^n) \le f_P \varepsilon^p + f_N \varepsilon^n = t_s \tag{5.17}$$

Therefore, a buffer sizing problem can be formulated to overcome the effects of process variations on skews and phase delays.

Formulation 3: *Given a buffered clock tree T with intermediate buffers, the problem of power minimization by device sizing (PMDS) while satisfying delay and skew constraints irrespective of process parameter variations is to determine the sizes of PMOS and NMOS devices of each buffer on T, such that, the total power dissipation, P_{tot} is minimized, subject to the phase delay constraint, t_p, pull-up path and pull-down path skew constraints, ε^p and ε^n:*

$$\max(d_i) \le t_p \tag{5.18}$$

$$\max(d_i^p - d_j^p) \le \varepsilon^p,\ \max(d_i^n - d_j^n) \le \varepsilon^n \tag{5.19}$$

where d_i^p and d_j^p are the pull-up path delays on the paths from s_0 to sinks s_i and s_j respectively; d_i^n and d_j^n are the pull-down path delays on the paths from s_0 to s_i and s_j respectively.

The above formulated buffer and device sizing problems, PMBS and PMDS can be transformed into posynomial programming problems for which the global minima can be obtained if a local minima is found [20].

Some experiments were done to construct buffered clock trees with the balanced buffer insertion algorithm. Buffer and device sizing were also done to minimize the total power dissipation. Table 5.1 gives the power dissipation results for four benchmarks. Significant power reduction can be seen when compared to an optimal wire sizing method which minimizes skew [23].

	Dynamic (W)		Short-circuit (mW)		Power Reduction (%)
	WS	BIS	WS	BIS	
Example1	1.334	0.389	75.8	42.9	326
Example2	1.623	0.792	96.3	48.4	200
Primary1	4.383	1.327	209.5	112.3	306
Primary2	4.578	2.218	458.3	241.2	200

Table 5.1 Comparisons of power dissipation in clock distribution. Buffer Insertion and Sizing method versus the Wire Sizing method.

In all the four examples, the parameters of a 0.65 μm CMOS process were used. The width of all branches was chosen as 1 μm as the result of buffer insertions. A 3.3 V supply voltage is used. The clock tree results are summarized in Table 5.2

Examples	Frequency (MHz)	Buffer levels	# of buffers	Skew (ns)
Example1	200	5	62	0.07
Example2	100	4	24	0.02
Primary1	300	6	86	0.23
Primary2	200	7	148	0.67

Table 5.2 Four clock tree examples, Primary 1 and Primary 2 are based on MCNC benchmarks.

5.4. Zero Skew vs. Tolerable Skew

Much research has been done in the area of performance driven clock distribution, mainly on the construction of clock trees to minimize clock skew [11]. Most techniques used for skew minimization are based on adjusting the interconnect lengths and widths: the *length adjustment technique* moves the balance points or elongates the interconnect length to achieve zero skew [19],[4],[22][2]; the *width sizing technique* achieves zero skew by adjusting the widths of wires in the clock tree [23],[16]. For power consideration, these techniques tend to be pessimistic in that they assume the skew between every pair of clock sinks has to be limited by the same amount. Due to the attempt to achieve the minimum skew for all clock sinks, wire lengths or widths of the clock tree may be increased substantially resulting in increased power dissipation.

Here, we discuss an alternative approach by taking advantage of tolerable skews. Tolerable skews are the maximum values of clock skew between each pair of clock sinks with which the system can function correctly at the desired frequency.

5.4.1. Derivation of Tolerable Skew

We use Figure 5.8 to model the operation of a synchronous digital system and to illustrate the concept of tolerable skew. A typical synchronous system such as one with a pipelined/parallel architecture can be considered as being formed by the basic building blocks shown in Figure 5.8. The Figure shows the sequential operation of the registers with one single-phase clock, assuming edge-triggered flip-flops are used. The flip-flops are characterized with setup time, d_{Setup}, hold time, d_{Hold} and flip-flop delay d_{FF}. The signal D_{01} is the input to the flip-flop ff_{01} and D_{02} the input of flip-flop ff_{02}, which is the output of the combinational logic. There are some phase delays due to interconnection delays between the clock source C_0 and C_{01} and C_{02} which are clock terminals of ff_{01} and ff_{02} respectively. Let d_{01} and d_{02} denote the arrival time of C_{01} and C_{02}. The combinational logic block is characterized with maximum delay $MAX(d_{logic})$ and minimum delay, $MIN(d_{logic})$, due to the variations of input values and input to output path delays through the combinational logic.

Figure 5.9 illustrates two cases of correct synchronous operations with tolerable skews. In Figure 5.9(a), the clock arrives at C_{02} later than the previous

---

2. Zero-skew in the sense that phase delays of all sinks calculated with a delay model, i.e. Elmore delay model, are equal under ideal process condition.

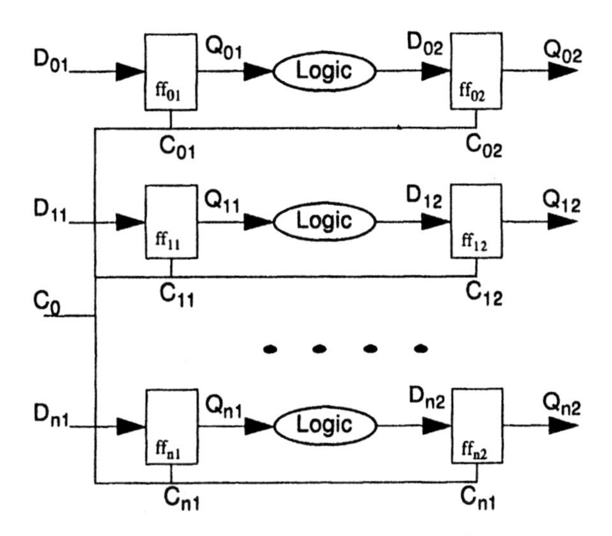

Figure 5.8 The basic building block in a parallel/pipelined synchronous system.

stage clock terminal C_{01} resulting in a negative clock skew. In Figure 5.9(b), the clock arrives at C_{02} earlier than C_{01} causing a positive clock skew. In both cases, the synchronous operation is correct since the correct data is latched in the next-stage flip-flop, ff_{02}. Thus in both cases the skews are tolerable.

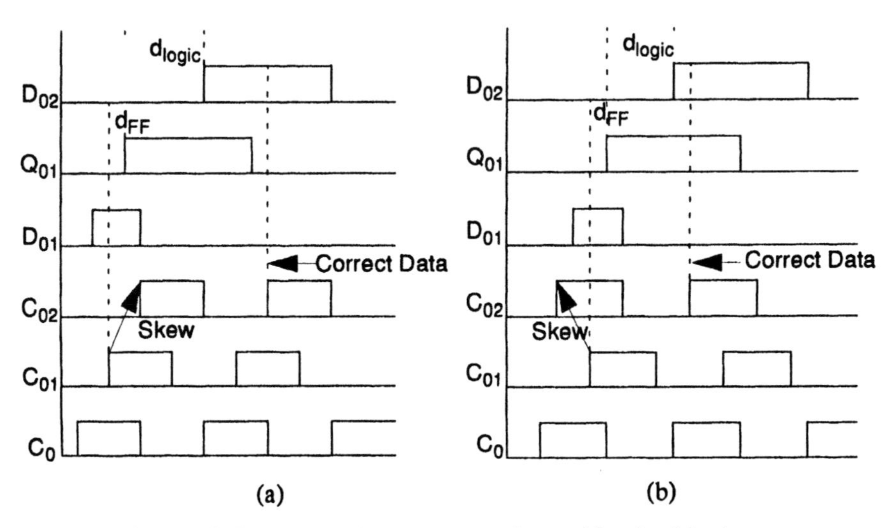

Figure 5.9 Correct synchronous operations with tolerable skews: (a) with negative clock skew; (b) with positive tolerable skew.

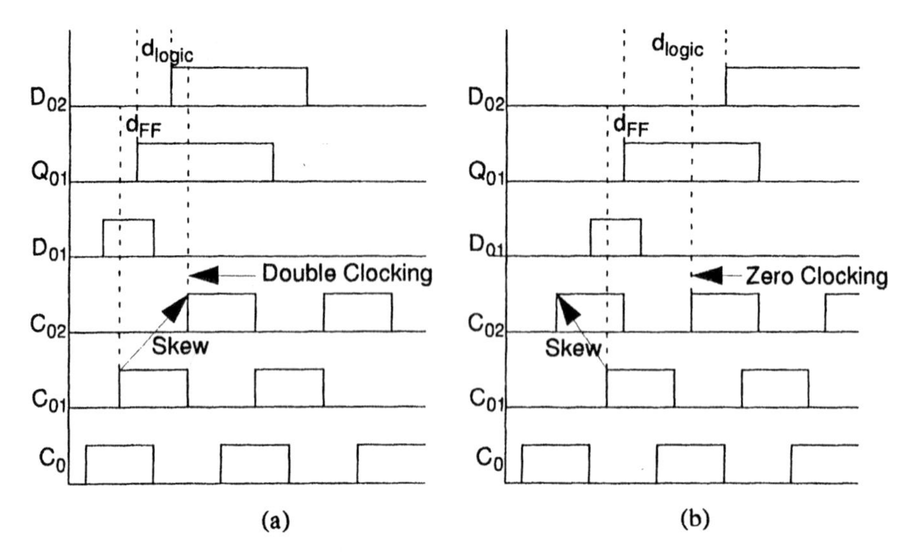

Figure 5.10 Incorrect synchronous operations with excessive clock skews: (a) Double-clocking with negative skew; (b) Zero-clocking with positive skew.

However, with excessive clock skew, incorrect operations may occur as shown in Figure 5.10. In the case shown in Figure 5.10(a), a phenomenon called *double-clocking* is caused by the late arrival of the clock signal at C_{02}. At the first clock transition, data is latched in by the first flip-flop, ff_{01}. The second stage flip-flop, ff_{02} latches in the data produced by the combinational logic at the same clock transition while missing the data at the next clock transition. In the case shown in Figure 5.10(b), a phenomenon called *zero-clocking* occurs when the clock arrives at C_{02} too early and the second stage flip-flop, ff_{02} is unable to latch in the data produced by the combinational logic at the next clock transition. To avoid these clock hazards under a given clock cycle time, P, the clock timing has to satisfy the following constraints [8],[15].

To avoid *double-clocking*:

$$d_{01} + d_{FF} + MIN(d_{logic}) \geq d_{02} + d_{Hold} \tag{5.20}$$

To avoid *zero-clocking*:

$$d_{01} + d_{FF} + MAX(d_{logic}) + d_{Setup} \leq d_{02} + P \tag{5.21}$$

From these constraints, the values of tolerable skew between clock sinks C_{01} and C_{02} can be derived:

- *Negative tolerable skew*, $d_{01} \le d_{02}$:

$$Skew_{tol} = d_{02} - d_{01} \le d_{FF} + MIN(d_{logic}) - d_{Hold} \tag{5.22}$$

- *Positive tolerable skew*, $d_{01} \ge d_{02}$:

$$Skew_{tol} = d_{01} - d_{02} \le P - (d_{FF} + MAX(d_{logic}) + d_{Setup}) \tag{5.23}$$

If level-sensitive latches are used instead of edge triggered flip-flops, the sequential operations of registers will allow larger positive tolerable skews and smaller negative skews [2]. Given the desired clock cycle time, P, clock pulse width, P_{pulse}, latch delay, d_{Latch}, latch setup time, d_{Setup} and hold time, d_{Hold},

- *Negative tolerable skew with latches*, $d_{01} \le d_{02}$:

$$Skew_{tol} = d_{02} - d_{01} \le d_{Latch} + MIN(d_{logic}) - d_{Hold} - P_{pulse} \tag{5.24}$$

- *Positive tolerable skew with latches*, $d_{01} \ge d_{02}$:

$$Skew_{tol} = d_{01} - d_{02} \le P + P_{pulse} - (d_{Latch} + MAX(d_{logic}) + d_{Setup}) \tag{5.25}$$

For a clock distributed to the parallel flip-flops or latches, such as ff_{01} and ff_{11} in Figure 5.9, the maximum tolerable skew between C_{01} and C_{11} can also be determined based on P, d_{Setup} and d_{Hold} [21].

5.4.2. A Two-level Clock Distribution Scheme

Given the desired clock frequency, the setup time, hold time, the delay of flip-flops and the maximum and minimum delays of the combinational logic between each pair of flip-flops, the tolerable skew between each pair of clock sinks is well defined. In designing a low power system, tolerable skew instead of minimum skew should be used during clock tree construction.

In high-performance VLSI processes with four or five metal layers, one layer, possibly the one with the smallest RC parameters can be dedicated to clock distribution. Placing clock tree on a single metal layer reduces delays and the attenuations caused by via's and decreases the sensitivity to process induced wire or via variations. The clock wiring capacitance is also substantially reduced. However, it is not always practical to embed the entire clock tree on a single layer.

We propose a two-level clock distribution scheme. Tolerable skew differs from one pair of clock sinks to another as logic path delays vary from one combinational block to another. The clock sinks that are close to each other and have very small tolerable skews among them are grouped into clusters. A global level clock tree connects the clock source to the clusters and is routed on a single layer with the smallest RC parameters by a planar routing algorithm [22]. For clock sinks that are located close to each other, tolerable skews among them can be easily satisfied. Little savings within a local cluster can be gained if the sinks within the cluster have large tolerable skews. Local trees may be routed on multiple layers since the total wiring capacitance inside each cluster is very small and has less impact on total power. The tolerable skews between two clusters can be determined from the smallest tolerable skew between a clock sink in one cluster and a clock sink in the other cluster. During clustering, we maximize the tolerable skews between clusters. This will give the global level clock tree construction more opportunity to reduce wire length, save buffer sizes, and reduce power consumption since the global level clock tree has much more impact on power.

The buffer insertion and sizing method described in the previous section can be used at the global level to minimize wire width and meet tolerable skew constraints. By taking advantage of the more relaxed tolerable skews instead of zero-skews, we can further reduce power dissipation. The buffer sizing problem (PMBS) and device sizing problem (PMDS) can be reformulated by replacing the minimum tolerable skew value, d_s, with the tolerable skew value for each pair of clock sinks, $t_{i,j}$. The constraint in **Formulation 2** becomes:

$$d_i - d_j \leq t_{i,j} \tag{5.26}$$

Accordingly, we set:

$$\varepsilon_{i,j}^{p} = \frac{t_{i,j}}{2f_P},\ \varepsilon_{i,j}^{n} = \frac{t_{i,j}}{2f_N} \tag{5.27}$$

the constraints in **Formulation 3** become:

$$d_i^p - d_j^p \leq \varepsilon_{i,j}^{\ p},\ d_i^n - d_j^n \leq \varepsilon_{i,j}^{\ n} \tag{5.28}$$

The new PMBS or PMDS problems remain constrained optimization problems which can again be solved by the posynomial programming approach used in [20].

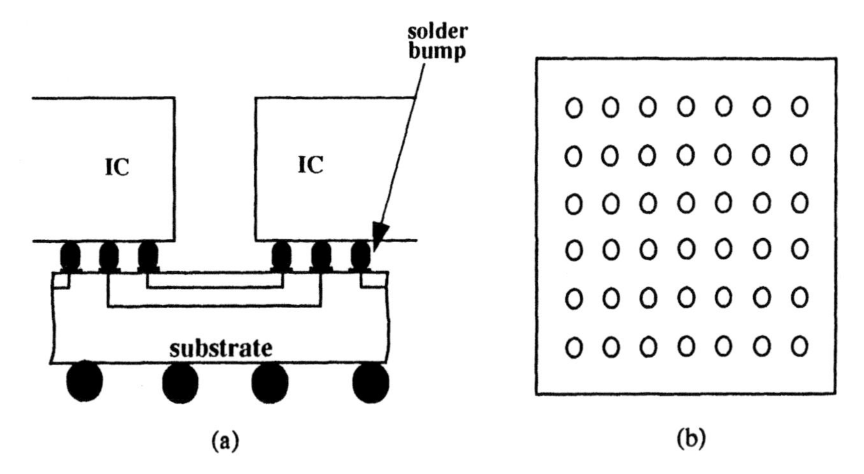

Figure 5.11 Flip-chip technology: (a) Direct attachment to substrate; (b) Area IO solder bumpers on chip surface.

5.5. Chip and Package Co-Design of Clock Network

With ever increasing IO counts, more and more chips will become "pad limited". At the same time, the simultaneous switching noise due to the wire bond inductance will limit the chip performance. This is more apparent to low power systems as supply voltage is scaled down to 3 V or less.

Area-IO provides an immediate and lasting solution to the IO bottleneck on chips. In area-IO, the IO pads are distributed over the entire chip surface rather than being confined to the periphery as shown in Figure 5.11(b). The flip-chip technology makes area IO scheme feasible. In this technology, the dice or bare chips are attached with pads facing down and solder bumps form the mechanical and electrical connections to the substrate as shown in Figure 5.11(a). Compared with the wire bonding and TAB, flip-chip has the highest IO density, smallest chip size, and lowest inductance.

With high density area-IO and low inductance solder bumps, it is possible to place global clock distribution on a dedicated layer off the chip, either on the chip carriers of single chips as shown in Figure 5.12 or on the substrates of multi-chip modules. This two-level clock-distribution scheme is depicted in Figure 5.13. The package level interconnect layer can be made to have far smaller RC parameters. This can be seen from the interconnect scaling properties.

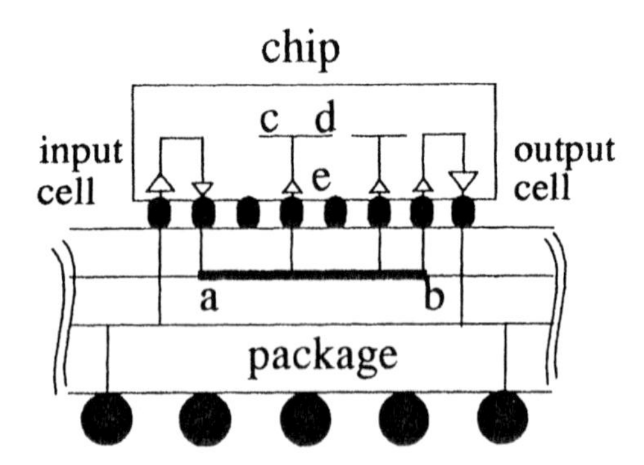

Figure 5.12 Chip and package co-design of interconnects.

For a multi-layer embedded microstrip-line, the capacitance per unit length and the inductance per unit length are scale invariant [9]:

$$C \approx (\varepsilon w) / t_i, L \approx (\mu t_i) / w \tag{5.29}$$

where w is the line width, t_i the dielectric thickness, ε the dielectric permittivity, and μ the dielectric permeability. Therefore, if we uniformly scale the interconnect cross-section of a line, the capacitance per unit length and inductance per unit length will remain the same. However, the resistance per unit length is inversely proportional to the area of the cross-section:

$$R \approx \rho / w t_m \tag{5.30}$$

where ρ is the metal resistivity, and t_m the metal thickness.

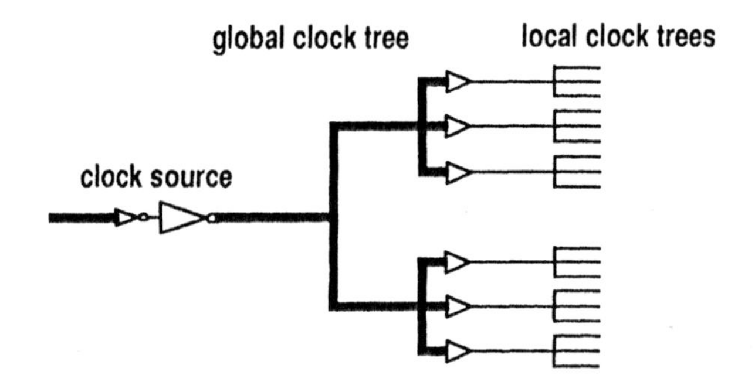

Figure 5.13 Two-level clock tree using chip and package co-design.

The interconnect scaling properties imply that with all other factors the same, the thicker film results in smaller signal delay and lower power. An experiment was designed to verify the benefits of the new chip and package co-design schemes for clock networks. The chip is designed to use a 3.0 V supply voltage and is 20×20 mm^2 in size. The clock net connects to 13440 loads. We redesigned the clock tree by putting the first level tree on the package layer. The package is a 40 × 20 mm^2 plastic ball grid array (BGA) using flip-chip assembly. HSpice simulation results showed the delays from the clock driver to loads are significantly reduced. The line capacitance on the package is 10 times less than that on the chip and the line resistance on the package is 500 times smaller than that on chip. In addition, by removing the global clock distribution from the chip, we can compact the layout and reduce the chip size. Furthermore, the clock driver can be scaled down from two parallel buffers to one buffer, since the *RC* delay of the clock net is significantly reduced. As supply voltage is scaled down, the logic level becomes sensitive to crosstalk noise between signals, especially noise generated by fast switching (or noisy) signals such as clocks. Moving the global clock off the chip will significantly improve the reliability of a low power system. Also saving on-chip wiring reduces the on-chip power dissipation.

This clock distribution scheme is also applicable to the design of MCMs. Figure 5.14 illustrates a two-level clock routing scheme for MCMs. We partition

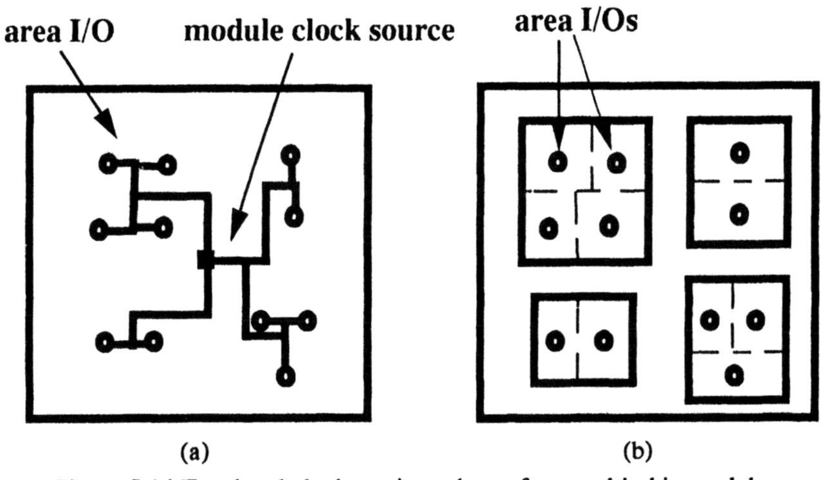

Figure 5.14 Two level clock routing scheme for a multi-chip module. (a) global clock tree on the substrate route module clock source to area clock pads; (b) A die is partitioned into isochronous bins with area pads as the roots of local clock trees.

each die into a number of regions, called *isochronous bins*, and assign an area IO as the root of the local clock tree in each bin. The global clock tree on the substrate connects the area IOs of local clock trees. MCMs bring chips much closer together so that the distances of clock distribution to multiple chips are shortened significantly thus reducing delay and total capacitance as well as power dissipation. Figure 5.15 shows the significant reduction in interconnect capacitance using silicon-on-silicon MCM with flip-chip assembly.

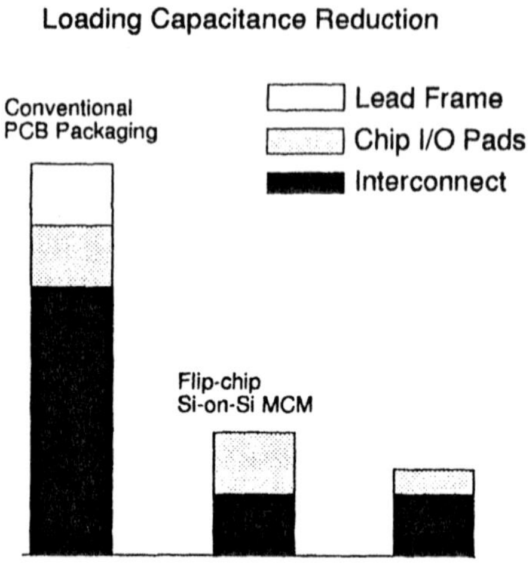

Loading Capacitance Reduction

Conventional PCB Packaging

Lead Frame
Chip I/O Pads
Interconnect

Flip-chip Si-on-Si MCM

Figure 5.15 Loading capacitance in flip-chip silicon-on-silicon MCM, IO pads optimized silicon-on-silicon MCM compared to conventional PCB packaging.

At present, noise considerations limit the speed of the IO buffers. Taking advantages of the low loading capacitance of MCM interconnect and the negligible series inductance of flip-chip attachment, AT&T has redesigned low voltage differential driver and receiver circuits [12]. Shown in Figure 5.16, the driver circuit uses a separate low-voltage power supply (V_{PWR}). The circuit uses n-channel pull-up and pull-down transistors to take advantage of their higher transconductance. The receiver circuit shown in Figure 5.16 is designed to restore the signals to the chip's regular operating level (V_{DD}) without dissipating excessive quiescent power. To improve the speed of the receiver, two p-channel pull-up transistors connected to V_{DD} are added to a cross-coupled latch. Interconnects which are internal to MCMs use low-voltage differential interconnection. For area IOs that do not connect to external pins, little electrostatic discharge (ESD) protection is needed. The active area required for the output driver is about one tenth that of

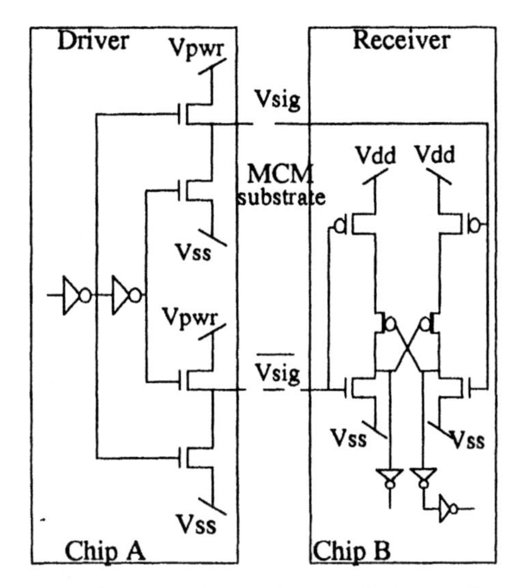

Figure 5.16 Low-voltage differential driver and receiver circuits.

conventional buffers. The optimized IO pads improve the speed by a factor of three and reduce the power consumption by a factor of six. Figure 5.16 also shows a further reduction of load capacitance due to the optimized IO pads.

5.6. Summary

Clock distribution is critical for determining system performance and power dissipation. In this chapter, several issues involved in low power clock distribution have been addressed. To minimize the power dissipated by clock, the *distributed buffering scheme* should be used instead of the *single driver scheme* because of its advantages in reducing wiring capacitance and driver power dissipation. In using the *distributed buffering scheme*, the process variation induced skews should be taken into account and an optimization method can be used to remedy this problem. For low power systems, tolerable skew instead of zero-skew should be considered as the goal of clock distribution to relax the delay balance constraints and reduce total capacitance. We can also apply a two-level clock distribution scheme by taking advantage of the flip-chip packaging and area-pad technology. By placing the global level clock tree on to the package level and lock clock trees on the chip, the clock delay and power can be substantially reduced in either a single-chip or an MCM system. Moreover, low-voltage differential driver and receiver circuits can be designed when using the flip-chip packaging technology to further reduce loading capacitance and power dissipation.

Acknowledgment

The authors would like to acknowledge Qing Zhu for his contributions on the chip/package co-design of clock network and his work on the planar equal path-length clock tree routing. The authors would also like to thank David Staepelaere for reviewing the manuscript and his valuable suggestions. This work was supported in part by the National Science Foundation Presidential Young Investigator Award under grant MIP-9058100 and in part by the Semiconductor Research Corporation under grant SRC-93-DJ-196.

References

[1] M. Afghahi and C. Svensson, "Performance of synchronous and asynchronous schemes for vlsi systems," In *IEEE Transactions on Computers*, pages 858–871, 1992.

[2] H. Bakoglu. *Circuits, Interconnections, and Packaging for VLSI*. Addison-Wesley Publishing Company, 1987.

[3] K-Y. Chao and D.F. Wong, "Low power considerations in floorplan design," In *Proc. Intl. Workshop on Low Power Design.*, pages 45–50, 1994.

[4] T.H. Chao, Y. Hsu, J. Ho, K. Boese, and A. Kahng, "Zero skew clock net routing," *IEEE Transactions on Circuits and Systems*, 39(11):799–814, November 1992.

[5] J.D Cho and M. Sarrafzadeh, "A buffer distribution algorithm for high-performance clock net optimization," *IEEE Transactions on VLSI Systems*, 3(1):84–97, March 1995.

[6] S. Dhar, M. A. Franklin, and D. F. Wann, "Reduction of clock delays in VLSI structures," In *Digest of Tech. Papers of IEEE Intl. Conf. on Custom Integrated Circuits*, pages 778–783, 1984.

[7] D. Dobberpuhl and R. Witek, "A 200mhz 64b dual-issue CMOS microprocessor," in *Proc. IEEE Intl. Solid-State Circuits Conf.*, pages 106–107, 1992.

[8] J. P. Fishburn, "Clock skew optimization," *IEEE Transactions on Computers*, 39(7):945–951, 1990.

[9] R. C. Frye, "Physical scaling and interconnection delay in multichip modules," *IEEE Transactions on Components, Packaging, And Manufacturing Technology. Part B: Advanced Packaging*, 17(1):30–37, February 1994.

[10] V. Hirendu and M. Pedram. Pcube, "A performance driven placement algorithm for lower power designs," In *Proc. of Euro-DAC*, pages 72–77, 1993.

[11] M. A. B. Jackson, A. Srinivasan, and E. S. Kuh, "Clock routing for high-performance ics," In *Proc. of 27th Design Automation Conf.*, pages 573–579, 1990.

[12] S.Knauer T.Gabara, W.Fischer. An i/o buffer set for silicon multichip modules. In *Proceedings of IEEE Multi-Chip Module Conference*, pages 147–152, 1993.

[13] C-K. Koh and J. Cong. Simultaneous driver and wire sizing for performance and power optimization. *IEEE Trans. on VLSI Systems*, 2(4):408–425, 1994.

[14] S. Kurosawa, M. Yamada, R. Nojima, N. Kojima, "A 200mhz 64b dual-issue CMOS microprocessor," In *Proc. IEEE Symp. on Low Power Electronics.*, pp. 50–51, 1994.

[15] J.L. Neves and E. Friedman, "Design methodology for synthesizing clock distribution networks exploiting non-zero localized clock skew," *To appear in IEEE Transaction on VLSI Systems*, 1995.

[16] S. Pullela, N. Menezes, J. Omar, and L. Pillage, "Skew and delay optimization for reliable buffered clock trees," in *Digest of Tech. Papers of IEEE Intl. Conf. on Computer Aided Design*, pages 556–562, 1993.

[17] S. Pullela, N. Menezes, and L. Pillage, "Reliable non-zero skew clock trees using wire width optimization," In *Proc. of 30th ACM/IEEE Design Automation Conference*, pp. 165–170, 1993.

[18] M. Shoji, "Elimination of process-dependent clock skew in CMOS VLSI," *IEEE Journal of Solid-State Circuits*, sc-21(1):875–880, 1986.

[19] K.S.Tsay, "An exact zero-skew clock routing algorithm," *IEEE Trans. on Computer-Aided Design*, 12(3):242–249, 1993.

[20] J. Xi and W. Dai, "Buffer insertion and sizing under process variations for low power clock distribution," *To appear in the 32nd Design Automation Conference, San Francisco.*, June 1995.

[21] J. Xi and W. Dai, "Clocking with tolerable skews for low power design," In *Technical Report, UCSC-CRL-95-21, University of California, Santa Cruz.*, 1995.

[22] D. Zhu and W. Dai, "Perfect-balance planar clock routing with minimal path-length. In *Digest of Tech. Papers of IEEE Intl. Conf. on Computer Aided Design, pages 473-476.*, Nov, 1992.

[23] D. Zhu, W. Dai and J. Xi, "Optimal sizing of high speed clock networks based on distributed rc and transmission line models," In *Digest of Tech. Papers of IEEE Intl. Conf. on Computer Aided Design, pages 628-633.*, Nov, 1993.

PART II

Logic and Module Design Levels

p.128
NH

Logic Synthesis for Low Power

Massoud Pedram

6.1. Introduction

Low power, yet high-throughput and computationally intensive, circuits are becoming a critical application domain. One driving factor behind this trend is the growing class of personal computing devices (portable desktops, audio- and video-based multimedia products) and wireless communications systems (personal digital assistants and personal communicators) which demand high-speed computations and complex functionalities with low power consumption. Another crucial driving factor is that excessive power consumption is becoming the limiting factor in integrating more transistors on a single chip or on a multiple-chip module. Unless power consumption is dramatically reduced, the resulting heat will limit the feasible packing and performance of VLSI circuits and systems. Furthermore, circuits synthesized for low power are also less susceptible to run time failures.

To address the challenge to reduce power, the semiconductor industry has adopted a multifaceted approach, attacking the problem on four fronts:

1. **Reducing chip and package capacitance:** This can be achieved through process development such as SOI with fully depleted wells, process scaling to submicron device sizes, and advanced interconnect

substrates such as Multi-Chip Modules (MCM). This approach can be very effective but is expensive.

2. **Scaling the supply voltage:** This approach can be very effective in reducing the power dissipation, but often requires new IC fabrication processing. Supply voltage scaling also requires support circuitry for low-voltage operation including level-converters and DC/DC converters as well as detailed consideration of issues such as signal-to-noise.
3. **Using power management strategies:** The power savings that can be achieved by various static and dynamic power management techniques are very application dependent, but can be significant.
4. **Employing better design techniques:** This approach promises to be very successful because the investment to reduce power by design is relatively small in comparison to the other two approaches and because it is relatively untapped in potential.

Low power VLSI design can be achieved at various levels of the design abstraction from algorithmic and system levels down to layout and circuit levels (see Figure 6.1). In the following, some of these optimization techniques will be briefly mentioned.

At the system level, inactive hardware modules may be automatically turned off to save power; Modules may be provided with the optimum supply voltage and interfaced by means of level converters; Some of the energy that is delivered from the power supply may be cycled back to the power supply; A given task may be partitioned between various hardware modules or programma-

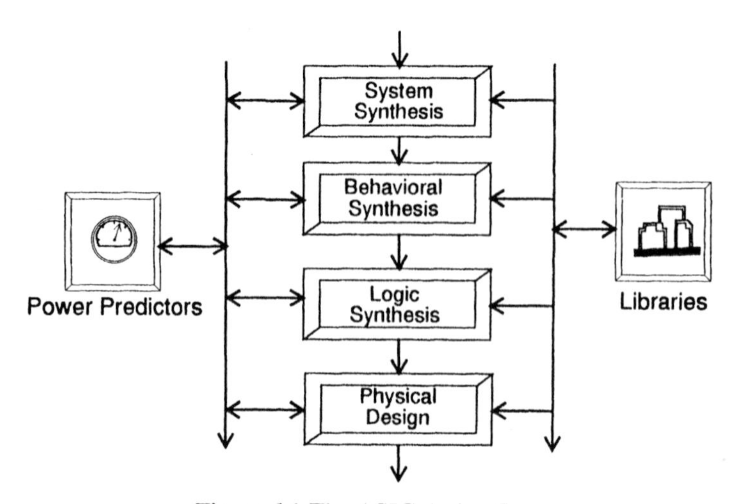

Figure 6.1 The ASIC design flow

ble processors or both so as to reduce the system-level power consumption.

At the architectural (behavioral) design level, concurrency increasing transformations such as loop unrolling, pipelining and control flow optimization as well as critical path reducing transformations such as height minimization, retiming and pipelining may be used to allow a reduction in supply voltage without degrading system throughput; Algorithm-specific instruction sets may be utilized that boost code density and minimize switching; A Gray code addressing scheme can be used to reduce the number of bit changes on the address bus; On-chip cache may be added to minimize external memory references; Locality of reference may be exploited to avoid accessing global resources such as memories, busses or ALUs; Control signals that are "don't cares" can be held constant to avoid initiating nonproductive switching.

At the register-transfer (RT) and logic levels, symbolic states of a finite state machine (FSM) can be assigned binary codes to minimize the number of bit changes in the combinational logic for the most likely state transitions; Latches in a pipelined design can be repositioned to eliminate hazardous activity in the circuit: Parts of the circuit that do not contribute to the present computation may be shut off completely; Output logic values of a circuit may be precomputed one cycle before they are required and then used to reduce the internal switching activity of the circuit in the succeeding clock cycle; Common sub-expressions with low transition probability values can be extracted; Network don't cares can be used to modify the local expression of a node so as to reduce the bit switching in the transitive fanout of the node; Nodes with high switching activity may be hidden inside CMOS gates where they drive smaller physical capacitances; Power dissipation may be further reduced by gate resizing, signal-to-pin assignment and I/O encoding.

At the physical design level, power may be reduced by using appropriate net weights during netlist partitioning, floorplanning, placement and routing; Individual transistors may be sized down to reduce the power dissipation along the non-critical paths in a circuit; Large capacitive loads can be buffered using optimally sized inverter chains so as to minimize the power dissipation subject to a given delay constraint; Wire and driver sizing may be combined to reduce the interconnect delay with only a small increase in the power dissipation; Clock trees may be constructed that minimize the load on the clock drivers subject to meeting a tolerable clock skew.

At the circuit level, power savings techniques that recycle the signal energies using the adiabatic switching principles rather than dissipating them as heat are promising in certain applications where speed can be traded for lower power. Similarly, techniques based on combining self-timed circuits with a mechanism for selective adjustment of the supply voltage that minimizes the power while satisfying the performance constraints show good signs. Techniques based on partial transfer of the energy stored on a capacitance to some charge sharing capacitance and then reusing this energy at a later time. Design of energy efficient level-converters and DC/DC converters is also essential to the success of adaptive supply voltage strategies.

The design for low power problem cannot be achieved without accurate power prediction and optimization tools or without power efficient gate and module libraries (see Figure 6.1). Therefore, there is a critical need for CAD tools to 1) estimate power dissipation during the design process to meet the power budget without having to go through a costly redesign effort, 2) enable efficient design and characterization of the design libraries and 3) reduce the power dissipation using various transformations and optimizations.

The remainder of this chapter describes the CAD methodologies and techniques required to effect power efficient design at the logic level. In Section 6.2, various power simulation and estimation techniques are reviewed. In Section 6.3, state of the art in sequential and combinational logic synthesis targeting low power consumption will be described. Concluding remarks will be given in Section 6.4.

6.2. Power Estimation Techniques

6.2.1. Sources of Power Dissipation

Power dissipation in CMOS circuits is caused by three sources: 1) the leakage current which consists of reverse bias current in the parasitic diodes formed between source and drain diffusions and the bulk region in a MOS transistor as well as the subthreshold current that arises from the inversion charge that exists at the gate voltages below the threshold voltage, 2) the short-circuit (rush-through) current which is due to the DC path between the supply rails during output transitions and 3) the charging and discharging of capacitive loads during logic changes.

The short-circuit and subthreshold currents in CMOS circuits can be made small with proper circuit and device design techniques. The dominant source of power dissipation is thus the charging and discharging of the node capacitances (also referred to as the dynamic power dissipation) and is given by:

$$P = \frac{V_{dd}^2}{2T_{cycle}} \sum_g C_g E_g(sw) \tag{6.1}$$

where V_{dd} is the supply voltage, T_{cycle} is the clock cycle time, C_g is the capacitance seen by gate g and $E_g(sw)$ (referred as the *switching activity*) is the expected number of transitions at the output of g per clock cycle. The product of C_g and $E_g(sw)$ is often referred as the *switched capacitance*. At the logic level, we assume that V_{dd} and T_{cycle} are fixed, and thus, minimize the total switched capacitance of the circuit.

Power dissipation is linearly dependent on the switched capacitance of the circuits. Calculation of the switched capacitances is however difficult as it depends on a number of circuit parameters and technology-dependent factors which are not readily available or precisely characterized. Some of these factors are described next.

Physical Capacitance

Power dissipation is dependent on the physical capacitances seen by individual gates in the circuit. Estimating this capacitance at the technology independent phase of logic synthesis is difficult and imprecise as it requires estimation of the load capacitances from structures which are not yet mapped to gates in a cell library; this calculation can however be done easily after technology mapping by using the logic and delay information from the library.

Interconnect plays an increasing role in determining the total chip area, delay and power dissipation, and hence, must be accounted for as early as possible during the design process. The interconnect capacitance estimation is however a difficult task even after technology mapping due to lack of detailed place and route information. Approximate estimates can be obtained by using information derived from a companion placement solution [33] or by using stochastic / procedural interconnect models [34].

Circuit Structure

The major difficulty in computing the switching activities is the reconvergent nodes. Indeed, if a network consists of simple gates and has no reconvergent fanout nodes (that is, circuit nodes that receive inputs from two paths that fanout from some other circuit node), then the exact switching activities can be computed during a single post-order traversal of the network. For networks with reconvergent fanout, the problem is much more challenging as internal signals may become strongly correlated and exact consideration of these correlations cannot be performed with reasonable computational effort or memory usage. Current power estimation techniques either ignore these correlations or approximate them, thereby improving the accuracy at the expense of longer run times.

Input Pattern Dependence

Switching activity at the output of a gate depends not only on the switching activities at the inputs and the logic function of the gate, but also on the spatial and temporal dependencies among the gate inputs. For example, consider a two-input **and** gate g with independent inputs i and j whose signal probabilities are 1/2, then $E_g(sw)$=3/8. This holds because in 6 out of 16 possible input transitions, the output of the two-input and gate makes a transition. Now suppose it is known that only patterns 00 and 11 can be applied to the gate inputs and that both patterns are equally likely, then $E_g(sw)$=1/2 (see Table 6.1 a). Alternatively, assume that it is known that every 0 applied to input i is immediately followed by a 1 while every 1 applied to input j is immediately followed by a 0, then $E_g(sw)$=4/9 (see Table 6.1 b). Finally, assume that it is known that i changes exactly if j changes value, then $E_g(sw)$=1/4 (see Table 6.1 c). The first case is an example of spatial correlations between gate inputs, the second case illustrates temporal correlations on gate inputs while the third case describes an instance of spatiotemporal correlations.

The straight-forward approach of estimating power by using a simulator is greatly complicated by this pattern dependence problem. It is clearly infeasible to

(a)

i	j	
0 → 0	0 → 0	0 → 0
0 → 1	0 → 1	0 → 1 *
1 → 0	1 → 0	1 → 0 *
1 → 1	1 → 1	1 → 1

(b)

i	j	
0 → 1	0 → 0	0 → 0
0 → 1	0 → 1	0 → 1 *
0 → 1	1 → 0	0 → 0
1 → 0	0 → 0	0 → 0
1 → 0	0 → 1	0 → 0
1 → 0	1 → 0	1 → 0 *
1 → 1	0 → 0	0 → 0
1 → 1	0 → 1	0 → 1 *
1 → 1	1 → 0	1 → 0 *

(c)

i	j	
0 → 0	0 → 0	0 → 0
0 → 0	1 → 1	0 → 0
0 → 1	0 → 1	0 → 1 *
0 → 1	1 → 0	0 → 0
1 → 0	1 → 0	1 → 0 *
1 → 0	0 → 1	0 → 0
1 → 1	0 → 0	0 → 0
1 → 1	1 → 1	1 → 1

Table 6.1 Effect of the input correlations

estimate the power by exhaustive simulation of the circuit. Recent techniques overcome this difficulty by using probabilities that describe the set of possible logic values at the circuit inputs and developing mechanisms to calculate these probabilities for gates inside the circuit. Alternatively, exhaustive simulation may be replaced by Monte-Carlo simulation with well-defined stopping criterion for specified relative or absolute error in power estimates for a given confidence level [8].

Delay Model

Based on the delay model used, the power estimation techniques could account for steady-state transitions (which consume power, but are necessary to perform a computational task) and/or hazards and glitches (which dissipate power without doing any useful computation). Sometimes, the first component of power consumption is referred as the *functional activity* while the latter is referred as the *spurious activity.* It is shown in [2] that the mean value of the ratio of hazardous component to the total power dissipation varies significantly with the considered circuits (from 9% to 38%) and that the spurious power dissipation cannot be neglected in CMOS circuits. Indeed, an average of 15-20% of the total power is dissipated in glitching. The spurious power dissipation is likely to become even more important in the future scaled technologies.

Current power estimation techniques often handle both zero-delay (non-glitch) and real delay models. In the first model, it is assumed that all changes at the circuit inputs propagate through the internal gates of the circuits instantaneously. The latter model assigns each gate in the circuit a finite delay and can thus account for the hazards in the circuit (see Figure 6.1). A real delay model significantly increases the computational requirements of the power estimation techniques while improving the accuracy of the estimates.

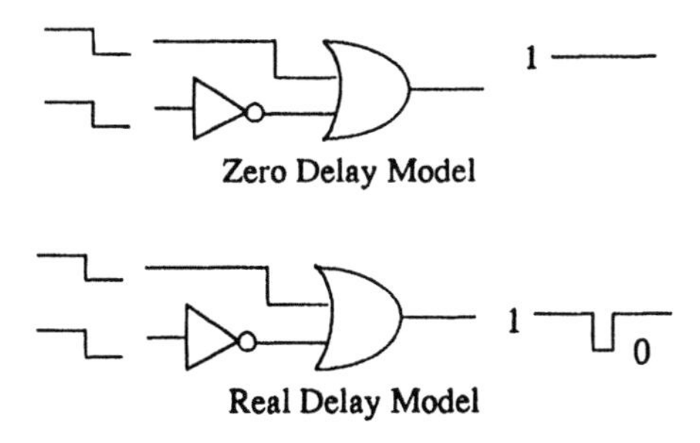

Figure 6.2 Effect of the delay model

Statistical Variation of Circuit Parameters

In real networks, statistical perturbations of circuit parameters may change the propagation delays and produce changes in the number of transitions because of the appearance or disappearance of hazards. It is therefore useful to determine the change in the signal transition count as a function of this statistical perturbations. Variation of gate delay parameters may change the number of hazards occurring during a transition as well as their duration. For this reason, it is expected that the hazardous component of power dissipation is more sensitive to IC parameter fluctuations than the power required to perform the transition between the initial and final state of each node.

In the following sections, various techniques for power estimation will be reviewed. These techniques are divided into two general categories: simulation based and probabilistic.

6.2.2. Simulation-Based Techniques

Circuit simulation based techniques ([20],[53]) simulate the circuit with a representative set of input vectors. They are accurate and capable of handling various device models, different circuit design styles, single and multi-phase clocking methodologies, tristate drives, etc. However, they suffer from memory and execution time constraints and are not suitable for large, cell-based designs. In general, it is difficult to generate a compact stimulus vector set to calculate accurate activity factors at the circuit nodes. The size of such a vector set is dependent on the application and the system environment [35].

A *Monte Carlo simulation* approach for power estimation which alleviates this problem has been proposed in [8]. This approach consists of applying randomly generated input patterns at the circuit inputs and monitoring the power dissipation per time interval T using a simulator. Based on the assumption that the power consumed by the circuit over any period T has a normal distribution, and for a desired percentage error in the power estimate and a given confidence level, the number of required power samples is estimated. The designer can use an existing simulator (circuit-level, gate-level or behavioral) in the inner loop of the Monte-Carlo program, thus trading accuracy for higher efficiency. The convergence time for this approach is fast when estimating the total power consumption of the circuit. However, when signal probability (or power consumption) values on individual lines of the circuit are required, the convergence rate is very slow. The method does not handle spatial correlations at the circuit inputs. The normal-ity assumption does not hold in many circuits (for example, small circuits or circuits with enable lines that control activity in a significant part of the logic), and thus the approach may converge to a wrong power estimate as a result of underestimating the number of required power samples.

Switch-level simulation techniques are in general much faster than circuit-level simulation techniques, but are not as accurate or versatile. Standard switch-level simulators (such as IRSIM [38]) can be easily modified to report the switched capacitance (and thus dynamic power dissipation) during a simulation run.

PowerMill [11] is a *transistor-level power simulator* and analyzer which applies an event-driven timing simulation algorithm (based on simplified table-driven device models, circuit partitioning and single-step nonlinear iteration) to increase the speed by two to three orders of magnitude over SPICE. PowerMill gives detailed power information (instantaneous, average and arms current values) as well as the total power consumption (due to steady-state transitions, hazards and glitches, transient short circuit currents, and leakage currents). It also tracks the current density and voltage drop in the power net and identifies reliability problems caused by EM failures, ground bounce and excessive voltage drops.

Entice-Aspen [14] is a power analysis system which raises the level of abstraction for power estimation from the transistor level to the *gate level.* Aspen computes the circuit activity information using the Entice power characterization data as follows. A stimulus file is to be supplied to Entice where power and timing delay vectors are specified. The set of power vectors discretizes all possible events in which power can be dissipated by the cell. With the relevant parameters set according to the user's specs, a SPICE circuit simulation is invoked to accurately obtain the power dissipation of each vector. During logic simulation, Aspen monitors the transition count of each cell and computes the total power consumption as the sum of the power dissipation for all cells in the power vector path.

6.2.3. Probabilistic Power Estimation Techniques

6.2.3.1 Combinational Circuits

Estimation under a Zero Delay Model

Most of the power in CMOS circuits is consumed during charging and discharging of the load capacitance. To estimate the power consumption, one has to calculate the (switching) activity factors of the internal nodes of the circuit. Methods of estimating the activity factor $E_n(sw)$ at a circuit node n involve estimation of signal probability $prob(n)$, which is the probability that the signal value at the node is one. Under the assumption that the values applied to each circuit input are temporally independent (that is, value of any input signal at time t is independent of its value at time t-1), we can write:

$$E_n(sw) = 2\,prob(n)\,(1 - prob(n)). \tag{6.2}$$

Computing signal probabilities has attracted much attention. In [32], some of the earliest work in computing the signal probabilities in a combinational network is presented. The authors associate variable names with each of the circuit inputs representing the signal probabilities of these inputs. Then, for each internal circuit line, they compute algebraic expressions involving these variables. These expressions represent the signal probabilities for these lines. While the algorithm is simple and general, its worse case time complexity is exponential.

If a network consists of simple gates and has no reconvergent fanout nodes, then the exact signal probabilities can be computed during a single post-order traversal of the network using the following equations [16]:

$$\text{not gate: } prob(o) = 1 - prob(i)$$

$$\text{and gate: } prob(o) = \prod_{i \in inputs} prob(i)$$

$$\text{or gate: } prob(o) = 1 - \prod_{i \in inputs} (1 - prob(i))$$

This simple algorithm is known as the *tree algorithm* as it produces the exact signal probabilities for a tree network. For networks with reconvergent fanout, the tree algorithm however yields approximate values for the signal probabilities.

In [43], a graph-based algorithm is used to compute the exact signal probabilities using Shannon's expansion. This algorithm relies on the notion of the super-gate of a node n which is defined as the smallest sub-circuit of the (transitive fanin) cone of n whose inputs are independent, that is, the cones of the inputs are mutually disjoint. Maximal super-gates are the super-gates which are not properly contained in any other super-gate. Maximal super-gates form a unique cover of circuit. Although this cover is disjoint for a single-output circuit, it is not necessarily disjoint for a multiple-output circuit [42]. The procedure, called the *super-gate algorithm*, essentially requires three steps: 1) Identification of reconvergent nodes, 2) Determination of the maximal super-gates, and 3) Calculation of signal probabilities at the super-gate outputs based on the Shannon's expansion with respect to all multiple-fanout inputs of the super-gate. In the worst case, a super-gate may include all the circuit inputs as multiple-fanout inputs. In such cases, the super-gate algorithm becomes equivalent to an exhaustive true-value simulation of the super-gate.

In [22] an extension to the tree algorithm, called the *weighted averaging algorithm*, is described. The new algorithm computes the signal probability at the output of a gate approximately using Shannon's expansion with respect to each multiple-fanout primary input in the support of the gate. The signal probability of the gate is then calculated as a weighted sum of these approximate signal probabilities. This approach attempts to take into account the first order effects of reconvergent fanout stems in the input variable support of the node. It is linear in the product of the number of circuit inputs and the size of the circuit.

In [39], an algorithm, known as the *cutting algorithm*, is presented that computes lower and upper bounds on the signal probability of reconvergent nodes by cutting the multiple-fanout reconvergent input lines and assigning an appropriate probability range to the cut lines and then propagating the bounds to all the other lines of the circuits by using propagation formulas for trees. The effectiveness of the cutting algorithm, however, depends on the non-deterministic choice of the cuts. Well-chosen cuts lead to better estimates of the signal probabilities while poorly chosen cuts results in poor estimates. The algorithm runs in polynomial time in terms of the size of the circuits.

In [9], an exact procedure based on Ordered Binary-Decision Diagrams (OBDDs) [6] is described which is linear in the size of the corresponding function graph (the size of the graph, however, may be exponential in the number of circuit inputs). In this method, which is known as the *OBDD-based* method, the signal probability at the output of a node is calculated by first building an OBDD corresponding to the *global function* of the node (i.e., function of the node in terms of the circuit inputs) and then performing a postorder traversal of the OBDD using equation:

$$prob(y) = prob(x) prob(f_x) + prob(\bar{x}) prob(f_{\bar{x}}) \tag{6.3}$$

This leads to a very efficient computational procedure for signal probabil-

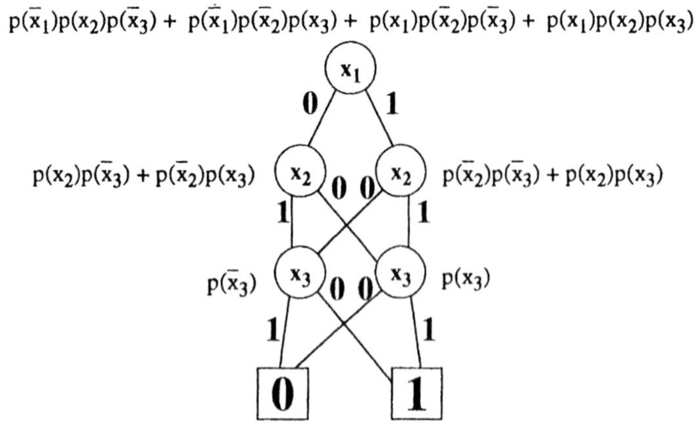

$p(\bar{x}_1)p(x_2)p(\bar{x}_3) + p(\bar{x}_1)p(\bar{x}_2)p(x_3) + p(x_1)p(\bar{x}_2)p(\bar{x}_3) + p(x_1)p(x_2)p(x_3)$

$p(x_2)p(\bar{x}_3) + p(\bar{x}_2)p(x_3)$ $p(\bar{x}_2)p(\bar{x}_3) + p(x_2)p(x_3)$

$p(\bar{x}_3)$ $p(x_3)$

Figure 6.3 Computing the signal probability using OBDDs

ity estimation. Figure 6.3 shows an example computation on the OBDD representation of a three-input **exor** gate.

In [12], a procedure for propagating signal probabilities from the circuit inputs toward the circuit outputs using only *pairwise correlations* between circuit lines and ignoring higher order correlation terms is described. The correlation coefficient of two signals i and j is defined as:

$$C(i,j) = \frac{prob(i \wedge j)}{prob(i)\,prob(j)} \tag{6.4}$$

The correlation coefficient of i and $\bar{j}$, $\bar{i}$ and j, etc. are defined similarly. Ignoring higher order correlation coefficients, it is assumed that $C(i,j,k) = C(i,j)\,C(i,k)\,C(j,k)$. The signal probability of g is thus approximated by:

not gate: $prob(g) = 1 - prob(i)$

and gate: $prob(g) = \prod_{i \in inputs} prob(i) \cdot \prod_{j>i} C(i,j)$

or gate: $prob(g) = 1 - \prod_{i \in inputs} (1 - prob(i)) \cdot \prod_{j>i} C(\bar{i},\bar{j})$

where $C(\bar{i},\bar{j})$ is calculated from $prob(i)$, $prob(j)$ and $C(i,j)$.

In [41] and [27], the temporal correlation between values of some signal x in two successive clock cycles is modeled by a time-homogeneous Markov chain which has two states 0 and 1 and four edges where each edge ij ($i,j = 0, 1$) is annotated with the conditional probability $prob_{ij}^{x}$ that x will go to state j at time t+1 if it is in state i at time t (see Figure 6.4). The transition probability $prob\ (x_{i \to j})$ is equal to $prob\ (x = i)\ prob_{ij}^{x}$. Obviously, $prob_{00}^{x} + prob_{01}^{x} = prob_{10}^{x} + prob_{11}^{x} = 1$ while $prob\ (x) = prob\ (x_{0 \to 1}) + prob\ (x_{1 \to 1})$ and $prob\ (\bar{x}) = prob\ (x_{0 \to 0}) + prob\ (x_{1 \to 0})$. The activity factor of line x can be expressed in terms of these transition probabilities as follows:

$$E_x(sw) = prob(x_{0 \to 1}) + prob(x_{1 \to 0}) . \tag{6.5}$$

The various transition probabilities can be computed exactly using the OBDD representation of the logic function of x in terms of the circuit inputs.

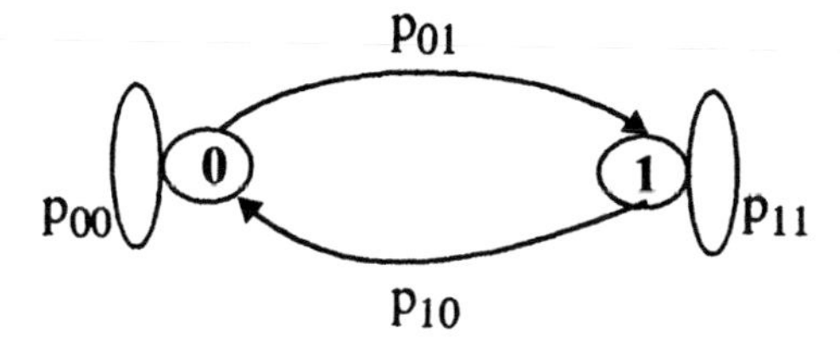

Figure 6.4 A Markov chain model for representing temporal correlations

The authors of [27] also describe a mechanism for propagating the transition probabilities through the circuit which is more efficient as there is no need to build the global function of each node in terms of the circuit inputs. The loss in accuracy is often small while the computational saving is significant. They then extend the model to account for spatio-temporal correlations. The mathematical foundation of this extension is a four state time-homogeneous Markov chain where each state represents some assignment of binary values to two lines x and y and each edge describes the conditional probability for going from one state to next. The computational requirement of this extension is however high as it is linear in the product of the number of nodes and number of paths in the OBDD representation of the Boolean function in question. A practical method using local OBDD constructions is described by the authors.

This work has been extended to handle highly correlated input streams using the notions of *conditional independence* and *isotropy of signals* [28]. Based on these notions, it is shown that the relative error in calculating the signal probability of a logic gate using pairwise correlation coefficients can be bounded from above.

Table 6.2 gives the various error measures (compared to exact values obtained from exhaustive logic simulation) for biased input sequences applied to the *f51m* benchmark circuit. It can be seen that accounting for either spatial or temporal correlations improves the accuracy while the most accurate results are obtained by considering both spatial and temporal correlations.

	With		Without	
	Spatial Correlations			
	With	Without	With	Without
	Temporal Correlations			
Max	0.0767	0.2893	0.1714	0.3110
Mean	0.0111	0.0939	0.0380	0.1049
RMS	0.0205	0.1164	0.0511	0.1239
STD	0.0174	0.0692	0.0344	0.0663

Table 6.2Effect of spatio-temporal correlations on switching activity estimation

Estimation under a Real Delay Model

The above methods only account for steady-state behavior of the circuit and thus ignore hazards and glitches. This section reviews some techniques that examine the dynamic behavior of the circuit and thus estimate the power dissipation due to hazards and glitches.

In [15], the exact power estimation of a given combinational logic circuit is carried out by creating a set of symbolic functions such that summing the signal probabilities of the functions corresponds to the average switching activity at a circuit line x in the original combinational circuit (this method is known as the *symbolic simulation* method). The inputs to the created symbolic functions are the circuit input lines at time instances 0^- and ∞. Each function is the exclusive or of the characteristic functions describing the logic values of x at two consecutive instances (see Figure 6.5 for an example symbolic network). The major disadvantage of this estimation method is its exponential complexity. However, for the circuits that this method is applicable to, the estimates provided by the method can serve as a basis for comparison among different approximation schemes.

The concept of a probability waveform is introduced in [7]. This waveform consists of a sequence of transition edges or events over time from the initial steady state (time 0^-) to the final steady state (time ∞) where each event is anno-

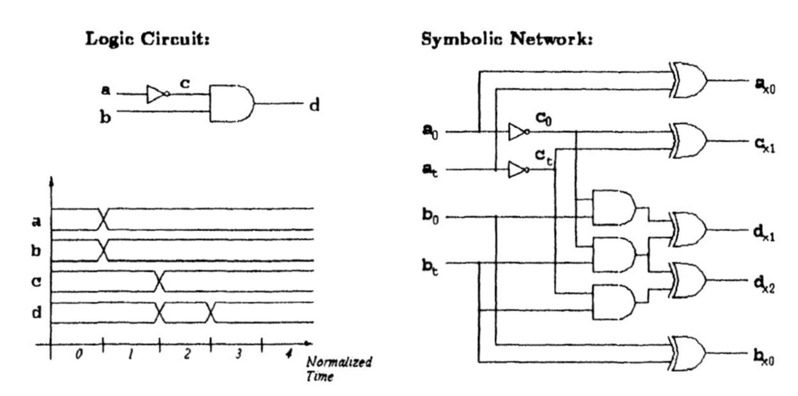

Figure 6.5 Symbolic simulation under a unit delay model

tated with an occurrence probability. The probability waveform of a node is a compact representation of the set of all possible logical waveforms at that node. Given these waveforms, it is straight-forward to calculate the switching activity of x which includes the contribution of hazards and glitches, that is:

$$E_x(sw) = \sum_{t \in eventlist(x)} \left(prob\left(x^t_{0 \to 1}\right) + prob\left(x^t_{1 \to 0}\right) \right). \tag{6.6}$$

Given such waveforms at the circuit inputs and with some convenient partitioning of the circuit, the authors examine every sub-circuit and derive the corresponding waveforms at the internal circuit nodes. In [31], an efficient *probabilistic simulation* technique is described that propagates transition waveforms at the circuit primary inputs up in the circuit and thus estimates the total power consumption (ignoring signal correlations due to the reconvergent fanout nodes).

A *tagged probabilistic simulation* approach is described in [49] that correctly accounts for reconvergent fanout and glitches. The key idea is to break the set of possible logical waveforms at a node n into four groups, each group being characterized by its steady state values (i.e., values at time instance 0^- and ∞). Next, each group is combined into a probability waveform with the appropriate steady-state tag (see Figure 6.6). Given the tagged probability waveforms at the input of a simple gate, it is then possible to compute the tagged probability waveforms at the output of the gate. The correlation between probability waveforms at the inputs is approximated by the correlation between the steady state values of

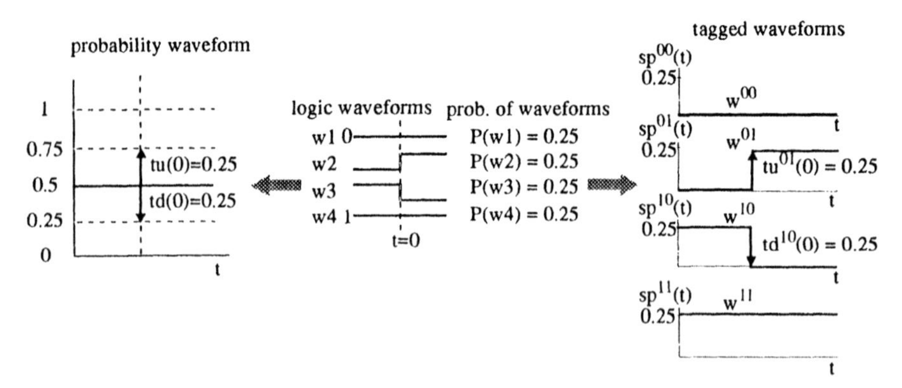

Figure 6.6 Probability waveforms

these lines. This is much more efficient than trying to estimate the dynamic correlations between each pair of events. This approach requires significantly less memory and runs much faster than symbolic simulation, yet achieves very high accuracy, e.g., the average error in aggregate power consumption is about 10%. In order to achieve this level of accuracy, detailed timing simulation along with careful *glitch filtering* and *library characterization* are needed. The first item refers to the scheme for eliminating some of the short glitches that cannot overcome the gate inertias from the probability waveforms. The second item refers to the process of generating accurate and detailed macro-modeling data for the gates in the cell library.

6.2.3.2 Sequential Circuits

Recently developed methods for power estimation have primarily focused on combinational logic circuits. The estimates produced by purely combinational methods can greatly differ from those produced by the exact method. Indeed, accurate average switching activity estimation for finite state machines (FSMs) is considerably more difficult than that for combinational circuits for two reasons: 1) The probability of the circuit being in each of its possible states has to be calculated; 2) The present state line inputs of the FSM are strongly correlated (that is, they are temporally correlated due to the machine behavior as represented in its State Transition Graph description and they are spatially correlated because of the given state encoding).

A first attempt at estimating switching activity in FSMs has been presented in [15]. The idea is to *unroll* the next state logic once (thus capturing the temporal

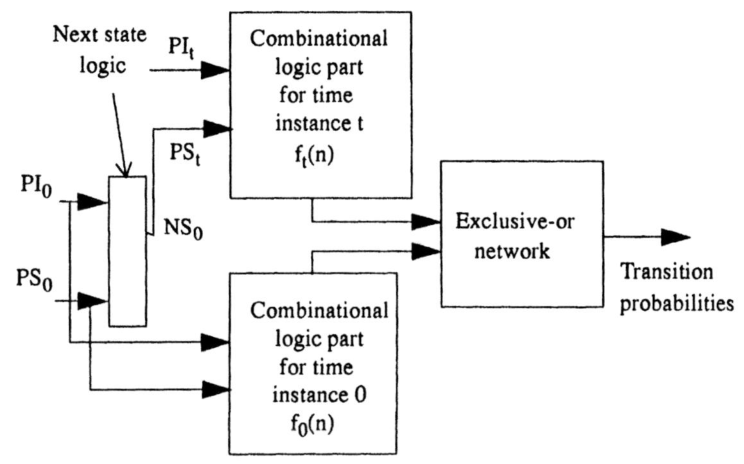

Figure 6.7 Power estimation for sequential circuits

correlations of present state lines) and then perform symbolic simulation on the resulting circuit (which is hence treated as a combinational circuit) as shown in Figure 6.7. This method does not however capture the spatial correlations among present state lines and makes the simplistic assumption that the state probabilities are uniform.

The above work is improved upon in [51] and [30] where results obtained by using the Chapman-Kolmogorov equations for discrete-time Markov Chains to compute the exact state probabilities of the machine are presented. We describe the method below.

For each state S_i, $1 \leq i \leq K$ in the STG, we associate a variable $prob(S_i)$ corresponding to the steady-state probability of the machine being in state S_i at $t = \infty$. For each edge e_{ij} in the STG, we have I_{ij} signifying the input combination corresponding to the edge. Given static probabilities for the primary inputs to the machine, we can compute $prob(S_j \mid S_i)$, the conditional probability of going from S_i to S_j. For each state S_j, we can write an equation:

$$prob(S_j) = \sum_{S_i \in instates(S_j)} prob(S_i)\, prob\langle S_j | S_i \rangle \qquad (6.7)$$

where $instates(S_i)$ is the set of fanin states of S_i in the STG. Given K states, we obtain K equations out of which any one equation can be derived from the remaining K - 1 equations. We have a final equation:

$$\sum_j prob(S_j) = 1 . \quad (6.8)$$

This linear set of K equations is solved to obtain the different $prob(S_j)$'s.

The Chapman-Kolmogorov method requires the solution of a linear system of equations of size 2^N, where N is the number of flip-flops in the machine. Thus, this method is limited to circuits with a small number of flip-flops, since it requires the explicit consideration of each state in the circuit.

The authors of [51] and [30] also describe a method for approximate switching activity estimation of sequential circuits. The basic computation step is the solution of a non-linear system of equations as follows:

$$\begin{aligned} ps_1 &= f_1(pi, ps_1, ps_2, \ldots, ps_n) \\ &\ldots \\ &\ldots \\ ps_n &= f_n(pi, ps_1, ps_2, \ldots, ps_n) \end{aligned} \quad (6.9)$$

where ps_j denotes the state bit probabilities of the ith next state bit at the output and the jth present state bit at the input of the FSM, respectively and f_i's are non-linear algebraic functions. The fixed point (or zero) of this system of equations can be found using the Picard-Peano (or Newton-Raphson) iteration [24].

Increasing the number of variables or the number of equations in the above system results in increased accuracy [47]. For a wide variety of examples, it is shown that the approximation scheme is within 1-3% of the exact method, but is orders of magnitude faster for large circuits. Previous sequential switching activity estimation methods exhibit significantly greater inaccuracies.

6.3. Power Minimization Techniques

Logic synthesis fits between the register transfer level and the netlist of gates specification. It provides the automatic synthesis of netlists minimizing some objective function subject to various constraints. Example inputs to a logic synthesis system include two-level logic representation, multi-level Boolean networks, finite state machines and technology mapped circuits. Depending on the input specification (combinational versus sequential, synchronous versus asynchronous), the target implementation (two-level versus multi-level, unmapped versus mapped, ASICs versus FPGAs), the objective function (area, delay, power, testability) and the delay models used (zero-delay, unit-delay, unit-fanout delay, or library delay models), different techniques are applied to transform and optimize the original RTL description.

Once various system level, architectural and technological choices are made, it is the switched capacitance of the logic that determines the power consumption of a circuit. In this section, a number of techniques for power estimation and minimization during logic synthesis will be presented. The strategy for synthesizing circuits for low power consumption will be to restructure or optimize the circuit to obtain low switching activity factors at nodes which drive large capacitive loads.

Both the switching activity and the capacitive loading can be optimized during logic synthesis. Research in this area is focusing on the following: (1) Developing low power version of various logic optimization and restructuring techniques and doing this in such a way that the timing and/or area constraints are met, (2) Developing more accurate, yet simple, computational models for estimating the parasitic wiring capacitances, signal correlations, short circuit currents, etc. In the following, we present a number of techniques for power reduction during sequential and combinational logic synthesis which essentially target dynamic power dissipation under a zero-delay or a simple library delay model.

Precomputation Logic

The basic idea is to selectively precompute the output logic values of the circuits one clock cycle before they are required, and then use the precomputed values to reduce internal switching activity in the succeeding clock cycle [1]. A precomputation architecture is shown in Figure 6.8. The inputs to block A have been partitioned into two sets, corresponding to registers R_1 and R_2. The output of the logic block A feeds register R_3. Two Boolean functions g_1 and g_2 are the *predictor functions*. It is required that:

$$g_1 = 1 \Rightarrow f = 1 \tag{6.10}$$

$$g_2 = 1 \Rightarrow f = 0 \tag{6.11}$$

Therefore, during clock cycle t if either g_1 or g_2 evaluates to 1, we set the load enable signal of register R_2 to 0. This implies that the outputs of R_2 during clock cycle $t+1$ do not change. However, since the outputs of R_1 are updated, function f will be calculated correctly. Power reduction is achieved because only a

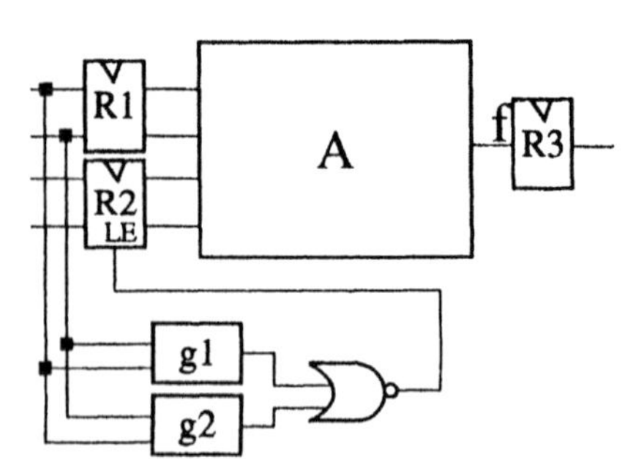

Figure 6.8 A precomputation architecture

subset of the inputs to block A change, implying reduced switching activity in block A. An example that illustrates the precomputation logic is the n-bit comparator that compares two n-bit numbers C and D and computes the function $C > D$. Assuming that each $C<i>$ and $D<i>$ has a 0.5 signal probability, the probability of correctly predicting the output result is 0.5, regardless of n. Thus, one can achieve a power reduction of 50%.

In a combinational circuit, it is also possible to identify subsets of gates which do not contribute to the computation initiated with some input stimulus. Power can thus be reduced by turning off these subsets of gates [46]. The overhead of detecting and disabling these sub-circuits may however be large. Power reductions of up to 40% have been reported in some examples.

Retiming

Retiming is the process of re-positioning the flip-flops in a pipelined circuit so as to either minimize the number of flip-flops or minimize the delay through the longest pipeline stage. In [29], it is noted that a flip-flop output makes at most one transition when the clock is asserted (see Figure 6.9). Based on this observation, the authors then describe a circuit retiming technique targeting low power dissipation. The idea is to identify circuit nodes with high hazard activity and high load capacitance as candidates for adding a flip-flop. The technique does not produce the optimal retiming solution as the retiming of a single node can dramatically change the switching activity of many other nodes in the circuit.

The authors report that the power dissipated by the 3-stage pipelined circuits obtained by retiming for low power with a delay constraint is about 8% less than that obtained by retiming for minimum number of flip-flops given a delay constraint.

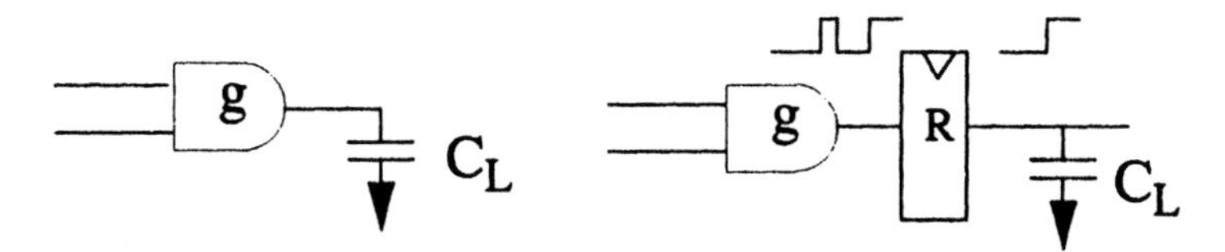

Figure 6.9 Flip-flop insertion to minimize hazard activity

State Assignment

State assignment of a finite state machine (which is the process of assigning binary codes to the states) has a significant impact on the area of its final logic implementation. In the past, many researchers have addressed the encoding problem for minimum area of two-level or multi-level logic implementations. These techniques can be modified to minimize the power dissipation. One approach is to minimize the switching activity on the present state lines of the machine by giving uni-distance codes to states with high transition frequencies to one another [37].

This formulation however ignores the power consumption in the combinational logic that implements the next state and output logic functions. A more effective approach [48] is to consider the complexity of the combinational logic resulting from the state assignment and to directly modify the objective functions used in the conventional encoding schemes such as NOVA [55]and JEDI [26] to achieve lower power dissipation. Experimental results on a large number of benchmark circuits show 10% and 17% power reductions for two-level logic and multi-level implementations compared to NOVA and JEDI, respectively.

Multi-Level Network Optimization

Network don't cares can be used for minimizing the intermediate nodes in a boolean network [40]. Two multi-level network optimization techniques for low power are described in [44] and [17]. One difference between these procedures and the procedure in [40] is in the cost function used during the two-level logic minimization. The new cost function minimizes a linear combination of the number of product terms and the weighted switching activity. In addition, [17] considers how changes in the global function of an internal node affects the switching activity (and thus, the power consumption) of nodes in its transitive fanout. The paper presents a greedy, yet effective, network optimization procedure as summarized below.

The procedure presented in [17] proceeds in a reverse topological fashion from the circuit outputs to the circuit inputs simplifying fanout's of a node before reaching that node. Once a node n is simplified, the procedure propagates those don't care conditions which could only increase (or decrease) the signal probability of that node if its current signal probability is greater than (less than or equal to) 0.5. This will ensure that as nodes in the transitive fanin of n are being simplified, the switching activity of n will not increase beyond its value when node n was optimized. Power consumption in a combinational logic circuit has been reduced by some 10% as a result of this optimization.

The above restriction on the construction of ODC may be overly constraining for the resynthesis process. In [23], a node simplification procedure is presented that identifies *good candidates* for resynthesis, that is, nodes where a local change in their activity plus the change in activity throughout their transitive fanout nodes, reduces the power consumption in the circuit. Both (delay-independent) functional activity and (delay-dependent) spurious activity are considered.

Common Sub-expression Extraction

The major issue in decomposition is the identification of common sub-expressions. Sharing of such expressions across the design reduces the complexity of the synthesized network and its implementation cost. Extraction based on algebraic division (using cube-free primary divisors or kernels) has proven to be very successful in creating an area-optimized multi-level Boolean network [4]. The kernel extraction procedure is modified in [37] to generate multi-level circuits with low power consumption. The main idea is to calculate the power savings factor for each candidate kernel based on how its extraction will affect the loading on its input lines and the amount of logic sharing. In [18], an alternate power-saving factor has been proposed which assumes that nodes in the multi-level network are in two-level logic form. This is consistent with the assumption made for calculating the literal-saving cost of a candidate divisors during algebraic operations.

An example decomposition is shown in Figure 6.10 where two network structures that compute the same function $f=ab+ac+bc$ are depicted. Note that the two configurations have the same number of literals in the factored form of their intermediate nodes. It can be also seen that if $E_a(sw) + E_g(sw) > E_c(sw) + E_h(\mathrm{sw})$, then $P^A > P^B$. For example, if $prob(a)=0.5$ and $prob(b)=prob(c)=0.25$, then $P^A - P^B = 13/128$. In general, power savings of about 10% (compared to a min-

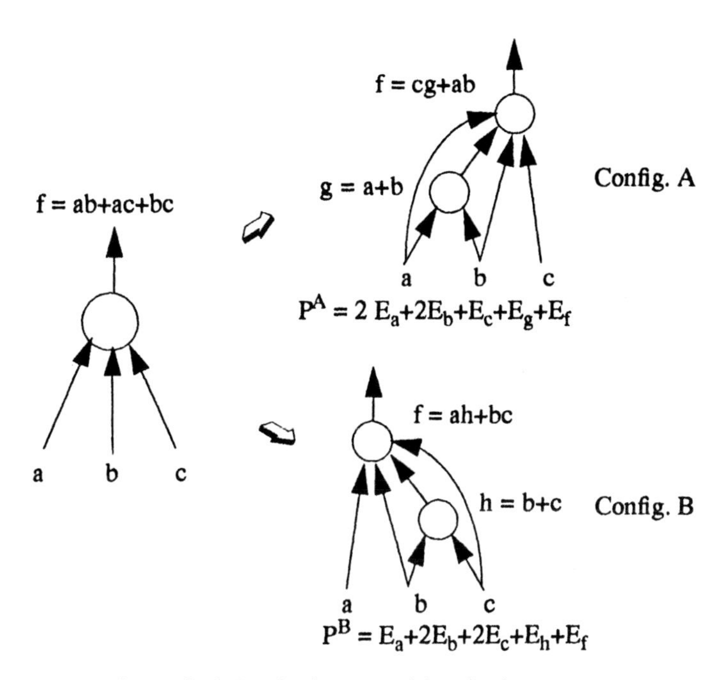

Figure 6.10 Logic decomposition for low power

imum-literal network) are expected.

As the active area (e.g., the number of literals) in a circuit strongly influences the power consumption, one must minimize a lexicographic cost (Δ_a, Δ_p) where Δ_a is the literal saving factor and Δ_p is the power saving factor. At the same time, the above power saving factor is expensive to compute, therefore, it is desirable to calculate it only for a subset of candidate divisors (say, the top 10% of divisors in terms of their literal saving factors).

Path Balancing

Balancing path delays reduces hazards/glitches in the circuit which in turn reduces the average power dissipation in the circuit. This can be achieved before technology mapping by selective collapsing and logic decomposition or after technology mapping by delay insertion and pin reordering.

The rationale behind selective collapsing is that by collapsing the fanins of a node into that node, the arrival time at the output of the node can be changed. Logic decomposition and extraction can be performed so as to minimize the level difference between the inputs of nodes which are driving high capacitive nodes.

The key issue in delay insertion is to use the minimum number of delay elements to achieve the maximum reduction in spurious switching activity. Path delays may sometimes be balanced by appropriate signal to pin assignment. This is possible as the delay characteristics of CMOS gates vary as a function of the input pin which is causing a transition at the output.

Technology Decomposition

This is the problem of converting a set of Boolean equations (or a Boolean network) to another set (or another network) consisting of only two-input NAND and inverter gates. It is difficult to come up with a NAND decomposed network which will lead to a minimum power implementation after technology mapping since gate loading and mapping information are unknown at this stage. Nevertheless, it has been observed that a decomposition scheme which minimizes the sum of the switching activities at the internal nodes of the network, is a good starting point for power-efficient technology mapping.

Given the switching activity value at each input of a complex node, a procedure for AND decomposition of the node is described in [50] which minimizes the total switching activity in the resulting two-input AND tree under a zero-delay model. The principle is to inject high switching activity inputs into the decomposition tree as late as possible. The decomposition procedure (which is similar to Huffman's algorithm for constructing a binary tree with minimum average weighted path length) is optimal for dynamic CMOS circuits and produces very good results for static CMOS circuits. An example is shown in Figure 6.11 where the input signal with the highest switching activity (that is, signal *d*) is injected last in the decomposition tree in configuration A, thus yielding lower power dissipation for this configuration.

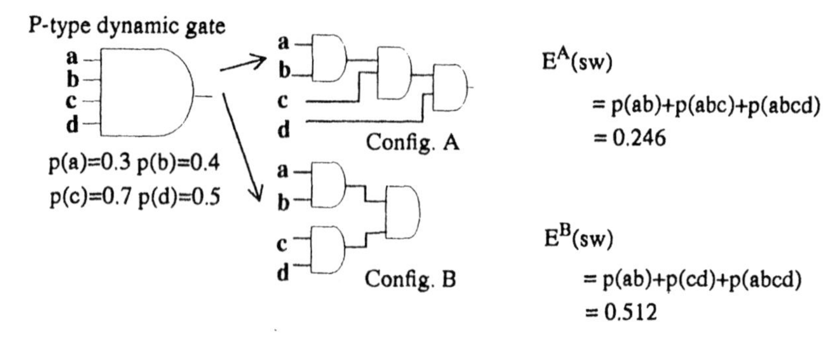

Figure 6.11 Technology decomposition for low power

In general, the low power technology decomposition procedure reduces the total switching activity in the networks by 5% over the conventional balanced tree decomposition method.

Technology Mapping

This is the problem of binding a set of logic equations (or a boolean network) to the gates in some target cell library. A successful and efficient solution to the minimum area mapping problem was suggested in [21] and implemented in programs such as DAGON and MIS. The idea is to reduce technology mapping to DAG covering and to approximate DAG covering by a sequence of tree coverings which can be performed optimally using dynamic programming.

The problem of minimizing the average power consumption during technology mapping is addressed in [50],[45] and [25]. The general principle is to hide nodes with high switching activity inside the gates where they drive smaller load capacitances (see Figure 6.12).

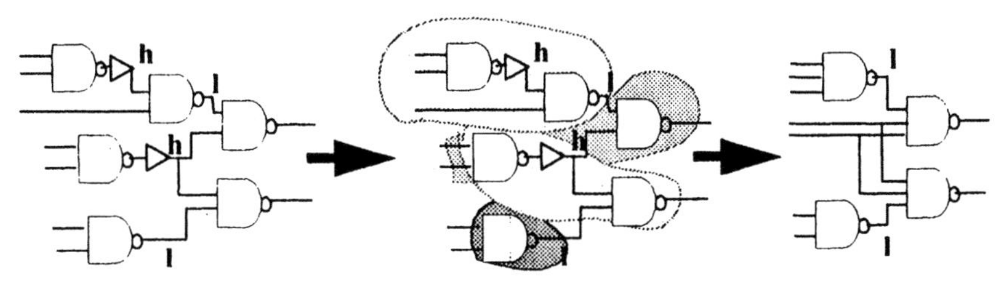

h: node with high switching activity
l: node with low switching activity

Figure 6.12 Technology mapping for low power

The approach presented in [50] consists of two steps. In the first step, power-delay curves (that capture power consumption versus arrival time trade-off) at all nodes in the network are computed. In the second step, the mapping solution is generated based on the computed power-delay curves and the required times at the primary outputs. For a NAND-decomposed tree, subject to load calculation errors, this two step approach finds the minimum area mapping satisfying any delay constraint if such a solution exists. Compared to a technology mapper that minimizes the circuit delay, this procedure leads to an average of 18% reduction in power consumption at the expense of 16% increase in area without any degradation in performance.

Generally speaking, the power-delay mapper reduces the number of high switching activity nets at the expense of increasing the number of low switching activity nets. In addition, it reduces the average load on the nets. By taking these two steps, this mapper minimizes the total weighted switching activity and hence the total power consumption in the circuit.

Under a real delay model, the dynamic programming based tree mapping algorithm does not guarantee to find an optimum solution even for a tree. The dynamic programming approach was adopted based on the assumption that the current best solution is derived from the best solutions stored at the fanin nodes of the matching gate. This is true for power estimation under a zero delay model, but not for that under a real delay model.

The extension to a real delay model is also considered in [49]. Every point on the power-delay curve of a given node uniquely defines a mapped subnetwork from the circuit inputs up to the node. Again, the idea is to annotate each such point with the probability waveform for the node in the corresponding mapped subnetwork. Using this information, the total power cost (due to steady-state transitions and hazards) of a candidate match can be calculated from the annotated power-delay curves at the inputs of the gate and the power-delay characteristics of the gate itself.

Gate Resizing

The treatment of gate sizing problem is closely related to finding a macro-model which captures the main features of a complex gate (area, delay, power consumption) through a small number of parameters (typically, width or transconductance). An efficient approach to gate sizing for low power is to linearize the path-based timing constraints and use a linear programming solver to find the global optimum solution [3]. The drawbacks of this approach are the omission of slope factor (input ramp time) of input waveforms from the delay model and use of simple power dissipation model that ignores short-circuit currents. This work can be easily extended to account for the slope factor and the short-circuit currents.

PLA Minimization

High speed PLAs are built by transforming the SOP representation of a two level logic to the NOR-NOR structure with inverting inputs and outputs and implementing it with two NOR arrays. Two common types of implementing the NOR arrays are pseudo-NMOS NOR gates and dynamic CMOS NOR gate.

The primary source of power consumption for a pseudo-NMOS NOR gate is the static power dissipation (see Figure 6.13). When the NOR gate evaluates to zero, both the PMOS and NMOS parts of the gate are on and there exists a direct current path. The charging and discharging energy is negligible compared with that dissipated by the direct current. Furthermore, the direct current I_{dc} is relatively constant irrespective of the number of NMOS transistors that are on. Therefore the power cost for a product (AND) term is given by:

$$V_{dd} \cdot I_{dc} \cdot prob^{0}_{AND} \tag{6.12}$$

where $prob^{0}_{AND}$ is the probability that the AND term evaluates to 0.

In a dynamic PLA circuit, dynamic power consumption is the major source of power dissipation (see Figure 6.13). The output of the product term is precharged to 1 and switches when it is evaluated to 0. Therefore the power cost for a product (AND) term is given by:

$$\frac{V_{dd}^2 \cdot f}{2} \cdot \left(\sum_{i=1}^{k} C_i E_i(sw) + C_{AND} prob^{0}_{AND} + 2C_{clock} \right) \tag{6.13}$$

where C_i is the gate capacitance seen by the ith input of the AND term, C_{AND} is the load capacitance that the AND term is driving, and C_{clock} is the load capacitance of the precharge and evaluate transistors that the clock drives and f is the clock frequency.

Fortunately, in both cases it has been shown that the optimum two-level cover will consist of only prime implicants [19]. The resulting minimization problems can be solved exactly by changing the cost function used in the

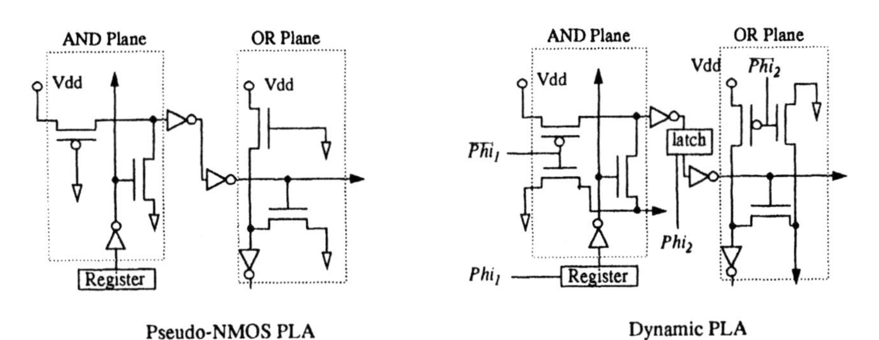

Figure 6.13 NOR-NOR PLAs

Quine-McClusky procedure or the Espresso heuristic minimizer [5]. In general, optimization for power resulted in a 5% increase in the number of cubes of the function while reducing the power by an average of 11%.

Signal-to-Pin Assignment

In general, library gates have pins that are functionally equivalent which means that inputs can be permuted on those pins without changing function of the gate output. These equivalent pins may have different input pin loads and pin dependent delays. It is well known that the signal to pin assignment in a CMOS logic gate has a sizable impact on the propagation delay through the gate.

If we ignore the power dissipation due to charging and discharging of internal capacitances, it becomes obvious that high switching activity inputs should be matched with pins that have low input capacitance. However, the internal power dissipation also varies as a function of the switching activities and the pin assignment of the input signals. To find the minimum power pin assignment for a gate g, one must solve a difficult optimization problem [52]. Alternatively, one can use heuristics, for example, a reasonable heuristic assigns the signal with largest probability of assuming a controlling value (zero for NMOS and one for PMOS) to the transistor near the output terminal of the gate. The rationale is that this transistor will switch off as often as possible, thus blocking the internal nodes from non-productive charge and discharge events.

6.4. Concluding Remarks

The need for lower power systems is being driven by many market segments. There are several approaches to reducing power, however the highest Return On Investment approach is through designing for low power. Unfortunately designing for low power adds another dimension to the already complex design problem; the design has to be optimized for Power as well as Performance and Area.

Optimizing the three axes necessitates a new class of power conscious CAD tools. The problem is further complicated by the need to optimize the design for power at all design phases. The successful development of new power conscious tools and methodologies requires a clear and measurable goal. In this context the research work should strive to reduce power by 5-10x in three years through design and tool development.

References

[1] M. Alidina, J. Monteiro, S. Devadas, A. Ghosh, and M. Papaefthymiou. " Precomputation-based Sequential Logic Optimization for Low Power. " In *Proceedings of the 1994 International Workshop on Low Power Design*, pages 57-62, April 1994.

[2] L. Benini, M. Favalli, and B. Ricco. " Analysis of hazard contribution to power dissipation in CMOS IC's. " In *Proceedings of the 1994 International Workshop on Low Power Design*, pages 27–32, April 1994.

[3] M. Berkelaar and J. Jess. " Gate sizing in {MOS} digital circuits with linear programming. " In *Proceedings of the European Design Automation Conference*, pages 217-221, 1990.

[4] R. K. Brayton, G. D. Hachtel, and A. L. Sangiovanni-Vincentelli. " Multilevel logic synthesis. " *Proceedings of the* IEEE, volume 78, pages 264–300, February 1990.

[5] R. K. Brayton, G. D. Hachtel, C. McMullen and A. L. Sangiovanni-Vincentelli. " Logic minimization algorithms for VLSI synthesis. " Kluwer Academic Publishers, Boston, Massachusetts, 1984.

[6] R. Bryant. " Graph-based algorithms for Boolean function manipulation. " IEEE *Transactions on Computers*, volume C-35, pages 677–691, August 1986.

[7] R. Burch, F. Najm, P. Yang, and D. Hocevar. " Pattern independent current estimation for reliability analysis of CMOS circuits. " In *Proceedings of the 25th Design Automation Conference*, pages 294–299, June 1988.

[8] R. Burch, F. N. Najm, P. Yang, and T. Trick. " A Monte Carlo approach for power estimation. " IEEE *Transactions on VLSI Systems*, 1(1):63-71, March 1993.

[9] S. Chakravarty. " On the complexity of using BDDs for the synthesis and analysis of boolean circuits. " In *Proceedings of the 27th Annual Allerton Conference on Communication, Control and Computing*, pages 730–739, 1989.

[10] K. Chaudhary and M. Pedram. " A near-optimal algorithm for technology mapping minimizing area under delay constraints. " In *Proceedings of the 29th Design Automation Conference*, pages~492-498, June 1992.

[11] C. Deng. " Power analysis for CMOS/BiCMOS circuits. " In *Proceedings of the 1994 International Workshop on Low Power Design*, pages 3-8, April 1994.

[12] S. Ercolani, M. Favalli, M. Damiani, P. Olivo, and B. Ricco. " Estimate of signal probability in combinational logic networks. " In *First European Test Conference*, pages 132–138, 1989.

[13] T. A. Fjeldly and M. Shur. " Threshold voltage modeling and the subthreshold regime of operation of short-channel MOSFET's. " IEEE *Transactions on Electron Devices*, 40(1):137–145, Jan. 1993.

[14] B. J. George, D. Gossain, S. C. Tyler, M. G. Wloka, and G. K. H. Yeap. " Power analysis and characterization for semi-custom design. " In *Proceedings of the 1994 International Workshop on Low Power Design*, pages 215–218, April 1994.

[15] A. Ghosh, S. Devadas, K. Keutzer, and J. White. " Estimation of average switching activity in combinational and sequential circuits. " In *Proceedings of the 29th Design Automation Conference*, pages 253–259, June 1992.

[16] H. Goldstein. " Controllability/observability of digital circuits. " IEEE *Transactions on Circuits and Systems*, 26(9):685–693, September 1979.

[17] S. Iman and M. Pedram. " Multi-level network optimization for low power. " In *Proceedings of the* IEEE *International Conference on Computer Aided Design*, pages 372–377, November 1994.

[18] S. Iman and M. Pedram. " Logic extraction and decomposition for low power. " In *Proceedings of the 32nd Design Automation Conference*, June 1995.

[19] S. Iman, M. Pedram and C. Y. Tsui. " PLA minimization for low power VLSI designs. " *CENG Technical Report*, Dept. of EE-Systems, University of Southern California, April 1995.

[20] S. M. King. " Accurate simulation of power dissipation in VLSI circuits. " IEEE *Journal of Solid State Circuits*, 21(5):889–891, Oct. 1986.

[21] K. Keutzer. " DAGON: Technology mapping and local optimization. " In *Proceedings of the 24th Design Automation Conference*, pages 341–347, June 1987.

[22] B. Krishnamurthy and I. G. Tollis. " Improved techniques for estimating signal probabilities. " IEEE *Transactions on Computers*, 38(7):1245–1251, July 1989.

[23] C. Lennard and A. R. Newton. " An estimation technique to guide low power resynthesis algorithms. " In *Proceedings of the* 1995 International Symposium on Low Power Design, pages 227-232, April 1995.

[24] H. M. Lieberstein. " *A Course in Numerical Analysis*. " Harper & Row Publishers, 1968.

[25] B. Lin and H. De Man. " Low-power driven technology mapping under timing constraints " In *Proceedings of the International Conference on Computer Design*, pages 421-427, October 1993.

[26] B. Lin and A. R. Newton. " Synthesis of multiple-level logic from symbolic high-level description languages. " In *Proceedings of IFIP International Conference on Very Large Scale Integration*, pages 187-196, August 1989.

[27] R. Marculescu, D. Marculescu, and M. Pedram. " Logic level power estimation considering spatiotemporal correlations. " In *Proceedings of the* IEEE *International Conference on Computer Aided Design*, pages 294–299, November 1994.

[28] R. Marculescu, D. Marculescu, and M. Pedram. " Efficient power estimation for highly correlated input streams. " In *Proceedings of the 32nd Design Automation Conference*, June 1995.

[29] J. Monteiro, S. Devadas, and A. Ghosh. " Retiming sequential circuits for low power. " In *Proceedings of the* IEEE *International Conference on Computer Aided Design*, pages 398–402, November 1993.

[30] J. Monteiro, S. Devadas, and A. Ghosh. " Estimation of switching activity in sequential logic circuits with applications to synthesis for low power. " In *Proceedings of the 31st Design Automation Conference*, pages 12-17 , June 1994.

[31] F. N. Najm, R. Burch, P. Yang, and I. Hajj. " Probabilistic simulation for reliability analysis of CMOS VLSI circuits. " IEEE *Transactions on Computer-Aided Design of Integrated Circuits and Systems*, 9(4):439–450, April 1990.

[32] K. P. Parker and J. McCluskey. " Probabilistic treatment of general combinational networks. " IEEE *Transactions on Computers*, C-24:668–670, Jun. 1975.

[33] M. Pedram and N. Bhat. " Layout driven technology mapping. " In *Proceedings of the 28th Design Automation Conference*, pages 99-105, June 1991.

[34] M. Pedram, B. T. Preas. " Interconnection length estimation for optimized standard cell layouts. " In *Proceedings of the IEEE International Conference on Computer Aided Design*, pages 390-393, November 1989.

[35] S. Rajgopal and G. Mehta. " Experiences with simulation-based schematic level current estimation. " In *Proceedings of the 1994 International Workshop on Low Power Design*, pages 9–14, April 1994.

[36] J. Rajski and J. Vasudevamurthy. " The testability-preserving concurrent decomposition and factorization of Boolean expressions. " IEEE *Transactions on Computer-Aided Design of Integrated Circuits and Systems*, 11(6):778–793, June 1993.

[37] K. Roy and S. C. Prasad. " Circuit activity based logic synthesis for low power reliable operations. " IEEE *Transactions on VLSI Systems*, 1(4):503–513, December 1993.

[38] A. Salz and M. A. Horowitz. " IRSIM: An incremental MOS switch-level simulator. " In *Proceedings of the 26th Design Automation Conference*, pages 173-178, June 1989.

[39] J. Savir, G. Ditlow, and P. Bardell. " Random pattern testability. " IEEE *Transactions on Computers*, 33(1):1041–1045, January 1984.

[40] H. Savoj, R. K. Brayton, and H. J. Touati. " Extracting local don't cares for network optimization. " In *Proceedings of the IEEE International Conference on Computer Aided Design*, pages 514–517, November 1991.

[41] P. Schneider and U. Schlichtmann. " Decomposition of boolean functions for low power based on a new power estimation technique. " In *Proceedings of the 1994 International Workshop on Low Power Design*, pages 123–128, April 1994.

[42] S. C. Seth and V. D. Agrawal. " *A new model for calculation of probabilistic testability in combinational circuits.* " Elsevier Science Publishers, 1989.

[43] S.C. Seth, L. Pan, and V.D. Agrawal. " PREDICT - Probabilistic estimation of digital circuit testability. " In *Proceedings of the Fault Tolerant Computing Symposium*, pages 220–225, June 1985.

[44] A. A. Shen, A. Ghosh, S. Devadas, and K. Keutzer. " On average power dissipation and random pattern testability of CMOS combinational logic networks. " In *Proceedings of the IEEE International Conference on Computer Aided Design*, November 1992.

[45] V. Tiwari, P. Ashar, and S. Malik. " Technology mapping for low power. " In *Proceedings of the 30th Design Automation Conference*, pages 74-79, June 1993.

[46] V. Tiwari, S. Malik and P. Ashar. " Guarded evaluation: Pushing power management to logic synthesis/design. " In *Proceedings of the 1995 International Symposium on Low Power Design*, pages 221-226, April 1995.

[47] C-Y. Tsui, J. Monteiro, M. Pedram, S. Devadas, A. M. Despain and B. Lin. " Power estimation in sequential logic circuits. " IEEE *Transactions on VLSI Systems, September 1995.*

[48] C-Y. Tsui, M. Pedram, C-H. Chen, and A. M. Despain. " Low power state assignment

targeting two- and multi-level logic implementations. " In *Proceedings of the* IEEE *International Conference on Computer Aided Design*, pages 82–87, November 1994.

[49] C-Y. Tsui, M. Pedram, and A. M. Despain. " Efficient estimation of dynamic power dissipation under a real delay model. " In *Proceedings of the* IEEE *International Conference on Computer Aided Design*, pages 224–228, November 1993.

[50] C-Y. Tsui, M. Pedram, and A. M. Despain. " Technology decomposition and mapping targeting low power dissipation. " In *Proceedings of the 30th Design Automation Conference*, pages 68–73, June 1993.

[51] C-Y. Tsui, M. Pedram, and A. M. Despain. " Exact and approximate methods for calculating signal and transition probabilities in fsms. " In *Proceedings of the 31st Design Automation Conference*, pages 18–23, June 1994.

[52] C-Y. Tsui, M. Pedram, and A. M. Despain. " Power efficient technology decomposition and mapping under an extended power consumption model. " IEEE *Transactions on Computer-Aided Design of Integrated Circuits and Systems*, 13(9), September 1994.

[53] A. Tyagi. " Hercules: A power analyzer of MOS VLSI circuits. " In *Proceedings of the* IEEE *International Conference on Computer Aided Design*, pages 530–533, November 1987.

[54] H. J. M. Veendrick. " Short-circuit dissipation of static CMOS circuitry and its impact on the design of buffer circuits. " IEEE *Journal of Solid State Circuits*, 19:468–473, August 1984.

[55] T. Villa and A. Sangiovanni-Vincentelli. " NOVA: State assignment of finite state machines for optimal two-level logic implementations. " IEEE *Transactions on Computer-Aided Design of Integrated Circuits and Systems*, 9: 905-924, September 1990.

7

Low Power Arithmetic Components

Thomas K. Callaway and Earl E. Swartzlander, Jr.

Minimizing the power consumption of circuits is important for a wide variety of applications, both because of the increasing levels of integration and the desire for portability. Since performance is often limited by the arithmetic components' speed, it is also important to maximize the speed. Frequently, the compromise between these two conflicting demands of low power dissipation and high speed can be accomplished by selecting the optimum circuit architecture.

7.1. Introduction

An important attribute of arithmetic circuits for most applications is maximizing the speed (for general purposes applications) or throughput (for signal processing applications). For a growing number of applications, minimizing the power consumption is also of great importance. The most direct way to reduce the power is to use CMOS circuits, which generally dissipate less power than their bipolar counterparts. Even for CMOS, the use of adders with minimum power consumption is attractive to increase battery life in portable computers, to avoid local areas of high power dissipation which may cause hot spots, and to reduce the need for a low impedance power and ground distribution network which may interfere with signal interconnections.

There are four factors which influence the power dissipation of CMOS circuits. Chandrakasan, *et al* [20] describe these as: technology, circuit design style, architecture, and algorithm. In the second section of this chapter, we will explore the effect that the circuit design style has on the power dissipation and delay. Ripple carry adders were designed and fabricated in a 2 micron CMOS process using 7 different circuit design styles. Using simulations and measurements of the chip, the effects of the circuit style on power dissipation and delay can be observed.

The third section investigates the effect of the algorithm. In a similar fashion to the previous section, 6 different types of adders were designed and fabricated in the same 2 micron CMOS process. These 6 adders were all designed using static CMOS circuits, and this highlights the effect that the choice of algorithm has on the power dissipation and delay.

The fourth section investigates the effect that the architecture has on the power dissipation and delay of a multiplier. Two different multiplier designs are examined, and a simple first order model is used to model their power dissipation and delay.

The next section includes a brief discussion of the power and delay characteristics of two different division algorithms. Again, this is done using a simple first order model. The final section draws some conclusions and summarizes the important points from the previous sections.

7.2. Circuit Design Style

There exist quite a number of CMOS circuit design styles, both static and dynamic in nature. In this section, we will examine seven of the more popular design styles to investigate their suitability for use in low-power, high-performance arithmetic units.

The full adder is the basis for almost every arithmetic unit, therefore any investigation into the suitability of circuit design style for use in arithmetic units must focus on the design of a full adder. In the next subsection, we will examine seven different full adders, each one designed using a different circuit design style. Circuit diagrams will be presented, and the operation of the circuit will be briefly discussed.

Following that, the performance of each full adder will be investigated, both in isolation and when used in a 16-bit ripple carry adder.

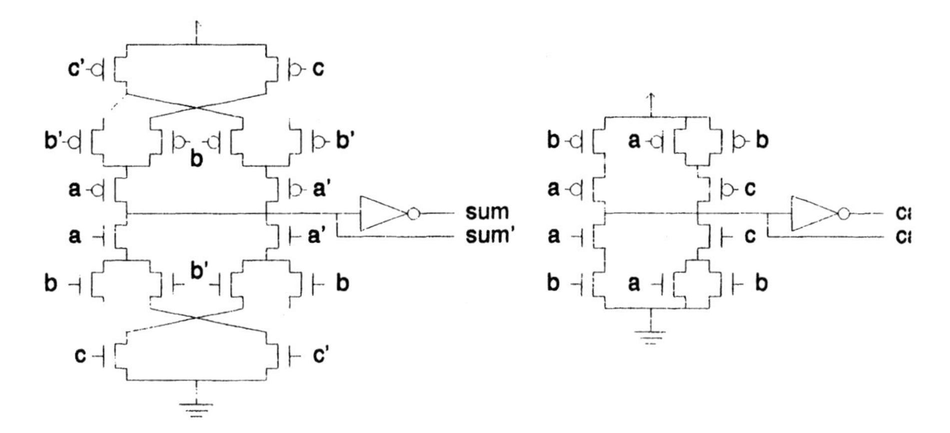

Figure 7.1 Static CMOS Full Adder

7.2.1. Full Adder Designs

An example of a full adder designed using static complementary MOS logic employing both P- and N-type logic trees [1] is shown in Figure 7.1. The P logic tree (upper half of the circuit) allows the output to be charged high, while the N tree (lower half of the circuit) allows the output to be discharged to ground. Both complemented and uncomplemented inputs are required, and both the true sum and carry and the complemented sum and carry are produced. The sum and carry functions are computed independently of each other. This full adder has transistor count of 30 (each of the inverting buffers takes 2 transistors).

Full adders constructed using NO RAce dynamic CMOS logic (NORA) [25] employ alternating stages of P- and N-type logic trees to form the carry and sum outputs as shown in Figure 7.2. The P-type stage that forms the carry output is dynamically precharged high, while the N-type transistor tree that computes the sum output is dynamically pre-discharged low. This precharging and pre-discharging process requires a two-phase complimentary clock, denoted as phi and phi'. NORA logic is unique among the CMOS families discussed here in that it does not require complemented inputs and does not compute both complemented and uncomplemented outputs. The NORA full adder requires 22 transistors.

Cascode Voltage Switch Logic (CVSL) [26] is a dynamic logic family. Like the NORA logic family, it requires a two-phase clock. Unlike NORA circuits, however, the complement of the clock signal is not necessary. In the CVSL full adder circuit shown in Figure 7.3, the outputs and their complements are all

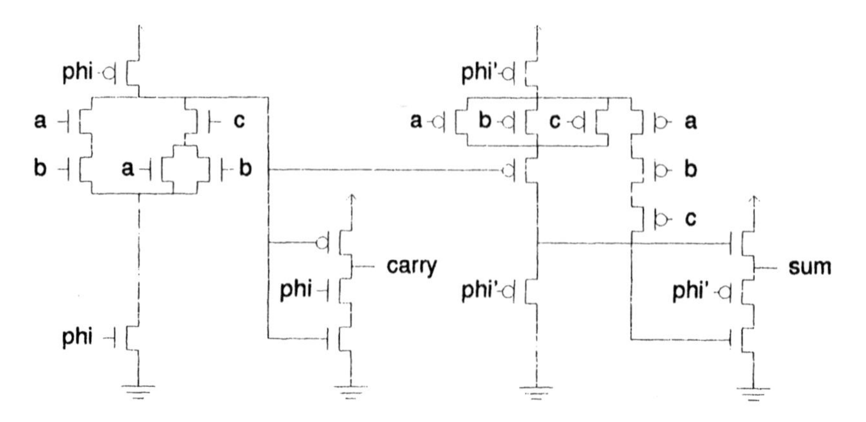

Figure 7.2 NO RAce dynamic CMOS logic (NORA) Full Adder

precharged high while the clock is low. When the clock signal goes high, the complementary cascoded differential N-type transistor trees pull either the output or its complement low. Again, the sum and carry are computed independently of each other.

Replacing the P-type transistors in CVSL with a cross coupled pair of P transistor yields a static version of that logic known as Differential Cascode Voltage Switch (DCVS) logic [27] as shown in Figure 7.4. The cross coupled P transistors act as a differential pair. When the output at one side gets pulled low, then the opposite P transistor will be turned on, and the output on that side will be pulled high.

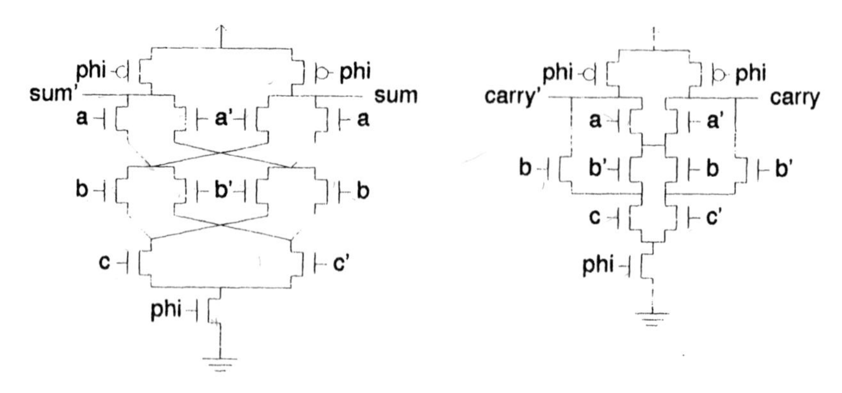

Figure 7.3 Cascode Voltage Switch Logic (CVSL) Full Adder

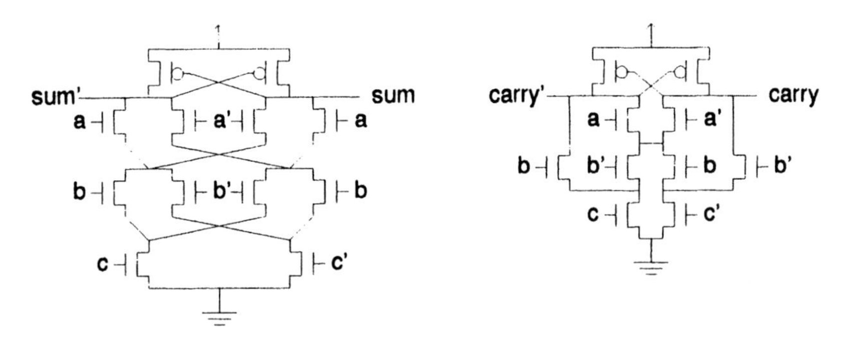

Figure 7.4 Differential Cascode Voltage Switch Logic (DCVSL) Full Adder

CMOS NonThreshold Logic (CNTL) [28] employs the same binary decision trees and cross coupled P transistors as DCVS, but extra N-type transistors are added as shown in Figure 7.5, in order to reduce the potential voltage swing at the output. Two N-type transistors are placed in between the cross coupled P transistors and the output in order to lower the voltage level of the high output. Another two N transistors are placed between the N transistor trees and ground in order to raise the voltage level of the low output. Shunt capacitors are also placed across these last two transistors in order to reduce negative feedback. The resulting reduced voltage swing allows the CNTL full adder to switch more quickly than the DCVS full adder. This full adder requires 30 transistors and 4 capacitors.

Another extension to the DCVS full adder is the addition of self-timing signals, as shown in Figure 7.6. Enable/disable CMOS Differential Logic (ECDL)

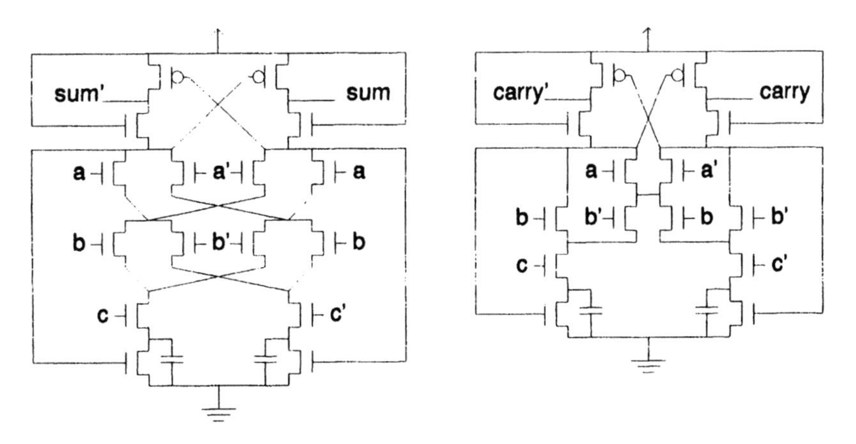

Figure 7.5 CMOS NonThreshold Logic (CNTL) Full Adder

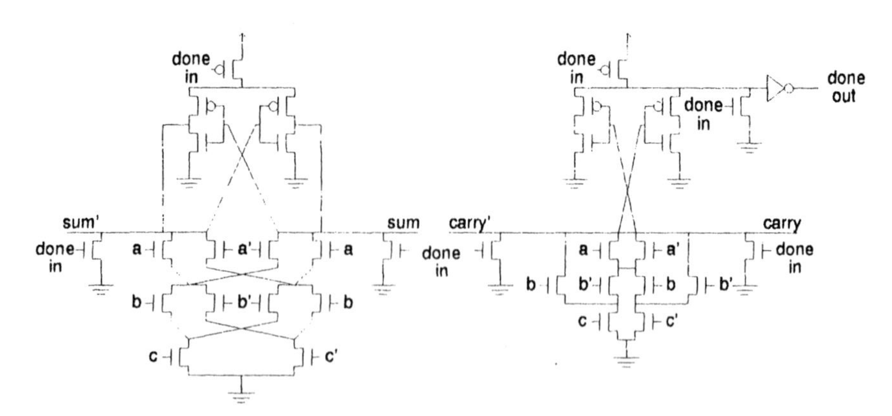

Figure 7.6 Enable/Disable CMOS Differential Logic (ECDL) Full Adder

[29] uses a completion signal, DONE, to pre-discharge the outputs of each stage. While the DONE input signal is held high, the sum and carry outputs are discharged. The DONE input signal goes low when the previous stage has finished its computation. Once the DONE input signal goes low, the full adder operates in the same manner as the standard DCVS full adder. This full adder requires 35 transistors.

Enhancement Source Coupled Logic (ESCL) [30] is another variation on the DCVS full adder. As shown in Figure 7.7, it replaces the P-type transistors with two biased N-type transistors, and replaces the cross coupled pair of P transistors with 2 N-type transistors, whose inputs are held at voltage *Vb*. The value for *Vb* is about a diode drop less than *Vdd*. This circuit is further biased by the insertion of a constant current source in between the bottom of the N-type transistor tree and ground. Inputs to the ESCL circuit steer current from one branch of the tree to another. The current source is a simple 2 transistor current mirror, bringing the total transistor count of this full adder to 24 transistors.

Now that we have introduced each of the full adder designs, we will evaluate their characteristics in the next subsection.

7.2.2. Ripple Carry Adder Characteristics

In this subsection, we will consider the characteristics of each of the full adder designs when used as the basic building block for a 16-bit ripple carry adder. One of the primary concerns in VLSI designs is minimizing the area,

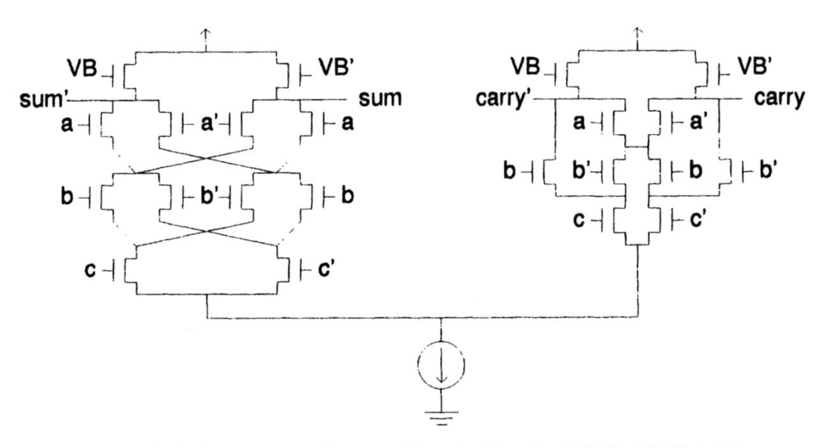

Figure 7.7 Enhancement Source Coupled Logic (ESCL) Full Adder

which is often approximated by the number of transistors, as shown in Table 7.1. The area given is the minimum bounding box for a 2 micron technology.

	Transistors	Rank	Area (μ^2)	Rank
CMOS	30	5	21,294	3
NORA	22	1	14,319	1
CVSL	24	3	25,740	4
DCVS	22	1	21,080	2
CNTL	34	6	40,020	7
ECDL	35	7	34,170	6
ESCL	24	3	26,522	5

Table 7.1 Full Adder Transistor Count and Area

It is interesting to note that even though the static CMOS full adder is the only one which has both full P and N transistor trees, it has one of the smaller areas. As expected, the adder with the largest and smallest numbers of transistors, occupy the largest and smallest areas, respectively. The NORA adder has an especially low area, likely due to the fact that it does not use either uncomplemented inputs or outputs. This reduces the amount of interconnect required, thus reducing the total area.

Table 7.1 gives both the transistor count and area for full adders implemented in each of the seven logic styles previously discussed. Because ripple carry adders are so simple, the area of an N bit ripple carry adder is almost exactly N times the area of a single full adder. Thus, multiplying every entry in Table 7.1 by 16 yields the transistor counts and area for the 16 bit ripple carry adders implemented in the seven circuit styles discussed above.

In studying the power dissipation, peak switching current, and delay, two different strategies will be followed. The first strategy consists of designing and laying out the adders in full custom CMOS, and then extracting information from the layouts for use in a circuit simulator.

The second method involves direct measurements from a test chip, which has been fabricated in order to verify the results of the circuit simulations. This test chip provides a separate power supply pin for each adder, thus simplifying the job of measuring the power dissipation of each adder. This chip is very similar in conception to the one described in detail in the next section.

Although ripple carry adders are not generally suitable for use in situations where a low latency is required, the delay of the seven ripple carry adders, as shown in Table 7.2, varies widely enough that the delay may be the deciding factor in the decision to use one circuit style. In this case, the slowest adder (ESCL) has a worst case delay more than twice that of the faster one (CVSL). The table includes both simulation results and actual measurements.

	Simulated (nsec)	**Rank**	**Measured (nsec)**	**Rank**
CMOS	46.3	3	60	3
NORA	45.9	2	47.2	1
CVSL	45.4	1	49.2	2
DCVS	61.5	5	72.6	4
CNTL	54.1	4	87.0	6
ECDL	79.7	6	85.0	5
ESCL	94.1	7	166.0	7

Table 7.2 16-bit Ripple Carry Adder Delays

One measure of the power consumption of a circuit is its peak switching current. The peak switching current is defined to be the largest difference between the steady state power supply current, and the value of the power supply current when the circuit is performing a computation. All of the full adders discussed in this section except for the ESCL design have an effective steady state power supply current of zero.

Because the test chip provides a separate power pin for each of the adders, measuring the peak switching current is relatively simple. Small (100 Ω) resistors are placed in series with the power supply and the power pins, along with some bypass capacitors to filter spikes. The current drawn by each adder can then be easily obtained. The measurements shown in Table 7.3 were made using a circuit simulator and a Tektronix digital oscilloscope while the adders were being presented with pseudo-random inputs at a frequency of 2 MHz.

	Simulated (mA)	Rank	Measured (mA)	Rank
CMOS	2.42	6	1.30	7
NORA	2.74	7	1.20	5
CVSL	1.08	2	1.06	3
DCVS	1.19	3	1.22	6
CNTL	1.25	4	1.18	4
ECDL	1.27	5	0.98	2
ESCL	0.04	1	0.23	1

Table 7.3 16 Bit Ripple Carry Adder Peak Switching Current

The ESCL adder switches its output by "steering" current from one branch of the transistor tree to another, in a fashion similar to Emitter Coupled Logic (ECL). Because there are always two open paths from power to ground, there is a steady state current of 0.3 mA. In comparison, the static CMOS adder has a steady state power supply current of 3 micro amps.

The ESCL logic family is therefore best suited to use in mixed-signal VLSI circuits, where power supply noise is a problem. It seems to be ill suited to use in low power applications.

In addition to the peak switching current, the average power dissipation is also of interest. In order to measure this average power dissipation, a digital multi-meter was placed in series with the power supply, and the average power supply current was recorded for each adder as shown in Table 7.4.

	Current (μA)	Rank
CMOS	98	1
NORA	948	5
CVSL	925	4
DCVS	116	2
CNTL	1320	6
ECDL	422	3
ESCL	4514	7

Table 7.4 Average Measured Power Supply Current

The results presented in this section indicate that although the ESCL circuit family has the lowest peak switching current of the seven studied, it is unsuitable for use in systems where low power dissipation is essential. Because it has a relatively large standby current, and offers little in the way of performance, it's use seems to be limited to use alongside sensitive analog circuits.

The other logic families all appear to have quite similar characteristics, with static CMOS and DCVS offering the best compromise between low standby current and low latency, while the NORA circuit style offers the smallest area penalty.

7.3. Adders

In static CMOS the dynamic power dissipation of a circuit depends primarily on the number of logic transitions per unit time [1]. As a result, the average number of logic transitions per addition can serve as the basis for comparing the efficiency of a variety of adder designs. If two adders require roughly the same amount of time and roughly the same number of gates, the circuit which requires fewer logic transitions is more desirable as it will require less dynamic power

This is only a first order approximation as the power also depends on switching speed, gate size (i.e., fan-in), fan-out, output loading, *etc.*

Previous attempts to estimate energy consumption (dissipation) for VLSI circuits have included attempts to estimate the worst-case energy consumption for general circuits. Kissin [2] calculated worst-case upper and lower bounds of acyclic circuits built out of 2-input and and or gates, and inverters where power is consumed by the wires connecting gates as well as the gates themselves. Cirit [3] attempts to measure the average power dissipation of CMOS circuits under a wide range of operating conditions by using statistical methods to calculate the probability that a gate will switch states. Jagau [4] attempts to find the worst-case power dissipation of static CMOS circuits by combining a logic simulator with results from the analog simulation of two switching reference gates.

More recently, Callaway and Swartzlander have investigated the average power dissipation of adders [5] and multipliers [6]. Devadas, *et al* [7] have attempted to estimate the maximum power dissipation of CMOS circuits using boolean function manipulation combined with a simple gate model. Nagendra, *et al* [19] investigated the effects of transistor sizing on power dissipation and delay for ripple carry, block carry lookahead, and signed digit adders designed using static CMOS.

The results presented in this section were obtained from three different sources. A gate level simulator was used to generate a first estimate of each adder's dynamic power consumption. For that gate level simulator, the adders are constructed using only and, or, and invert gates. The circuits are then subjected to 50,000 pseudo-random inputs. For each input, the number of gates transitions that occurred during the addition is counted and the average number of logic transitions per addition is computed.

The second estimate is derived from detailed circuit simulation using a program called CAzM, which is similar in behavior to SPICE. A 16 bit version of each adder was designed in 2 micron static CMOS using MAGIC, and the resulting layout was the basis for the circuit simulation. Transistor parameters are those of a recent 2 micron MOSIS fabrication run. Each adder is presented with 1,000 pseudo-random inputs, and the average power supply current waveform is plotted. Each addition is allowed a time of 100 nanoseconds, so that the voltages will stabilize completely before the next addition is performed. Since the worst case delay of the longest adder is approximately 55 nsec, this provides adequate time for the circuits to stabilize.

The third method for obtaining results relies on direct physical measurement. A chip has been designed, fabricated, and tested, containing 16 bit implementations of the six different adders. Each of the six adders has its own power pin, which enables direct measurement of each adder's dynamic power dissipation without having to estimate other power sinks.

The resulting chip occupies an area of 6400 × 4800 square microns, and has 65 pins. There are 33 adder input pins (16 bits + 16 bits + 1 carry in), 17 output pins (16 bits + 1 carry out), 3 multiplexer control pins (an 8:1 mux is used to select either one of the six adder outputs or either of the 2 inputs), 6 adder power pins, 1 mux power pin, 1 adder and mux ground pin, and 2 power and 2 ground pins for the pads.

7.3.1. Adder Types

The following types of adders were simulated: Ripple Carry [8], Constant Block Width Single-level Carry Skip [8], Variable Block Width Multi-level Carry Skip [12], Carry Lookahead [9], Carry Select [10], and Conditional Sum [11]. Each is described briefly below.

The N-bit Ripple Carry adder consists of N full adders (shown in Figure 7.8) with the carry output of the full adder connected to the carry input of its most significant neighbor as shown in Figure 7.9.

This is the smallest possible parallel adder. Unfortunately, it is also the slowest. Since the adders are connected by the carry chain, a worst case addition

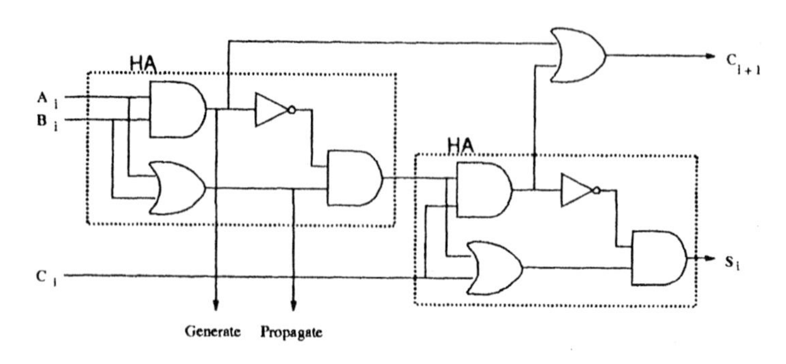

Figure 7.8 Full Adder

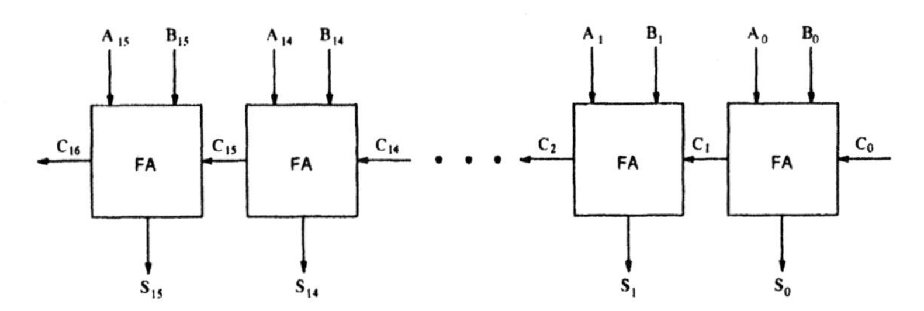

Figure 7.9 16 Bit Ripple Carry Adder

will require that a carry ripple from the least significant full adder through N - 2 intermediate full adders (each with 2 gate delays from carry input to carry output) to the most significant full adder. If all of the carries were known ahead of time, the addition could be completed in only 6 gate delays. That is the amount of time it takes for the two input bits to propagate to the sum output of a single full adder.

Since the sum delay is very small compared to the carry delay (6 vs. 68 delays for a 32 bit Ripple Carry adder), all of the other adders discussed in this subsection use additional logic to speed up carry generation and/or propagation.

For the Carry Lookahead adder shown in Figure 7.10, the OR gate used in the full adder to generate the carry out is eliminated, resulting in an adder with only eight gates. The carry input for each full adder is computed by a tree of lookahead logic blocks. Each block receives p_i and g_i signals from the full adders, and computes the carries. lookahead logic is implemented with four-bit modules, each of which has 14 gates.

The carry select adder consists of pairs of k-bit wide blocks, each block consisting of a pair of ripple carry adders (one assuming a carry input and one assuming no carry input), and a multiplexer controlled by the incoming carry signal. For further details on the circuits involved, see [10]. The carry generation is done using the same carry lookahead modules used in the carry lookahead adder.

The Carry Skip adder uses the idea that if corresponding bits in the two words to be added are not equal, a carry signal into that bit position will be passed to the next bit position. In the Constant Block Width Carry Skip adder shown in Figure 7.12, the optimal width of each block is determined by a formula based on the number of bits to be added [8]. This implementation uses only a single level

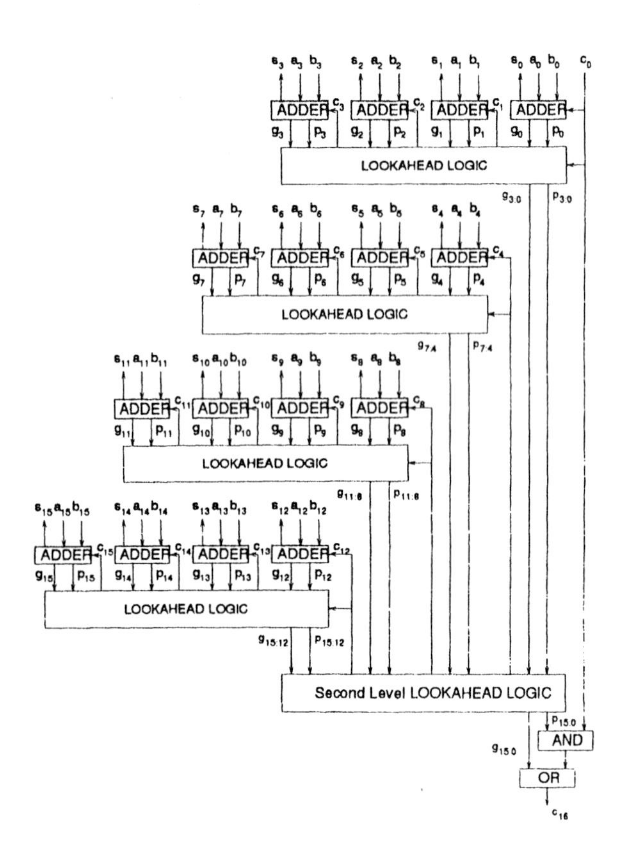

Figure 7.10 16-Bit Carry Lookahead Adder

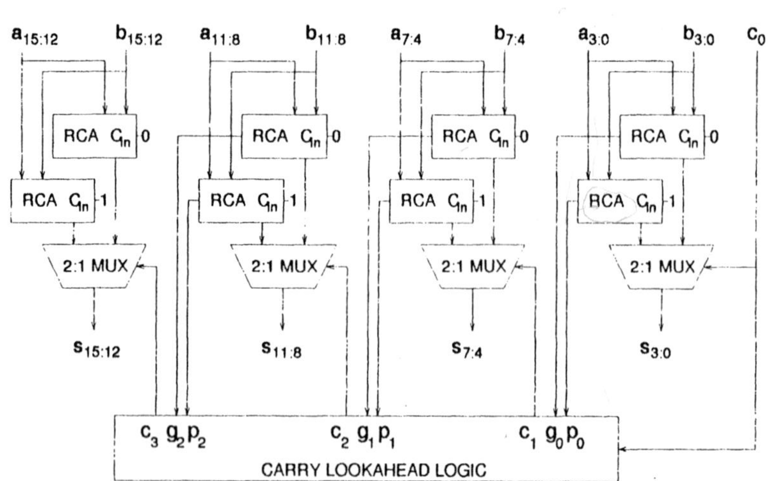

Figure 7.11 16-Bit Carry Select Adder (CSA)

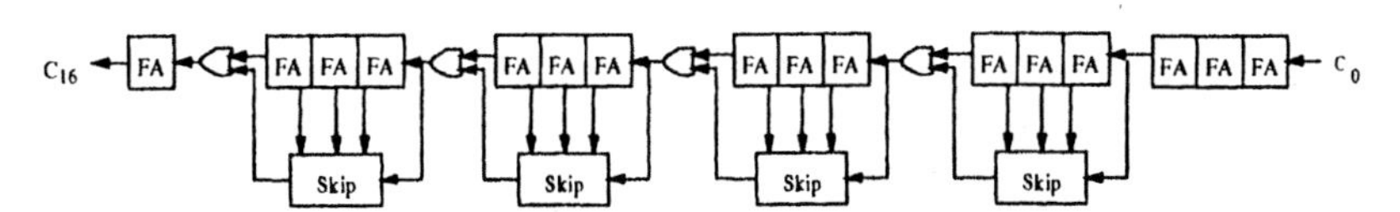

Figure 7.12 16 Bit Constant Block Width Carry Skip Adder

of skip logic. The optimal block size for a 16 bit adder is 3 bits, so that the 16 bit adder is realized by dropping the two most significant bits of an 18 bit adder.

For the Variable Block Width Carry Skip adder shown in Figure 7.13, multiple levels of skip logic (two or three in this chapter) and variable block sizes are used in order to create a faster circuit. The Turrini algorithm [12] is used to generate the optimal grouping of skip groups and block sizes. Application of this algorithm generally results in an adder larger than the desired size. The unnecessary full adders and skip logic on the most significant end can be eliminated without affecting the delay.

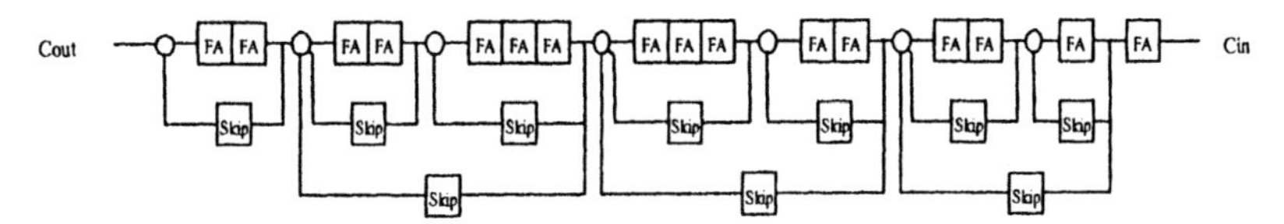

Figure 7.13 16 Bit Variable Block Width Carry Skip Adder

The Conditional Sum adder [11] shown in Figure 7.14 uses 2:1 multiplexers to recursively combine successively larger blocks of conditional sum and carry bits. The basic cell of this adder is based on the Sklansky "H" cell [11]. This circuit accepts as inputs the two bits to be added, and produces four outputs: the sum and carry for an assumed carry in of zero, and the sum and carry for an assumed carry in of one.

In choosing an adder for a particular application, several things must be considered, including speed, size, and dynamic power consumption. The speed is usually considered to be the worst case number of gate delays from input to the slowest output (typically the most significant sum bit). Table 7.5 presents the worst case number of gate delays for the six adders For the purposes of this chapter, all gates are assumed to have the same delay, regardless of the fan-in or fan-out.

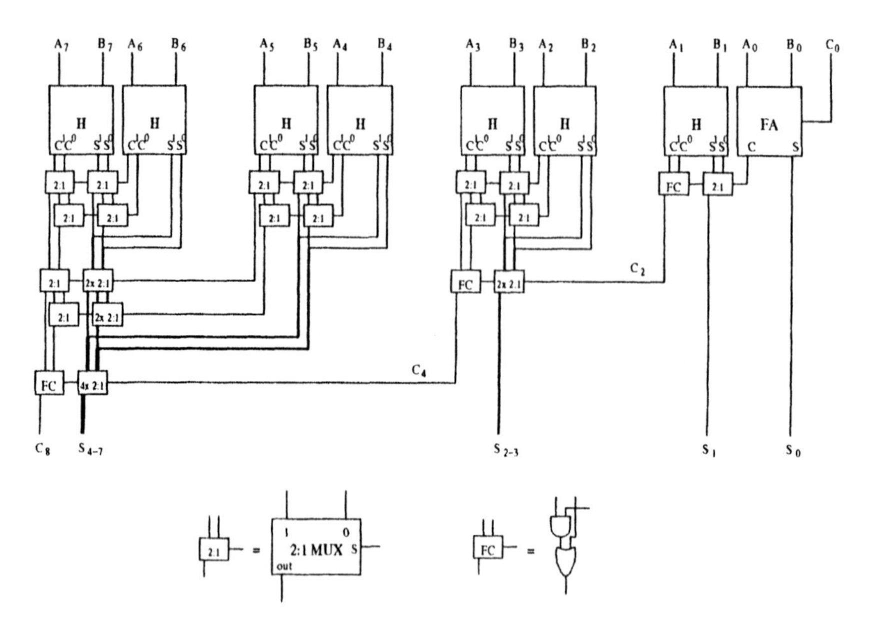

Figure 7.14 8-Bit Conditional Sum Adder

Adder Type	Adder Size		
	16	32	64
Ripple Carry	36	68	132
Constant Width Carry Skip	23	33	39
Variable Width Carry Skip	17	19	23
Carry Lookahead	10	14	14
Carry Select	14	14	14
Conditional Sum	12	15	18

Table 7.5 Worst Case Delay (in gate units)

The size of each adder is approximated to a first order by the number of gates, which is given in Table 7.6. The actual area required for a given adder will depend on the types of gates (i.e., three or four input gates require more area than two input gates), the amount of wiring area, and the available form factor, as will be seen in the next subsection.

Adder Type	Adder Size		
	16	32	64
Ripple Carry	144	288	576
Constant Width Carry Skip	156	304	608
Variable Width Carry Skip	170	350	695
Carry Lookahead	200	401	808
Carry Select	284	597	1228
Conditional Sum	368	857	1938

Table 7.6 Number of Gates

7.3.2. Gate Level Simulation

The gate-level simulator used to measure the average number of gates that switch during an addition accepts as its input a linked list of gates, with each gate pointing to its inputs, and also the next gate to be evaluated. The circuits themselves are built in C subroutines, and a pointer to the first gate in the circuit is passed to the simulator. The circuits then perform a user-specified number of additions with pseudo-random input patterns. For each pattern, the number of gates that switch during the addition is counted.

Before the first pseudo-random pattern is applied, the adder is initialized by applying zero inputs and allowing it to stabilize. Any gate transitions during this initialization are not counted. After that, pseudo-random inputs are presented one after the other, with no zero inputs in between. This is analogous to the adder being used in a vector processor with a vector length of 50,000. Table 7.7 gives the average number of transitions per addition observed for each adder in 50,000 randomly distributed input patterns with a random distribution of carry inputs.

Figures 7.15 and 7.16 show the probability distributions of the number of gate transitions for 16-bit and 64-bit adders of the six different types, based on the simulations with 50,000 input patterns. In both sizes, the ripple carry, carry skip, and carry lookahead adders have roughly the same distribution, while the carry

Adder Type	Adder Size		
	16	32	64
Ripple Carry	90	182	366
Constant Width Carry Skip	102	199	392
Variable Width Carry Skip	108	220	437
Carry Lookahead	100	202	405
Carry Select	161	344	711
Conditional Sum	218	543	1323

Table 7.7 Average Number of Logic Transitions

select adder peaks at about twice the ripple carry adder, and the conditional sum peaks at three times that value.

7.3.3. Circuit Simulation Results

The simulator used to obtain the results in the previous section uses a rather simple model, and provides no time information. In order to verify those results, 16 bit versions of the six different adders were custom designed in 2 micron CMOS using Magic. Netlists were extracted from the layout, and the

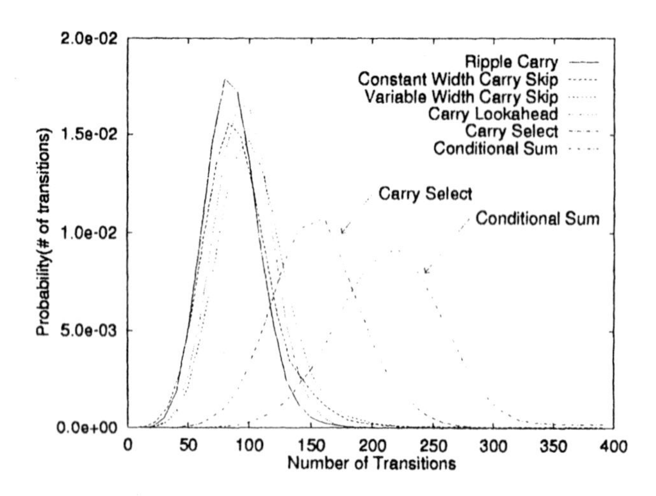

Figure 7.15 16-Bit Adder Logic Transition Histogram

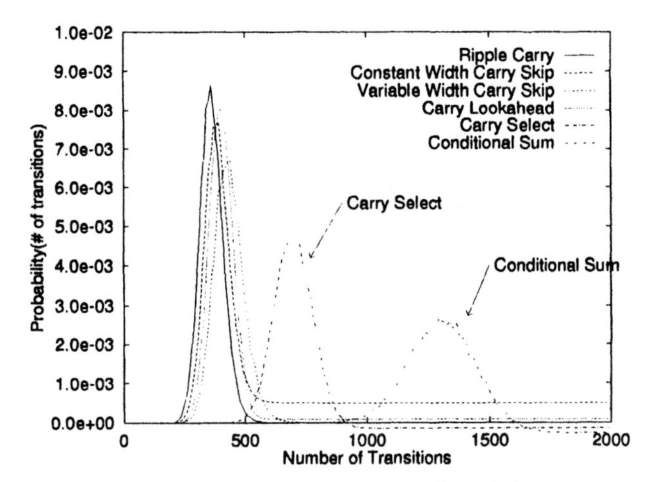

Figure 7.16 64-Bit Adder Logic Transition Diagram

CAzM circuit simulator was then used to estimate the worst case delay and the average power dissipation.

Table 7.8 shows the worst case delay of the six adders as estimated with CAzM. This delay was obtained by setting all of the inputs to zero initially, and then changing the carry in and one of the inputs words to all ones. The delay is then the length of time that the circuit takes to stabilize.

Type of Adder	Delay from CAzM (nsec)	Delay from Unit Delay Model (nsec)	Difference
Ripple Carry	54.27	52.2	2.1
Constant Width Carry Skip	28.38	33.3	-4.9
Variable Width Carry Skip	21.84	24.6	-2.8
Carry Lookahead	17.13	14.5	2.6
Carry Select	19.56	20.3	-0.7
Conditional Sum	20.05	17.4	2.7

Table 7.8 Worst case delay of a 16-bit Adder Estimated with CAzM

The table also shows an approximate delay based on the unit gate delay model (with the delays shown in Table 7.5) for an assumed 1.45 nsec for the "unit" delay. Δ = 1.45 nsec. The delay from CAzM for the constant block width carry skip adder is about 15% faster than estimated by the "unit delay" model, while the CAzM delay estimates for the carry lookahead and the carry select adders are about 20% to 25% slower than predicted by the unit delay model. In the case of the carry lookahead adder, the greater delay is probably due to its use of three and four input gates and a fan out of greater than two in the carry lookahead logic. For the conditional sum adder, the greater delay is probably due to the long path lengths.

Table 7.9 shows the number of gates and the area occupied by each adder in a 2 micron MOSIS technology. The areas indicated in Table 7.9 exhibit a wider

Type of Adder	Area (mm^2)	Gate Count
Ripple Carry	0.2527	144
Constant Width Carry Skip	0.4492	156
Variable Width Carry Skip	0.5149	170
Carry Lookahead	0.7454	200
Carry Select	1.0532	284
Conditional Sum	1.4784	368

Table 7.9 Size of 16-Bit Adders

range (approximately 6:1) than the gates counts from Table 7.6 (range of approximately 2.5:1), but the relative sizes are consistent with the gate counts. Part of the explanation for the wider range in sizes is that the ripple carry adder is laid out in a single row, while all of the others require multiple rows with varying amounts of "wasted space" for signal routing between the rows. With additional effort, the larger adders (i.e., carry lookahead, carry select, and conditional sum) could be laid out more compactly.

The average power dissipation per addition shown in Table 7.10 is obtained by simulating the addition of 1,000 pseudo-random inputs, and averaging the results. The average power dissipation is lower for the constant width carry skip adder than for the ripple carry adder because the power supply current falls to zero faster, even though it is larger at the peak for the constant width carry skip adder.

Type of Adder	Power (mW)
Ripple Carry	0.117
Constant Width Carry Skip	0.109
Variable Width Carry Skip	0.126
Carry Lookahead	0.171
Carry Select	0.216
Conditional Sum	0.304

Table 7.10 Average Power Consumption of 16-Bit Adders Calculated with CAzM

Because CAzM also provides timing information, it is useful to look at the distribution in time of the average power dissipation. This easily done by storing the curve of power supply current versus time for each individual addition. Then after all 1,000 additions, the curves are averaged together. This one curve then represents the average power dissipation (power supply current times power supply voltage) with respect to time of the adder, as shown in Figure 7.17 for each of the six adders.

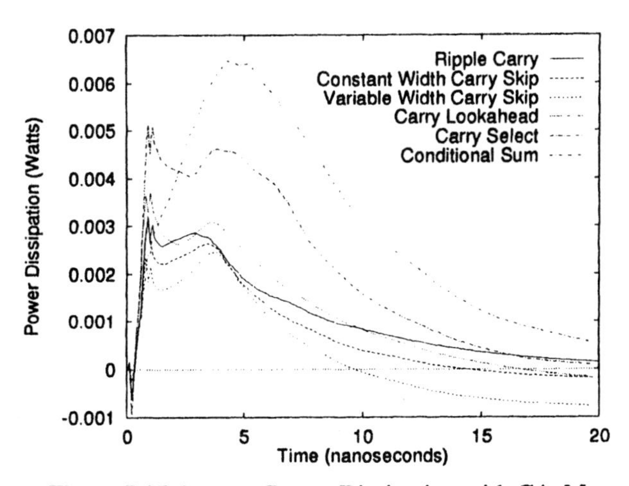

Figure 7.17 Average Power Dissipation with CAzM

7.3.4. Physical Measurement

The third method of obtaining results is to directly measure the power dissipation. The test chip shown in Figure 7.18 was constructed so that each adder has a separate power pin. Only the pads and the output multiplexer share a power net. There is a common ground for all devices.

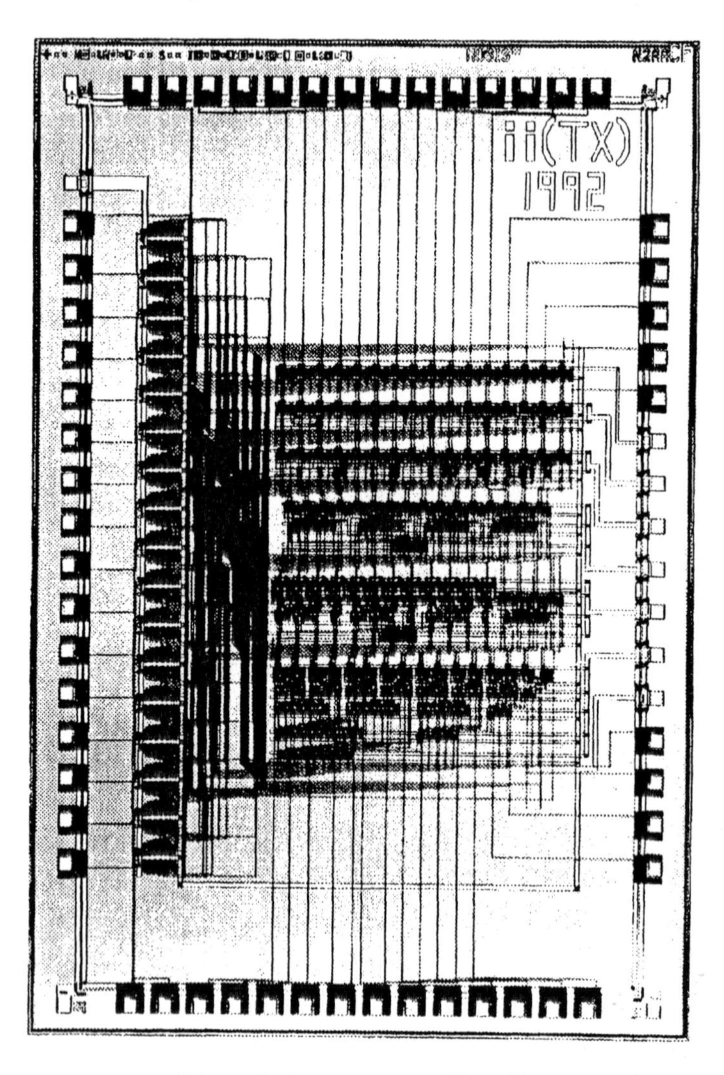

Figure 7.18 Die Photo of Test Chip

All six of the adders in Figure 7.18 are contained in the center of the chip. The adders run horizontally, and are laid out from top to bottom in the same order as in the tables, i.e. the topmost adder is the ripple carry adder, and the bottommost adder is the conditional sum adder. The vertical row of cells to the left of the adders is a 17-bit wide 8:1 output multiplexer.

Because each adder has its own power pin, measuring the power dissipation is simple. The mux is set to select the correct adder output for test purposes, and two power supplies are connected, one for the pads and the mux, and one for the adder. A simple test board was constructed to provide inputs and functionality testing for the chips. Linear Feedback Shift Registers on the test board were used to generate the pseudo-random operands, and a separate adder on the test board was used to verify that the adders were functional at the 2 MHz clock rate of the test board.

The power supply current was measured by inserting a 100 Ω resistor in series with the power supply, and a small (about 1 nanofarad) bypass capacitor to provide some spike filtering. The power supply current can then be determined by measuring the voltage across the resistor. A Tektronix TDS 460 digital Oscilloscope was used to collect and average the voltage across that resistor over a moving window of 1,000 additions. Figures 7.19 - 7.24 show photos of the average voltage across that 100 Ω resistor for the six different adders at a clock frequency of 2 MHz.

As seen in the scope photos in Figures 7.19 - 7.24, the test board adds quite a bit of delay. The power supply current indicates that the operands are not even received by the adder inputs until about 50 nsec have passed. The long tail of the

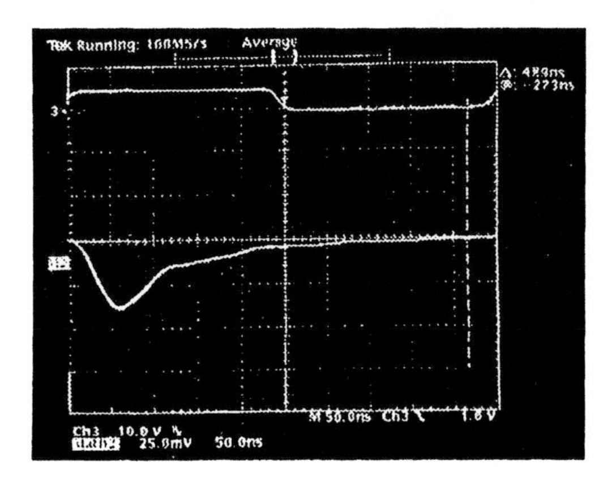

Figure 7.19 Ripple Carry Adder

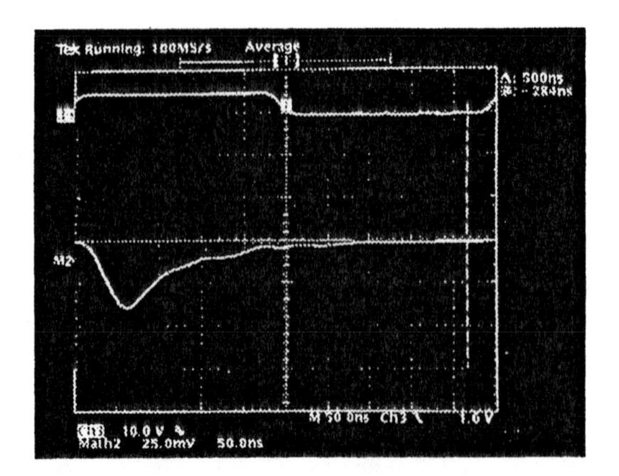

Figure 7.20 Constant Width Carry Skip Adder

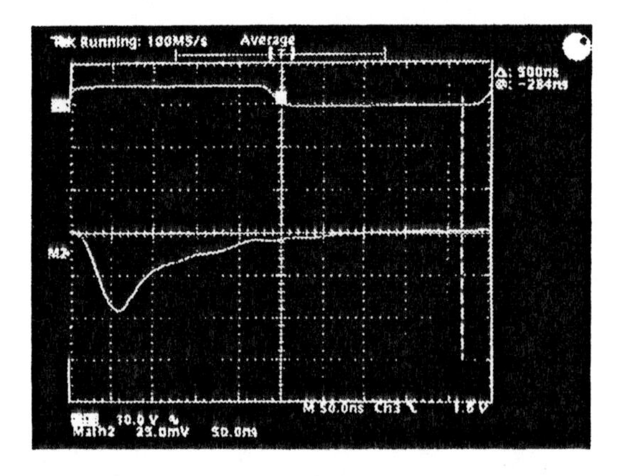

Figure 7.21 Variable Width Carry Skip Adder

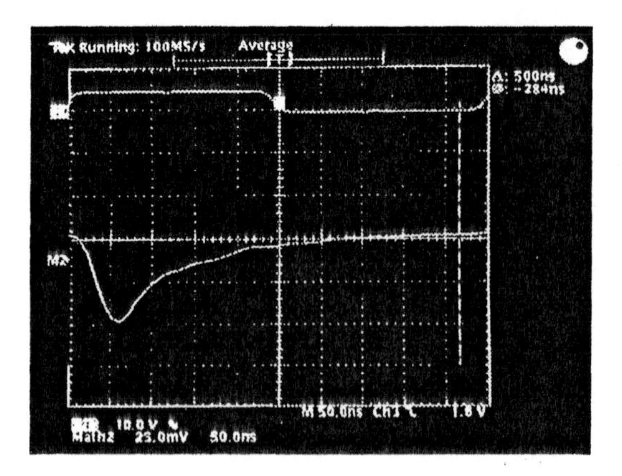

Figure 7.22 Carry Lookahead Adder

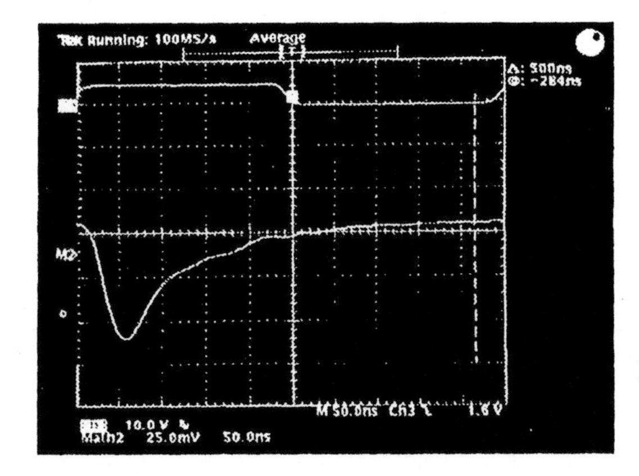

Figure 7.23 Carry Select Adder

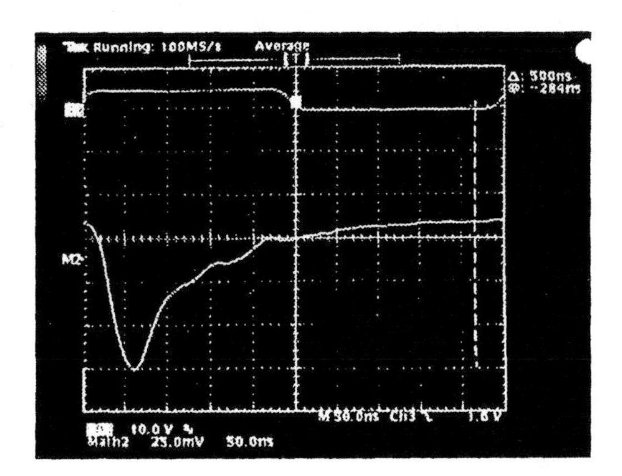

Figure 7.24 Conditional Sum Adder

power supply current is most likely due to the oscilloscope being set for 20 MHz band limiting.

Another set of measurements was made by connecting a multi-meter in series with the power supply and measuring the current while the adders are accepting pseudo-random inputs. The average current drawn by each adder is presented in Table 7.11. The problem with this measurement is that it is difficult to tell exactly what period of time the multi-meter is measuring the current over. These measurements are in close agreement with the results from the simple gate leve! model and the results from CAzM.

Type of Adder	Current (μA)
Ripple Carry	0.2527
Constant Width Carry Skip	0.4492
Variable Width Carry Skip	0.5149
Carry Lookahead	0.7454
Carry Select	1.0532
Conditional Sum	1.4784

Table 7.11 Average Measured Power Supply Current

7.3.5. Conclusions

In this section we have examined 6 types of adders in an attempt to model their power dissipation. We have shown that the use of a relatively simple model provides results that are qualitatively accurate, when compared to more sophisticated models, and to physical implementations of the circuits.

The main discrepancy between the simple model and the physical measurements seems to be the assumption that all gates will consume the same amount of power when they switch, regardless of their fan-in or fan-out. Because the carry lookahead adder has several gates with a fan-out and fan-in higher than 2, the simple model underestimates its power dissipation. The next step in this research is the development of a more accurate gate-level model which will include the ability to provide timing information and will model the power dissipation of each gate according to its fan-in and fan-out.

7.4. Multipliers

In this section, we investigate the importance of the choice of algorithm on the power and delay of multipliers. Although most previous research has focused on adders, there has been some work directed towards multipliers. Chau and Powell [15] investigate the power dissipation of VLSI processor arrays suitable for use as matrix-matrix multipliers. They focus primarily on the optimum partitioning of an algorithm onto a smaller number of processing elements. Recently, the power dissipation of adders was examined [5] using a method similar to the one presented here. Lemonds and Shetti [18] demonstrate that the use of latches to prevent glitches can drastically reduce power dissipation in a 16 by 16 array multiplier.

7.4.1. Multiplier Types

In general, multiplication can be viewed as repeated shifts and adds. Multiplication can be implemented very easily and simply using only an adder, a shift register, and a small amount of control logic. The advantage of this approach is that it is small. The obvious disadvantage is that it is slow. Not only is it a serial circuit, which will result in delays due to latching, but the number of additions it requires is equal to one minus the number of bits in the operands.

One fairly simple improvement to this is to form the matrix of partial products in parallel, and then use a 2-dimensional array of full adders to sum the rows of partial products. The only difficulty in such an approach is that the matrix of partial products is rhomboidal in shape due to the shifting of the partial products. This can be overcome by simply skewing the matrix into a square and then propagating the sum and carry signals of the full adders accordingly. This structure shown in Figure 7.25 is known as an array multiplier.

Each cell in the N by N array includes an and gate to form the appropriate partial product bit. The cells labeled "G" consist of just the and gate. The cells labeled "FA" are an and gate and a full adder as shown in Figure 7.26. The cells labeled "HA" are similar to the full adder cells, but with a half adder. The bottom row of one half adder and $N - 2$ full adders is simply a ripple carry adder. None of the cells in the bottom row include the and gate.

The advantages of the array multiplier are that it has a regular structure and a local interconnect. Each cell is connected only to its neighbors. This translates into a small, dense layout in actual implementations. The disadvantage is that the worst case delay path of an N by N array multiplier goes from the upper left corner diagonally down to the lower right corner and then across the ripple carry adder. This means that the delay is linearly proportional to the operand size.

One method which can be employed to decrease the delay of the array multiplier is to replace the ripple carry adder at the bottom with a carry lookahead adder. By doing this, the delay can be reduced by 10% for the 8 bit multiplier and 20% for the 32 bit multiplier while increasing the number of gates by only 6% for the 8 bit multiplier and 1% for the 32 bit multiplier. In this section, such a multiplier is referred to as a Modified Array multiplier.

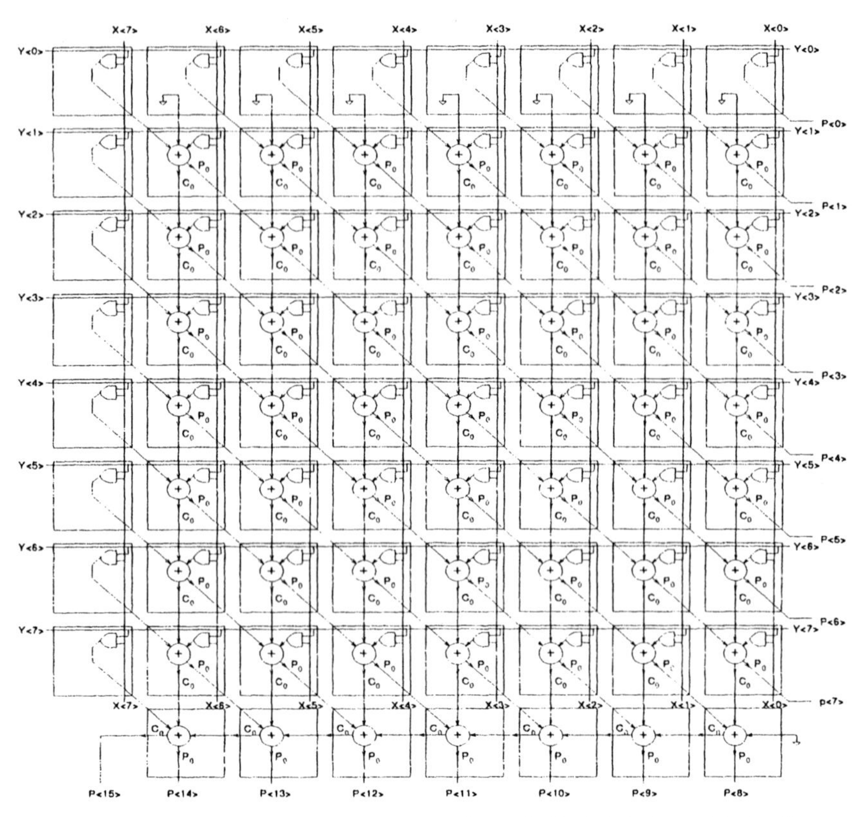

Figure 7.25 8-Bit Array Multiplier

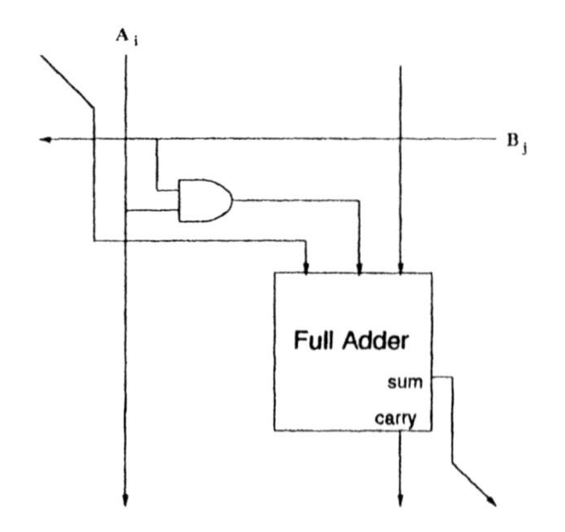

Figure 7.26 One Cell of an Array Multiplier

The area and the power dissipation of the two types of array multipliers do not differ by much. The carry lookahead adder has roughly the same power dissipation as the ripple carry adder it replaces, and only increases the gate count very slightly, while reducing the delay quite a bit. Because of this combination, only the Modified Array Multiplier will be discussed.

Wallace [16] showed that it was possible to improve the delay to logarithmic time. He noted that by using a pseudo-adder (a row of N full adders with no carry chain), it is possible to sum three operands into a two operand result with only a single full adder delay. The two resulting words are added together using a carry propagate adder, and the result is the sum of the three initial operands. He proposed using pseudo-adders repeatedly in a tree structure for summing the partial products into 2 larger partial products, and then using a fast carry propagate adder to sum them and produce the product. The structure for such a Wallace tree is shown in Figure 7.27.

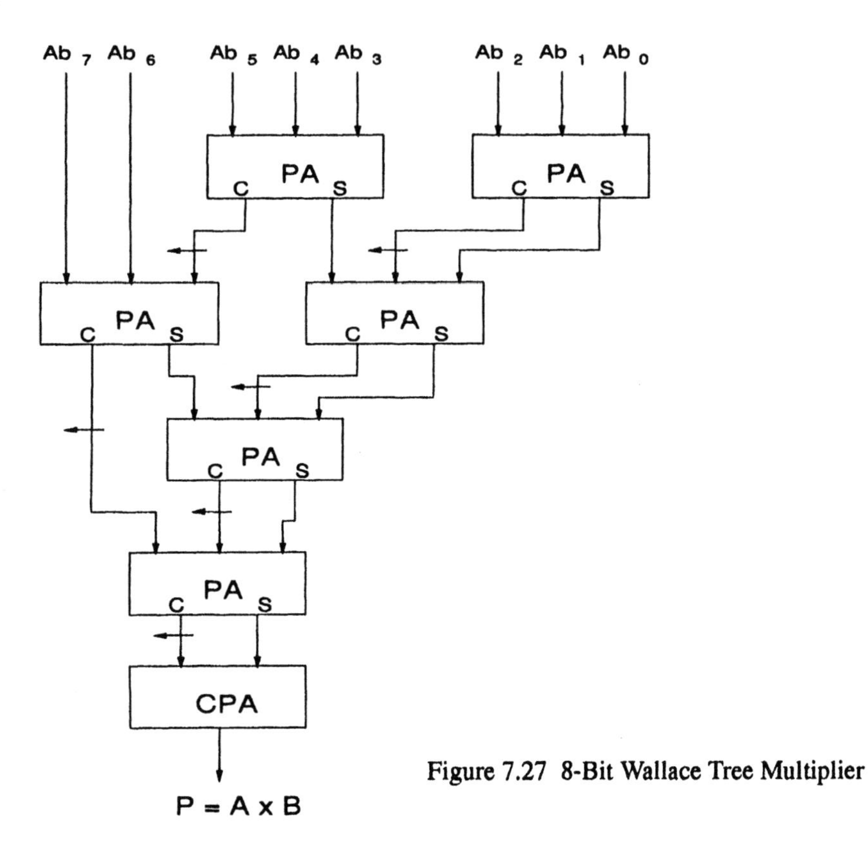

Figure 7.27 8-Bit Wallace Tree Multiplier

Each block labeled "PA", called a pseudo-adder, is an appropriately-sized block of full adders with the carry in and carry out used as inputs and outputs. Each of the 8 inputs Ab_j is the appropriately-shifted bit-wise and of A with the j-th bit of operand B. Basically, each of these is one row in the matrix of partial products. The arrows on each carry out line indicate that all of those bits will be shifted one position to the left, since they have a higher weight. These are then reduced to two wider rows in 4 full adder delays. The final addition is then done with a fast carry propagate adder. That fast adder is assumed to be a Carry Lookahead adder which is nearly $2N$ bits wide.

The method proposed by Wallace has a delay proportional to $log\ N$ for an N by N multiplier, making it faster than the array multiplier. It also requires slightly fewer gates. It's disadvantage is a fairly irregular structure. This means that it will require a larger area than the array multiplier and will be much more difficult to implement in VLSI.

Dadda [17] generalized and extended Wallace's results by noting that a full adder can be thought of as a circuit which counts the number of ones in the input and outputs that number in 2-bit binary form. Using such a counter, Dadda realized that the height of the matrix at each stage could be reduced by at most a factor of 1.5. This results in the following sequence for the reduction of the partial product matrix to two rows:

$$\{2, 3, 4, 6, 9, 13, 19, 28, 42, 63\} \tag{7.1}$$

He then postulated that at each stage, only the minimum amount of reduction should be done in order to reduce the height of the partial product matrix to the next number in the sequence.

For an 8 by 8 multiplier, this results in the following stages of reduction: 8, 6, 4, 3, 2. This is shown in Figure 7.28 using a Dadda dot diagram. Each dot represents a single bit in the matrix of partial products. Ovals indicate that those two or three bits will be the inputs to a half or full adder. Lines in the next level are used to indicate that those two dots are the sum and carry out of an adder (crossed lines indicate half adder outputs). The two final rows are then added together using a fast carry propagate adder (CPA) to obtain the final sum.

Compared to the Wallace multiplier, Dadda's method generally results in a design with fewer full and half adders, but with a slightly wider fast CPA. The worst case delay for both will be the same. The disadvantage of Dadda's method

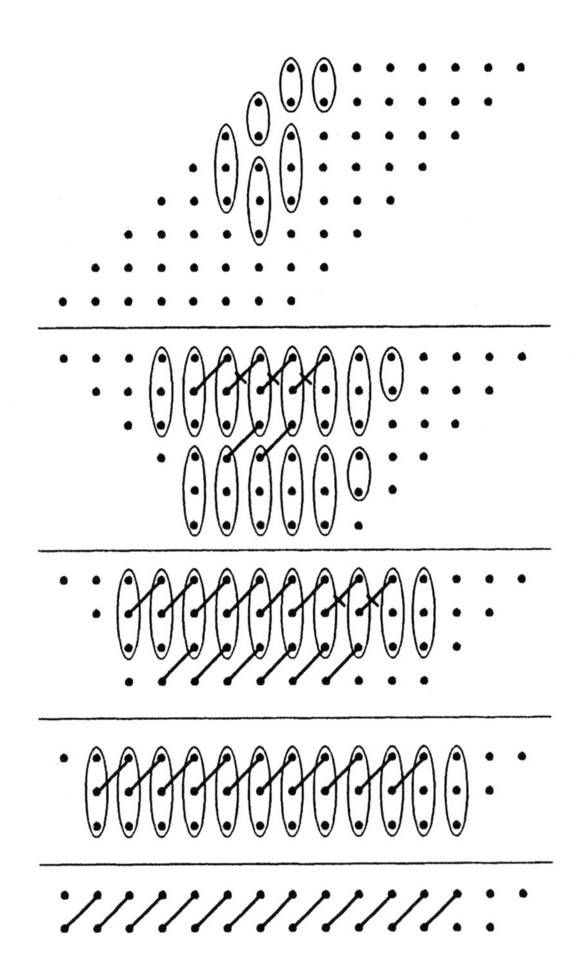

Figure 7.28 8-Bit Dadda Multiplier

is that it results in a structure which is less regular than Wallace's, making it more difficult to lay out in VLSI. Because the Wallace and Dadda multipliers are so similar, they will be treated as identical for the purposes of this discussion.

Table 7.12 gives the worst case delay in gate delays for both of the above multiplier types for operand sizes of 8, 16, and 32. The number of gates required for each circuit is shown in Table 7.13. For a random distribution of inputs, the average number of logic transitions per multiplication, computed on the basis of 50,000 trials, is shown in Table 7.14.

Multiplier Type	Multiplier Size (bits)		
	8	16	32
Modified Array	50	98	198
Wallace/Dadda	35	51	63

Table 7.12 Worst Case Delay

Multiplier Type	Multiplier Size (bits)		
	8	16	32
Modified Array	567	2405	9918
Wallace/Dadda	613	2569	10,417

Table 7.13 Number of Gates

Multiplier Type	Multiplier Size (bits)		
	8	16	32
Modified Array	583	7348	99,102
Wallace/Dadda	573	3874	19,548

Table 7.14 Average Number of Logic Transitions per Multiplication

7.4.2. Results

Figures 7.29, 7.30 and 7.31 show the probability distributions of the average number of logic transitions per multiplication for the 8-, 16-, and 32-bit multipliers.

At 8 bits, both of the multiplier types are close, but as the word length increases, the array multiplier becomes increasingly unattractive. This is due both to the higher average number of logic transitions and to its much higher (linear) delay. The one advantage of the array multiplier is regularity, which will lead to a compact and simple layout. On the other hand, the Wallace and Dadda type multipliers will tend to have an irregular layout, making them larger than the gate count indicates.

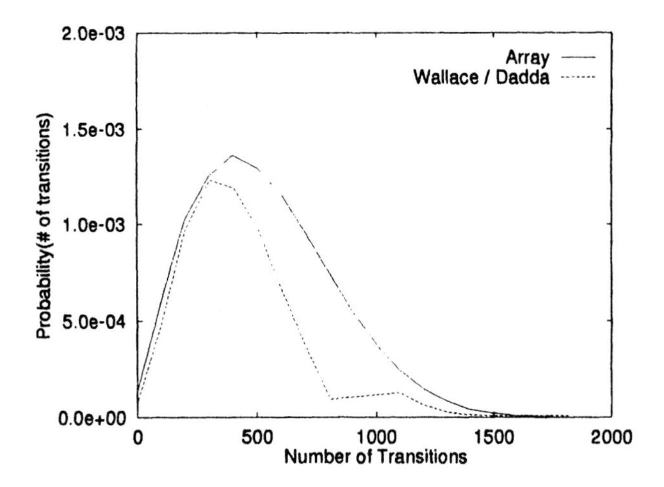

Figure 7.29 8-Bit Multiplier Logic Transition Histogram

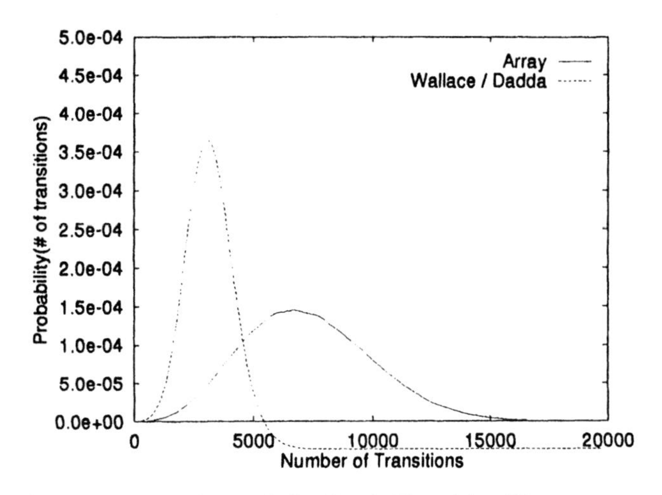

Figure 7.30 16-Bit Multiplier Logic Transition Histogram

7.4.3. Multiplier Conclusions

This section has examined several parallel multipliers to determine their utility for low power, high speed applications and presented a figure of merit that can be used to compare the multipliers based on their delay and dynamic power consumption. This figure of merit provides a common ground for ranking the multipliers in terms of their utility for low power applications. By this figure of

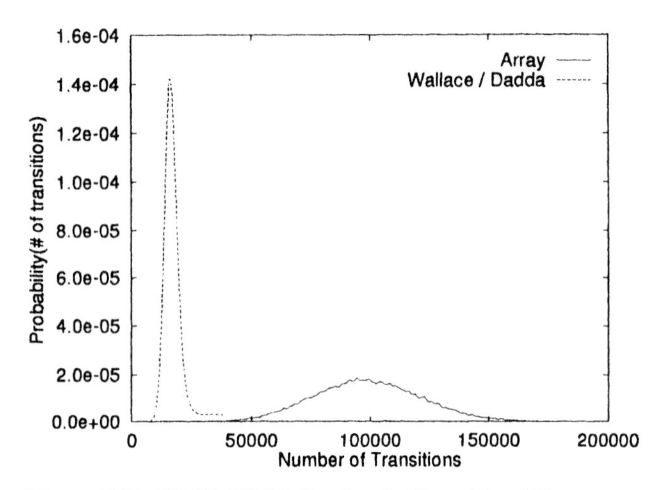

Figure 7.31 32-Bit Multiplier Logic Transition Histogram

merit, either the Wallace or the Dadda multipliers are the best for word sizes between 8 and 32 bit inclusive.

7.5. Division

In this section, we will develop some general estimates for the number of logic transitions per division operation. Division is inherently a serial operation, unlike addition and multiplication, which are usually implemented with large combinational circuits. There are generally two methods for high-speed division: SRT [21] and Newton-Raphson [22]. In both of these cases, the division operation takes the form of successive approximation until an accurate enough answer is obtained.

7.5.1. SRT Division

SRT division takes the form of repeated shift and add/subtract operations. The amount of hardware required is rather modest: an N-bit adder, some gates for the quotient prediction logic, and a state machine.

SRT division converges linearly, and in its simplest form requires N cycles, resulting in a delay of $N \times$ (adder delay + predict delay), and a total average number of logic transitions equal to $N \times$ (adder transitions + predict transitions), ignoring the delay and power dissipation of the latches or flip-flops.

As an example, let us now discuss the design of a radix-2 SRT divider for 64-bit numbers. We require one 64-bit adder, and about 10 gates for the quotient selection logic, plus some small control overhead. We know that each cycle, the adder will be responsible for about 405 gate output transitions (from Table 7.7). Simulating the quotient prediction gates with random inputs reveals that they will account for about 5 logic transitions per cycle. This gives a total of 410 logic transitions per cycle. An average of 64.5 cycles are required to perform the division ($N = 64$, so we need at least 64 cycles, and half the time the remainder will be negative, requiring an additional corrective step). This results in an average of $64.5 \times 410 = 26{,}445$ logic transitions per division.

The worst case delay is 65 times the delay of the adder plus the delay of the quotient prediction logic. The quotient logic has a delay of 3 in the worst case, so the worst case delay for the divider is $65 \times (14 + 3) = 1105$ gate delays.

It is possible to reduce the number of cycles required by using a radix larger than 2. As an example, we will examine the use of radix 4 SRT division. Using radix 4 means that two bits of the answer are produced each cycle, so that we need do only half as many cycles. However, the quotient prediction logic becomes more complicated, having a delay of 6 gates, and an average of 37 logic transitions per cycle.

This means that the average number of logic transitions is $32.5 \times (405 + 37) = 14365$, and the worst case delay is $33 \times (14 + 6) = 660$ gate delays. Tables 7.15 and 7.16 summarize these results, and also includes figures for 16- and 32-bit operands, again ignoring the effects of the latches on both the delay and number of logic transitions.

Tables 7.15 and 7.16 indicate that radix 4 SRT division is better than radix 2 SRT division for word lengths between 16 and 64 bits, inclusive.

Divider Type	Per Cycle	Operand Size		
		16	32	64
Radix 2	17	221	561	1105
Radix 4	20	180	340	660

Table 7.15 Worst Case Delay for SRT Division

Divider Type	Per Cycle	Operand Size		
		16	32	64
Radix 2	410	6765	13,325	26,445
Radix 4	442	3757	7293	14,365

Table 7.16 Average Logic Transitions for SRT Division

7.5.2. Newton-Raphson Division

A popular alternative to SRT division is Newton-Raphson division. In one of its common forms, the division $q = \frac{a}{b}$ is obtained first by computing the reciprocal of b, and then multiplying that reciprocal by a to obtain q. The basic iteration for obtaining $\frac{1}{b}$ is: $x_{i+1} = x_i (2 - b\, x_i)$. Each iteration, therefore, requires 2 multiplications and a subtraction. Because each iteration requires two multiplications, a high-speed multiplier is required in order to provide acceptable performance.

However, in contrast to the linear convergence of the SRT division algorithm, Newton-Raphson division converges quadratically, roughly doubling the number of correct bits each iteration. This means that fewer cycles are required than in SRT division, although each cycle is longer.

One area of active research is selection of the initial approximation [23], [24]. One approach presented by Schulte *et al* demonstrates that a simple linear approximation will provide an initial approximation accurate to at least 4 bits. This means that after one Newton-Raphson iteration, the answer will be accurate to 8 bits, after two iterations we will have 16 bits, and after 4 iterations, we will have 64 bits, which is sufficient in this case.

Calculating the initial approximation requires one 8-bit multiplication, and one 8-bit subtraction. This represents a worst case delay of 45 gate delays, and an average of 625 logic transitions.

In order to form an approximation of the average number of logic transitions, we will follow the same procedure as with SRT division, and ignore the effects of latches and the control logic. Extrapolating from Table 7.14, a 64-bit Wallace/Dadda multiplier will have an average of about 90,000 logic transitions per multiplication. However, only the final iteration needs to be done in full precision. The other iterations can be done in much smaller precision. As an example, the result after the first iteration is accurate to only 8 bits, so that the arithmetic in the first iteration needs to performed to an accuracy of only 8 bits.

By doing this, the total number of logic transitions can be greatly reduced, although the worst case delay is unlikely to be affected, unless some sort of self-timed logic is used. The last iteration will require full precision operations, and will thus determine the cycle time for a synchronous system.

A 64 by 64-bit multiplier has a worst case delay of 80 gate delays, and a 64-bit carry lookahead adder has a worst case delay of 14 gate delays, so the cycle time looks to be 80 + 14 + 80 = 174 gate delays. However, the operation $2 - b\,x_i$ simply computes the 2's complement of $b\,x_i$ and can be incorporated into the multiplier with very little overhead. So, the cycle time is actually the time needed to perform two multiplications, or 160 gate delays.

Again, the total number of iterations needed to form $1/b$ is 4, so the total number of gate delays is 4×160, plus an additional 45 gate delays to form the initial approximation, plus an additional 80 gate delays to multiply $1/b$ by a to get q, Therefore, the total delay is 765 gate delays. For smaller operands, a smaller multiplier is needed, so not only will fewer iterations be required, but each iteration will be faster as well. Table 7.17 presents the delay for Newton-Raphson division for 64, 32, and 16 bit operands.

Operand Size	Number of Iterations	Total Delay
16	3	367
32	4	581
64	5	894

Table 7.17 Worst Case Delay for Newton-Raphson Division

In order to form an estimate of the average number of logic transitions per cycle, we will use Table 7.14 as a basis. The estimates for the average number of logic transitions for Newton-Raphson division are shown in Table 7.18. The first column shows the average number of logic transitions in the initial approximation and the iterations needed to obtain $1/b$. The second column shows the total average number of logic transitions, which includes the final multiplication needed to obtain $q = a/b$.

Operand Size	Iterations	Total
16	14,000	18,000
32	74,300	95,000
64	341,000	428,000

Table 7.18 Average Number of Logic Transitions for Newton-Raphson Division

Comparing the average number of logic transitions required indicates that SRT division is clearly superior. This is not unexpected, because the basic operations in SRT division are addition and shifting, which tend to have fewer logic transitions than multiplication, which is the basis for Newton-Raphson division. In terms of the worst-case delay, the comparison is not quite so simple. Although looking at Table 7.15 and Table 7.17 indicates that SRT division will be faster, this may not always be the case. In particular, because both are sequential operations, if the clock period is long enough to allow one Newton-Raphson iteration every other cycle, then the SRT division will take more time.

7.6. Summary

We have seen that both circuit design styles and the choice of algorithm can have a significant impact on the area, delay, and power dissipation of arithmetic units. There is still quite a lot of research to be done in this area, including examinations of the interactions between the 4 factor mentioned above (circuit design style,technology,algorithm, and architecture).

Acknowledgments

The authors would like to thank Crystal Semiconductor, and in particular David Medlock, for help in testing the chips. The authors would also like to express their appreciation to Mel Hagge and Jeff Dorst for their help in designing and testing the different full adder types.

References

[1] N. Weste and K. Eshraghian, *Principles of CMOS VLSI Design: A Systems Perspective*, Addison-Wesley Publishing Company, 1988.

[2] G. Kissin, "Measuring Energy Consumption in VLSI Circuits: A Foundation," *Proceedings of the 14th ACM Symposium on the Theory of Computing*, pp. 99–104, 1982.

[3] M. Cirit, "Estimating Dynamic Power Consumption of CMOS Circuits," *ICCAD*,

pages 534-537, 1987.

[4] U. Jagau, "SIMCURRENT - An Efficient Program for the Estimation of the Current Flow of Complex CMOS Circuits," *DAC*, pp. 396-399, 1990.

[5] T. K. Callaway and E. E. Swartzlander, Jr., "Optimizing Adders for WSI," In *IEEE International Conference on Wafer Scale Integration*, pp. 251-260, 1992.

[6] T. K. Callaway and E. E. Swartzlander, Jr., "Optimizing Arithmetic Elements for Signal Processing," *VLSI Signal Processing, V*, pp. 91-100, 1992.

[7] S. Devadas, K. Keutzer, and J. White, "Estimation of Power Dissipation in CMOS Combinational Circuits Using Boolean Function Manipulation," *IEEE Transactions on CAD*, vol. 11, pp. 373-380, 1992.

[8] D. Goldberg. "Computer Arithmetic," In D. A. Patterson and J. A. Hennessey, *Computer Architecture: A Quantitative Approach*, Morgan Kaufmann Publishers, 1990, pp. A2, A3, A31-A39.

[9] O. L. MacSorley, "High-Speed Arithmetic in Binary Computers," *IRE Proceedings*, vol. 49, pp. 67-91, 1961.

[10] O. J. Bedrij, "Carry-Select Adder," *IRE Transactions on Electronic Computers*, vol. EC-11, pp. 340-346, 1962.

[11] J. Sklansky, "Conditional-Sum Addition Logic," *IRE Transactions on Electronic Computers*, vol. EC-9, pp. 226-231, 1960.

[12] S. Turrini, "Optimal Group Distribution in Carry-Skip Adders," *Proceedings of the 9th Symposium on Computer Arithmetic*, pp. 96-103, 1989.

[13] P. K. Chan, *et al*, "Delay Optimization of Carry-Skip Adders and Block Carry-Lookahead Adders Using Multidimensional Dynamic Programming," *IEEE Transactions on Computers*, vol. EC-41, pp. 920-930, August 1992.

[14] A. Aggarwal, A. Chandra, and P. Raghavan, "Energy Consumption in VLSI Circuits," *Proceedings of the 20th Annual ACM Symposium on the Theory of Computing*, pp. 205-216, 1988.

[15] P. M. Chau and S. R. Powell, "Estimating Power Dissipation of VLSI Signal Processing Chips: The PFA Technique," *Journal of VLSI Signal Processing, 4*, pp. 250-259, 1990.

[16] C. S. Wallace, "A Suggestion for a Fast Multiplier," *IEEE Transactions on Electronic Computers*, EC-13, pp. 14-17, 1964.

[17] L. Dadda, "Some Schemes for Parallel Multipliers," *Alta Frequenza*, 34, pp. 349-356, 1965.

[18] C. Lemonds and S. S. Shetti, "A Low Power 16 by 16 Multiplier Using Transition Reduction Circuitry," *Proceedings of the 1994 International Workshop on Low Power Design*, pp. 139-142, 1994.

[19] C. Nagendra, R. M. Owens, and M. J. Irwin, "Power-Delay Characteristics of CMOS Adders," *IEEE Transactions on VLSI Systems*, vol. 2, pp. 377-381, September 1994.

[20] A. Chandrakasan, S. Sheng, and R. Broderson, "Low Power CMOS Digital Design," *IEEE Journal of Solid-State Circuits*, SC-27, pp. 685–691, 1992.

[21] J. E. Robertson, "A New Class of Digital Division Methods," *IRE Transactions on Electronic Computers*, EC-7, pp. 681–687, 1958.

[22] M. J. Flynn, "On Division by Functional Iteration," *IEEE Transactions on Computers*, C-19, pp. 702–706, 1970.

[23] D. L. Fowler and J. E. Smith, "An Accurate High Speed Implementation of Division by Reciprocal Approximation," *Proceedings of the 9th Symposium on Computer Arithmetic*, pp. 60–67, 1989.

[24] D. D. Sarma and D. W. Matula, "Measuring the Accuracy of ROM Reciprocal Tables," *Proceedings of the 11th Symposium on Computer Arithmetic*, pp. 95–102, 1993.

[25] N. F. Goncalves and H. J. De Man, "NORA: A Racefree Dynamic CMOS Technique for Pipelined Logic Structures," *IEEE Journal of Solid-State Circuits*, SC-18, pp. 261–266, 1983.

[26] L. G. Heller, W. R. Griffen, J. W. Davis, and N. G. Thoma, "Cascode Voltage Switch Logic: A Differential CMOS Logic Family," *International Solid State Circuits Conference*, pp. 16–17, 1984.

[27] K. M. Chu and D. L. Pulfrey, "A Comparison of CMOS Circuit Techniques: Differential Cascode Voltage Switch Logic Versus Conventional Logic," *IEEE Journal of Solid-State Circuits*, SC-22, pp. 528–532, 1987.

[28] J. Wang, C. Wu, and M. Tsai, "CMOS Nonthreshold Logic (NTL) and Cascode Nonthreshold Logic (CNTL) for High-Speed Applications," *IEEE Journal of Solid-State Circuits*, SC-24, pp. 779–786, 1989.

[29] S. L. Lu and M. D. Ercegovac, "Evaluation of Two Summand Adders Implemented in ECDL CMOS Differential Logic," *IEEE Journal of Solid-State Circuits*, SC-26, pp. 1152–1160, 1991.

[30] M. Maleki and S. Kiaei, "Enhancement Source-Coupled Logic for Mixed-Mode VLSI Circuits," *IEEE Transactions on Circuits and Systems II: Analog and Digital Signal Processing*, vol. 39, pp. 399–402, 1992.

8

Low Power Memory Design

K. Itoh

8.1. Introduction

Low power LSI memory technology is becoming an increasingly important and growing area of electronics. In particular, low power RAM technology is a major area of this technology. Rapid and remarkable progress has been made in power reduction for RAM subsystems and there is strong potential for future improvement. This progress has even formed the basis for power reduction for other important memory types such as flash memory and ROM. Low power RAM technology includes low-power chip technology, multi-data-bit chip-configurations in which a large number of data-bits are processed simultaneously, small package technology, and low voltage chip-to-chip interfaces. Low power chip technology has contributed mainly to subsystem power reduction. Multi-data-bit chip-configuration has become more popular as chip memory capacity increases, since it effectively reduces subsystem power with resultant lower chip count for a fixed subsystem memory capacity. In addition to lowered chip counts, small package technology combined with high-density chip technology has reduced AC power, with less capacitance for address lines, control lines and data input/output bus lines on memory boards. Lowering voltage swing on data-bus lines reduces the AC power which is always increasing, in line with the strong requirement for higher data-throughput for memory subsystems. In any event, RAM chip power is a primary concern of the subsystem designer, since it dominates subsystem power.

In large memory capacity RAM chips, active power reduction is vital to realizing low-cost, high-reliability chips because it allows plastic packaging, low operating current, and low junction temperature. Hence, various low power circuit technologies [1] concerning reductions in charging capacitance, operating voltage, and static current have been developed. As a result, active power has been reduced at every generation despite a fixed supply voltage, increased chip size, and improved access time, as exemplified by the DRAM chips shown in Figure 8.1(a).

Recent exploratory achievements in the movement towards low voltage operation (departing from the standard power supply of 5 V) seemingly give promise of future improvements. Low voltage operation has already been one of the most important design issues for 64-Mb or larger DRAM chips, since it is essential not only to further reduce power dissipation, but also to ensure reliability for miniaturized devices. This movement implies that the state-of-the-art device technology in miniaturization has at last progressed to the extent of using operating voltages of 3 V or less. In fact, a 64-Mb DRAM chip with an operating

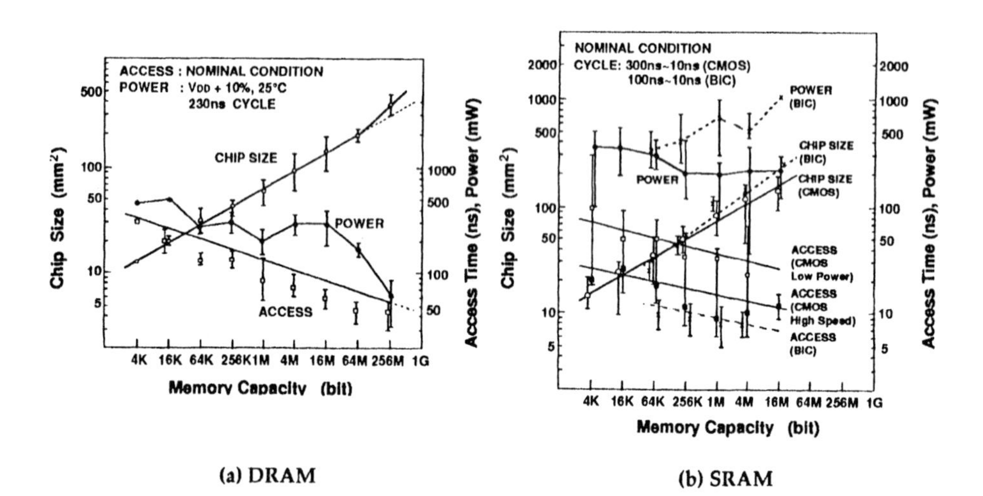

(a) DRAM (b) SRAM

Figure 8.1 Trends in RAM chip performance [1]. Bit-width for I/O pins is mostly 1-bits or 4-bits for DRAMs and high-speed SRAMs, and mostly 8-bits for low-power SRAMS.

voltage of 1.5 V has been reported [2], affording an active current of 23 mA at a 230 ns cycle time. Reducing the data-retention power in DRAMs has been increasingly important for battery backup applications (where SRAMs are normally used), since DRAMs are inherently less expensive to produce than SRAMs. A data-retention current as low as 3 mA at 2.6 V has been reported for 4-Mb chip [3], as shown in Figure 8.2. Such low current finally allows the design of DRAM battery operation, even under active operation mode, which is a key in battery-based hand-held equipment such as mobile communication systems. The ever-decreasing supply voltage requirements could result in ultra-low power and extended battery life. One target is 0.9 V, the minimum voltage of a NiCd cell. Even in this case, higher performance will eventually be required due to the ever-increasing demand for digital signal processing capability in such mobile equipment. To simultaneously achieve low voltage and high speed operation, however, threshold voltage (V_T) scaling is highlighted as an emerging issue. This is because the MOSFET subthreshold DC current [1], even in CMOS chips, increases with decreasing V_T. The resultant DC current eventually dominates both active and retention currents, thereby losing the low power advantage of CMOS circuits that we take for granted today. The issue is essential not only for designing multi-gigabit DRAM chips with a feature size of 0.1 μm or less in the future, but also for designing ultra-low voltage multi-megabit DRAM chips with an existing fabrication process tailored to scaled V_T.

For SRAMs, there has also been a strong requirement for low power. In early days, SRAM chip development was focused on low-power applications

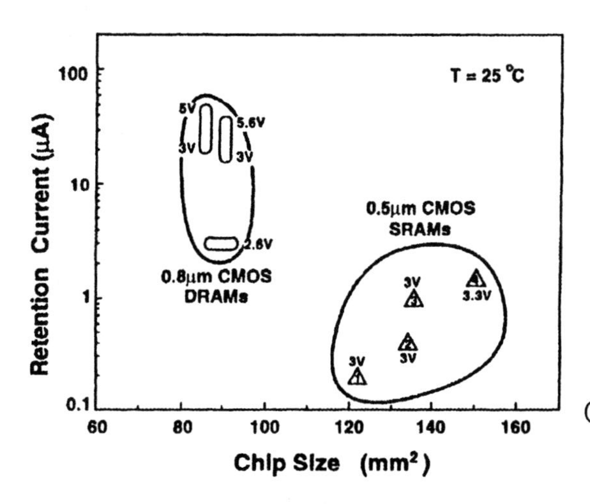

Figure 8.2 Data retention current versus chip size for 4-Mb RAM chips [1].
(4-Mb SRAM: 1,2,3:TFT load cell, 4:Poly load cell).

especially with very low standby and data-retention power, while increasing memory capacity with high-density technology. Nowadays, however, more emphasis has been placed on high-speed rather than large memory capacity [24][25], primarily led by cache applications in high speed microprocessors. ECL BiCMOS SRAM chips with access times of 5 ns to 10 ns are good examples. In this case, power and chip size are of less concern, as shown in Figure 8.1(b). For low power applications, however, the primary concern is for CMOS technology to realize high-speed with minimum power. Tremendous efforts have been made to minimize power with more emphasis on static current reduction and low voltage operation on long signal transmission lines. Consequently, there has been a decrease in CMOS SRAM power dissipation, in spite of a rapid increase in chip size and improved operating speed. However, this reduction trend does not look very significant compared with the trend in DRAMs. This is because high-speed characteristics have been more emphasized in the recent development of 1-Mb or larger SRAM chips. A typical example of a low-power high-density SRAM chip is a 16-Mb CMOS SRAM that achieves an access time of 15 ns and power dissipation of 165 mW at 30 MHz with 3-V power supply [4]. As for data-retention current, recent 4-Mb CMOS SRAM chips achieve sub-mA data-retention currents, as shown in Figure 8.2, which are still about one order of magnitude less than those of 4-Mb DRAM chips, because they don't require a refresh operation. Ultra-low voltage operation is also essential in SRAM design. This poses serious concerns for a subthreshold current and a V_T mismatch developed in a memory cell.

DRAMs and SRAMs both have reduced power dissipation through the use of common low power circuit designs. However, there are few papers clarifying which circuits are common or different based on a systematic comparison. This chapter consists of two parts: low-power RAM subsystem design and low-power standard CMOS RAM chip design. In the first, power sources and power reductions in RAM subsystems are discussed, with emphasis given to the importance of RAM-chip power reduction. In the second, low-power RAM chip designs are discussed in terms of three key issues: charging capacitance, operating voltage, and DC static current, differentiating between DRAMs and SRAMs. First, in section 8.3., power sources for active and data retention modes in a chip are described and essential differences in power sources between DRAMs and SRAMs are clarified. Next, the power reduction circuits for each power source are separately reviewed in the following chapters, first for DRAMs, then for SRAMs. In section

8.4., DRAM circuits are discussed, with emphasis on charging capacitance for the memory array. In section 8.5., SRAM circuits are intensively discussed with emphasis on DC current reduction for memory array and column circuitry. Perspectives on ultra-low voltage design are also described in each section, mainly in terms of subthreshold current, although this style of design is still in its infancy.

8.2. Sources and Reductions of Power Dissipation in Memory Subsystem

Major sources of active power in a modern CMOS memory subsystem are the active chips, whose power is a product of the activated chip power and the number of simultaneously activated chips. Other minor sources are the capacitances of data-bus lines, address lines and control lines on a memory board. In particular, data-bus lines are of special concern. They cause a heavy capacitance compared with other lines because of their inherently large values in number and capacitance. Active chip power and AC capacitive power have both been reduced by low-power circuits, miniaturized device technology, multi-data-bit chip-configuration, small package technology and low voltage interfaces. On the other hand, sources of subsystem retention power are all the memory chips in the subsystem, and thus the retention power has also been reduced by low power circuits and the improvement of device technologies.

In this section, sources and reductions of power dissipation in a typical DRAM subsystem are described in terms of multi-data-bit chip-configuration, small package technology and low voltage interface. Here, chip power data, as a major determinant of subsystem power, is from modern catalog specifications [5] which have resulted from low-power chip-technology advancement, as discussed in the succeeding sections.

8.2.1. Multi-data-bit Configuration Chip

The multi-data-bit configuration is becoming important and more favorable for power reduction, as well as design flexibility, as chip memory capacity increases. This is due to the resultant lower chip count, combined with low-power chip technology, for a fixed memory capacity of a subsystem. For example, a 4-MByte (B), 32-bit data-bus subsystem necessitates 32 memory chips for 1-Mb chip of 1-Mword × 1 bit configuration, as shown in Figure 8.3(a). With multi-data-bit configuration, however, the number of chips necessary is 8 for a 4-Mb chip (1-Mword × 4-bit) and 2 for a 16-Mb chip (1-Mword × 16-bit). Gener-

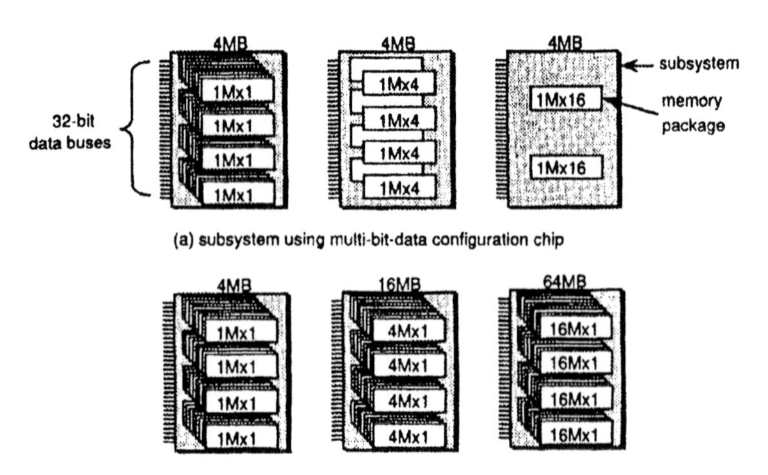

Figure 8.3 Impact of multi-bit data configuration on a 32-bit data-bus system.
1M×1: 1-M word × 1-bit configuration.

ally, in addition to increasing memory capacity of a chip, widening data-bit width increases chip power because of the resultant increased number of column and sense circuits. However, chip power has been suppressed with low power circuits, as shown later. Thus, multi-data-bit configuration combined with larger memory capacity dramatically reduces subsystem power with less chip count. Less chip count also allows design flexibility so subsystem memory capacity can be added on with an increment of as small as 4MB with a minimum memory capacity of 4MB. These features are essential for small systems such as personal computers and hand-held equipment in which power, physical size and price are critical. On the contrary, a fixed data-bit configuration gives disadvantages; such as larger power as well as design inflexibility, as shown in Figure 8.3(b). Table 1 shows more detailed advantages of the multi-data-bit configuration. Data is from a catalog specification [5] which guarantees the power at the worst combination within variations of V_{DD}, temperature and fabrication process. Obviously, drastic reduction in subsystem power is achieved with less chip count resulting from multi-data-bit configuration combined with large memory capacity, and low-power chip-technology advancement, as shown in Table 2. For example, the chip component of the normal mode active power for a 110 ns cycle is reduced from 17.4 W for a 1-Mb chip subsystem down to 1.1 W for a 16-Mb chip subsystem. Note that inactive (non-selected) chips consume a negligibly small power compared with active (selected) chips. Thus, chip power is almost determined by

Table 1 5V DRAM subsystems with 32 data-bit bus lines

memory chip		1-Mb (1-Mword ×1-bit)	4-Mb (1-Mword ×4-bit)	16-Mb (1-Mword ×16-bit)
memory subsystem		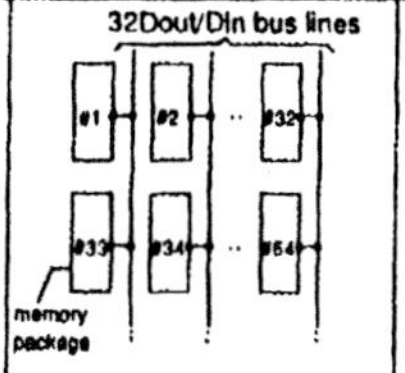 32Dout/Din bus lines; #1 #2 .. #32; #33 #34 .. #64; memory package	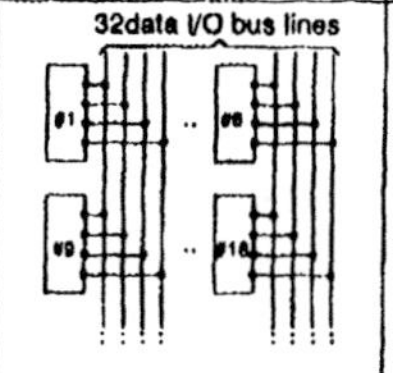 32data I/O bus lines; #1 .. #8; #9 .. #16	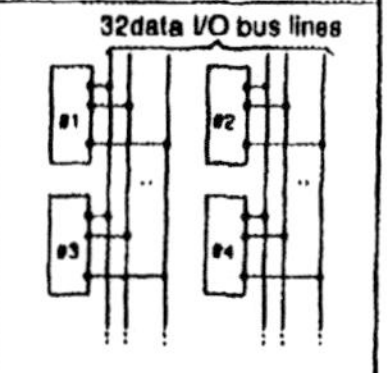 32data I/O bus lines; #1 #2; #3 #4
number of chips simultaneously activated		32	8	2
normal mode active power (t_{RC} =110ns)		17.45W	4.85W	1.15W
	chip / AC *	17.4/0.05W	4.8/0.05W	1.1/0.05W
page mode active power (t_{PC} =40ns)		15.63W	4.93W	1.23W
	chip / AC *	15.5/0.13W	4.8/0.13W	1.1/0.13W
data-retention power for 8MB subsystem		108.8mW	17.6mW	11.2mW
memory density on memory board (ratio)		1	4	8

* $32 \times 1/2C(\Delta V)^2 f$ for TTL interface: C=20pF for 8MB subsystem, $\Delta V = V_{DD} - V_T = 4V$ for 5V V_{DD}, f≈9MHz or 25MHz

Table 2 5V DRAM chip and package advancement [5]

memory capacity		1-Mb (1-Mword×1-bit)	4-Mb (1-Mword×4-bit)	16-Mb (1-Mword×16-bit)
power supply (V_{DD})		5V	5V	5V (internal 3.3V)
active power	normal mode (t_{RC} = 110ns)	545 mW max.	605 mW max.	550 mW max.
	page mode (t_{PC}= 40ns)	484 mW max.	605 mW max.	550 mW max.
data retention power		1.7 mW max.	1.1 mW max.	2.8 mW max.
refresh cycle (n)		512	1024	4096
feature size		1.3 μm	0.8 μm	0.5 μm
chip size (ratio)		1	1.5	2.5
package size		$8.5^{mm} \times 16.9^{mm}$ (1.0)	$8.5^{mm} \times 16.9^{mm}$ (1.0)	$11.2^{mm} \times 27.1^{mm}$ (2.1)

the number of chips simultaneously activated, independently of the subsystem memory capacity.

8.2.2. Small Package

In addition to the lower chip count described above, small package technology reduces the AC capacitive power by reducing the capacitance on the memory board. Common I/O pin assignment in which the data input (Din) pin and the data output (Dout) pin are common enables the use of small packages by halving the number of data pins that increase with multi-data-bit configuration. Furthermore, remarkable progress in package structure accommodates ever-enlarged chips at every successive generation in a small package. A typical example is the LOC (Lead-On-Chip) package [6] featuring a vertical structure of chip and inner leads, as shown in Figure 8.4. Figure 8.5 shows the resultant small packages achieved despite multi-data-bit configuration and large memory capacity. The memory density on a memory board is almost doubled when 4-Mb chips are replaced by 16-Mb chips, as shown in Tables 1 and 2, reducing the parasitic capacitances of data-bus lines.

8.2.3. Low-voltage Data-bus Interface

Recently, further AC power reduction has been required because the ever-increased data-throughput of memory subsystems, resulting from the rapidly improved performance of microprocessors, increases AC power. The reduction is especially vital for DRAM column mode which offers data-throughput inherently higher than normal read-write mode.

In a traditional address-multiplexed DRAM chip, as shown in Figure 8.6, a selected word line, WL, is activated after strobing the corresponding set of row addresses with an external clock called Row Address Strobe (/RAS). As a result,

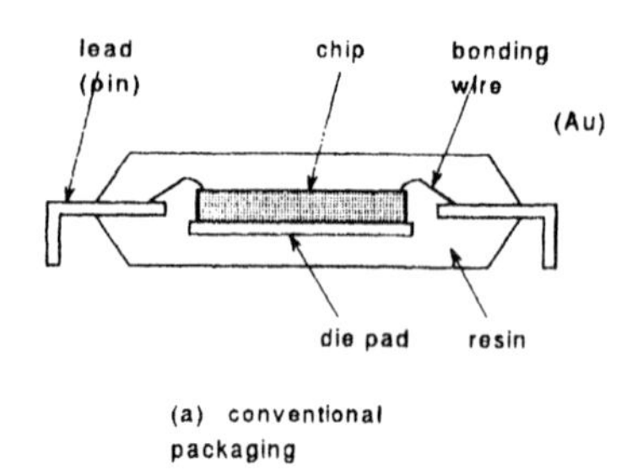

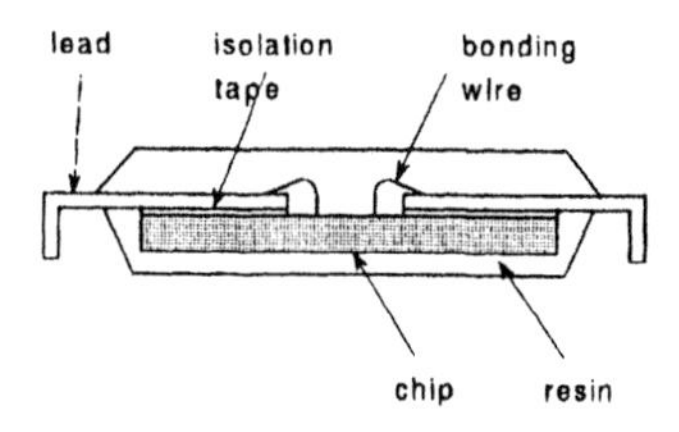

Figure 8.4 Advancements in packages [6].

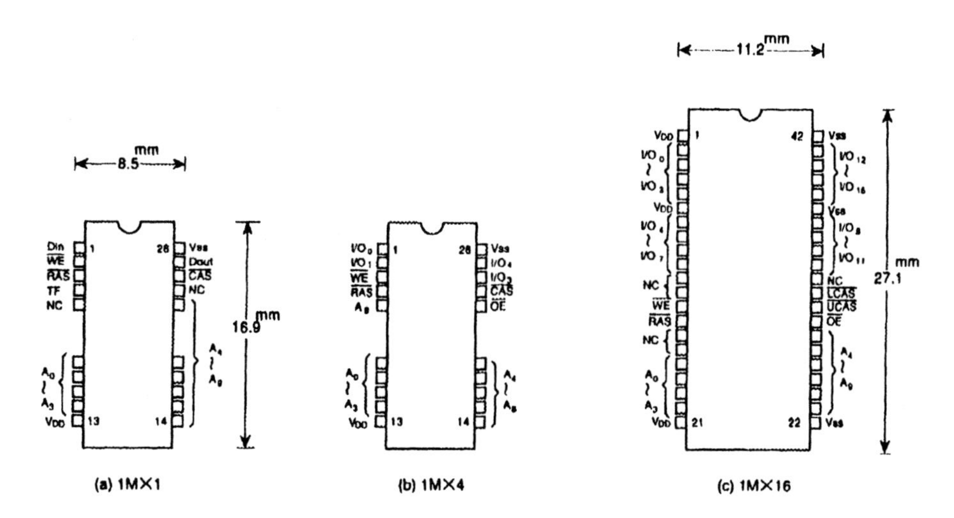

Figure 8.5 Top View of SOJ package and pin assignment [5].

all the stored information of memory calls along the selected WL are read on the corresponding data lines, and then amplified by the corresponding amplifiers. Each amplified signal is held on the data line as read information. Any read information on the selected data line can be read out to the data output (*Dout*) pin by strobing the corresponding set of column addresses with another clock called Col-

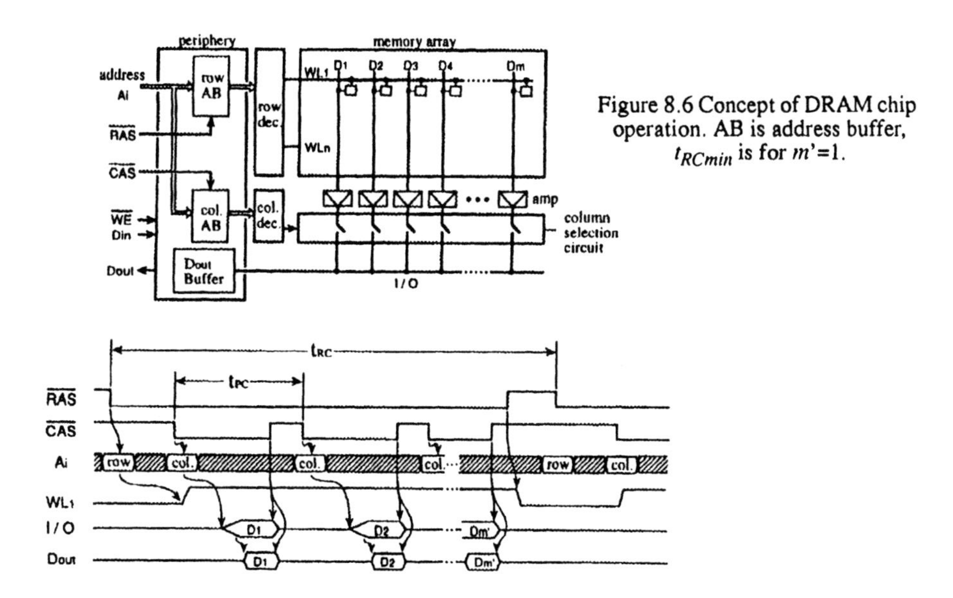

Figure 8.6 Concept of DRAM chip operation. AB is address buffer, t_{RCmin} is for $m'=1$.

umn Address Strobe (/CAS). A different word line is activated by the succeeding /RAS strobing of the corresponding row addresses. This is the so-called normal mode always entailing word-line activations. Here, during a /RAS activation (/RAS: Low), any amplified signal held on the data line can be successively read out to the Dout pin by successive data-line selection (column mode) that is performed only by /CAS and address activations. Column mode (/CAS) cycle (t_{PC}) is inherently much faster than normal mode (/RAS) cycle (t_{RC}). The difference in speed originates from the fact that a /CAS cycle is accomplished only by column circuits operation while a /RAS cycle entails both row and column circuits operations. The inherently slow operation of row circuits, which include relatively heavy loads such as word and data lines responsible for the large physical size of memory array, makes the difference larger. Thus, the minimum t_{PC} is about 40 ns while the minimum t_{RC} is about 110 ns in a modern DRAM chip. Page mode, nibble mode and static column mode are typical examples of column mode addressing [5]. Note that for 4 data-bit configuration, for example, a memory array is divided into 4 sub-arrays which are simultaneously accessible. Thus, a 4-bit parallel read operation from 4 sub-arrays can be performed.

Now, we will evaluate the AC capacitive power of data-bus lines at the minimum column cycle. For 5 V TTL (Transistor Transistor Logic) shown in Figure 8.7, each bus line power is expressed by

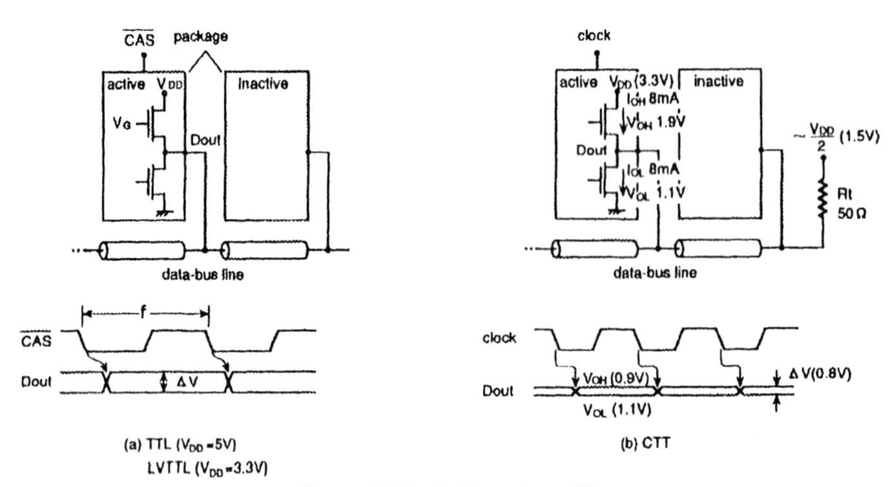

Figure 8.7 Typical interfaces [7].
$\Delta V = V_{DD} - V_T = 4$ V for TTL, $= V_{DD} = 3.3$ V for LVTTL, $= 0.8$ V for CTT.

$$P = \frac{1}{2}C(\Delta V)^2 f \tag{8.1}$$

where C, ΔV and f are the loading capacitance, voltage swing and frequency, respectively. It is found that the AC power of an 8-Mb, 32-bit data-bus-line subsystem becomes relatively large as the width of data I/O increases, although it is still small portion of total power as shown in Table 1. In fact, it is as small as 0.13 W even for a 40 ns page mode cycle, occupying 10% of the total power for a 16-Mb chip subsystem. In this subsystem two 16-Mb chips are connected to each data-bus line with a resulting capacitance of 20 pF, and ΔV is 4 V (= V_{DD}-V_T,V_T: transistor threshold voltage), since the gate voltage of the Dout transistor usually equals V_{DD}. For a large system connecting many chips on each data-bus line, however, the AC power component is more prominent with large capacitance. For the example of a 16-Mb chip subsystem, it occupies about 37% of total power for C = 100 pF.

Low V_{DD} operation enabling low voltage interfaces reduces not only the chip power, but also the AC power, as shown in Table 3. The LVTTL (Low Voltage TTL) reduces the subsystem power by 34% by reducing V_{DD} and ΔV to 3.3 V. However, even LVTTL can manage no further power increase due to more

Table 3 Impact of low voltage 16-Mb DRAM operation on subsystem power [5]

power \ chip		5V TTL 1-Mword×16-bit	3.3V LVTTL 1-Mword×16-bit
active chip power	normal mode (t_{RC} = 110ns)	550mW max.	360mW max.
	page mode (t_{pc} = 40ns)	550mW max.	360mW max.
data-retention chip power		2.8mW max.	1.4mW max.
normal mode power of 8MB subsystem (t_{RC} = 110ns)		1.15W	0.75W
	chip/AC	1.1/0.05W	0.72/0.03W
page mode power of 8MB subsystem (t_{pc} = 40ns)		1.23W [1]	0.81W [0.66]
	chip/AC	1.1/0.13W	0.72/0.09W
data-retention power of 8MB subsystem		11.2mW	5.6mW

enhanced cycle time. Chip power also increases since the Dout transistor, enlarged according to high speed driving of the bus capacitance, increases the internal node capacitance of the chip. Recently, new interfaces [7] called GTL (Gunning Transceiver Logic), CTT (Center Tapped Termination) and Rambus interfaces featuring small-amplitude impedance-matched bus lines have been proposed. Small amplitude minimizes AC capacitive power while impedance matching allows extremely high-speed transmission. They enable high-speed column cycles of 2 to 20 ns with low power. This is when high-speed DRAM chip operation is achieved using pipelined column-circuit operation synchronous to an external clock. The power consumption at each CTT interface, as shown in Figure 8.7, is difficult to express because of formation of a complicated transmission line [7][30]. However, it is roughly estimated by using a lumped circuit approximation as

$$\begin{aligned} P &= \left(I_{OL}V_{OL} + R_T I_{OL}^2\right)(1 - duty) \\ &+ \left(I_{OH}(V_{DD} - V_{OH}) + R_T I_{OH}^2\right) duty + \frac{1}{2} C (\Delta V)^2 f \end{aligned} \tag{8.2}$$

where I_{OL} and I_{OH} are currents for the low signal voltage (V_{OL}) and the high signal voltage (V_{OH}), respectively, and R_T is a termination resistance. The signal swing ΔV is 0.8 V for 8 mA I_{OL}, 8 mA I_{OH} and 50 Ω R_T. Total power consumed at 32 bus lines is about 500 mW even for 100 pF and 10 ns cycle, as shown in Figure 8.8. Thus, the interface provides low power with high speed compared with TTL

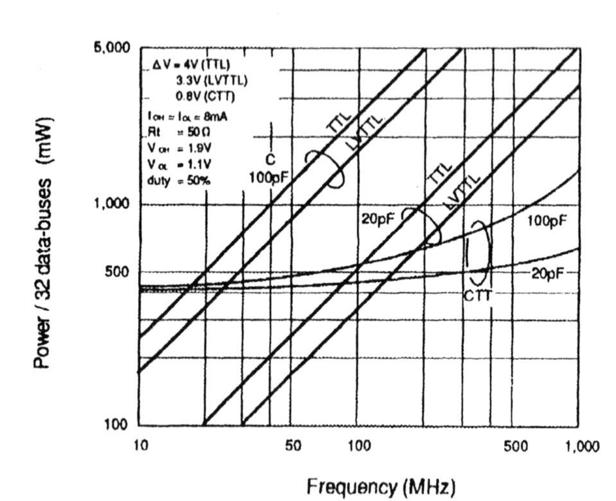

Figure 8.8 Interface power for various interfaces.

and LVTTL even in the high speed region of less than 40 ns cycle. Note that even for new interfaces, the chip power component dominates the total power. This is because it increases due to additional on-chip interface circuits while the interface power component is still low enough in high speed region. With regard to the data retention power of the memory subsystem, the main determinant is also the chip power component. Multi-bit configuration combined with low-power chip technology reduces data retention power, as shown in Table 1.

8.3. Sources of Power Dissipation in DRAM and SRAM

8.3.1. Active Power Sources

Figure 8.9 shows a simplified memory chip architecture [1] for investigating its power dissipation. The chip comprises three major blocks of power sources: memory cell array, decoders (row and column), and periphery. Note that all the m cells on one word line are simultaneously activated in this logical array model. A unified active power equation for modern CMOS DRAMs and SRAMs is approximately given by

$$P = V_{DD}\, I_{DD} \tag{8.3}$$

$$I_{DD} = m\, i_{act} + m(n-1)i_{hld} + (n+m)C_{DE}V_{INT} + C_{PT}V_{INT} + I_{DCP} \tag{8.4}$$

for normal read cycle, where V_{DD} is the external supply voltage, I_{DD} is the current of V_{DD}, i_{act} is the effective current of active or selected cells, i_{hld} is the effec-

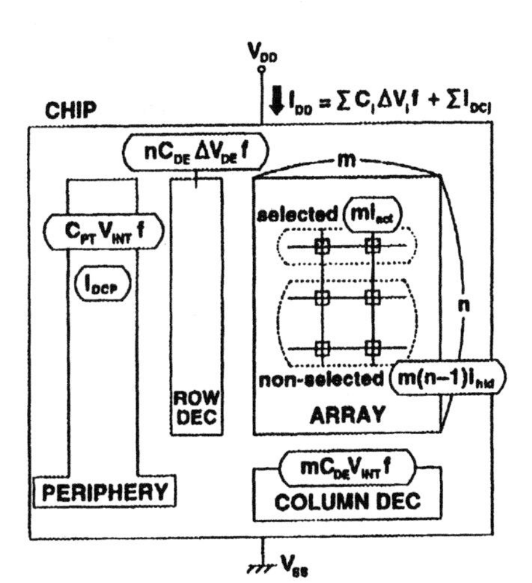

Figure 8.9 Sources of power dissipation in a RAM chip [1]

tive data retention current of an inactive or non-selected cell, C_{DE} is the output node capacitance of each decoder, V_{INT} is the internal supply voltage, C_{PT} is the total capacitance of the CMOS logic and the driving circuits in periphery, I_{DCP} is the total static (DC) or quasi-static current of the periphery, and f is the operating frequency (= $1/t_{RC}$: t_{RC} is the cycle time). Major sources of I_{DCP} are the column circuitry and the differential amplifiers on the I/O lines. Other contributors to I_{DCP} are the refresh-related circuits and on-chip voltage converters essential to DRAM operation; these include a substrate back-bias generator, a voltage-down converter, a voltage-up converter, a voltage reference circuit, and a half-V_{DD} generator. The DC currents of these circuits are virtually independent of the operating frequency. Hence, for high speed operation I_{DCP} becomes relatively small compared to other AC current components. At high frequencies, the data retention current, $m\,(n - 1)\, i_{hld}$, is negligible evidenced by the small cell leakage current and small periphery current necessary for the refresh operation in DRAMs. In SRAMs, i_{hld} is also quite small, as described later. The decoder charging current, $(n+m)C_{DE}\, V_{INT}$, is also negligibly small in modern RAMs incorporating a CMOS NAND decoder [8], because only one out of n or m (nodes) is charged at every selection: $n+m = 2$. The active current for both DRAMs and SRAMs is shown in Figure 8.10. Note that I_{DD} increases with increasing memory capacity, that is, with increased m and n.

DRAM — Destructive read-out characteristics of a DRAM cell necessitate successive operations of amplification and restoration for a selected cell on every data line. This is performed by a latch-type CMOS sense amplifier on each data line. Consequently, a data line is charged and discharged with a large voltage swing of ΔV_D (usually 1.5 ~ 2.5 V) and with charging current of $C_D \Delta V_D$ where C_D is the data line capacitance. Hence, the current is expressed as

$$I_{DD} \cong [m\, C_D\, \Delta V_D + C_{PT}\, V_{INT}] + I_D C_P \tag{8.5}$$

Eq. (8.3) and Eq. (8.5) show that the following issues are the keys to reducing active power for a fixed cycle time: (1) reducing the charging capacitance ($m\, C_D$, C_{PT}), (2) lowering the external and internal voltages (V_{DD}, V_{INT}, ΔV_D), and (3) reducing the static current (I_{DCP}). In particular, emphasis must be placed on reduction of the total data-line dissipation charge ($mC_D\Delta V_D$), since it dominates the total active power, as described in section 8.4. However, the dissipation charge must be reduced while maintaining an acceptable signal-to-noise-ratio (S/N), since they are closely related to each other. S/N is an extremely important

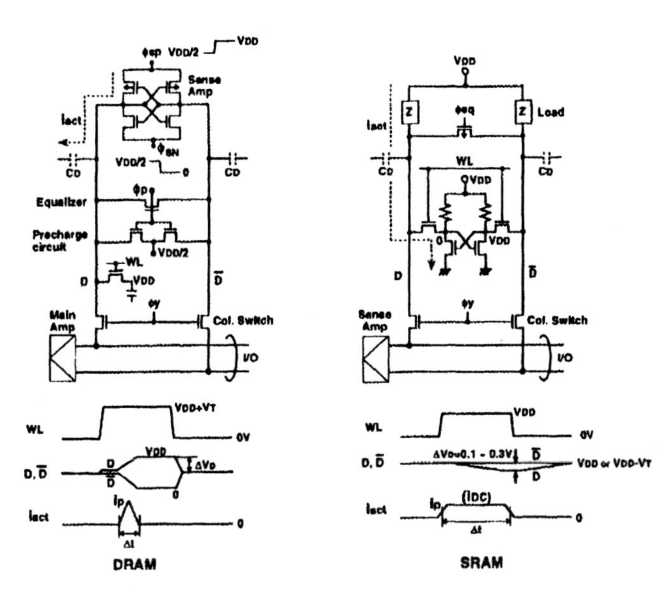

Figure 8.10 Comparison between DRAM and SRAM read operation [1]. The half-V_{DD} precharging scheme is assumed in DRAMs.

issue [9] for stable operation. The DRAM cell l is not only very small, but is furthermore read out onto a floating data line, that is susceptible to noise. The signal, v_s, is approximately expressed as

$$v_s \cong (C_S/C_D)V_{DD}/2 = (C_S/C_D)\Delta V_D = Q_S/C_D \tag{8.6}$$

for the half-V_{DD} precharging scheme ($\Delta V_D = V_{DD}/2$) described in section 8.4., where C_S and Q_S are the cell capacitance and cell signal charge, respectively. It is obvious from Eq. (8.5) and Eq. (8.6) that reducing C_D is effective for both reducing I_{DD} and increasing v_s, while reducing ΔV_D degrades v_s, despite the I_{DD} reduction. This implies the importance of increasing C_S and/or decreasing noise instead.

SRAM — The non-destructive read-out characteristics of SRAM never require restoration of cell data, allowing the elimination of a sense amplifier on each data line. To obtain a fast read, the cell signal on the data line is made as small as possible, transmitted to the common I/O line through the column switch, and amplified by a sense amplifier. Since the cell signal is developed as the ratio of the data-line load impedance to the cell transistors, a ratio current i_{DC} flows along the data line during the word-line activation time, Δt. Here, the data-line charging current is negligibly small due to a very small ΔV_D (= 0.1 V ~ 0.3 V), although it is prominent for write operation, as described in section 8.5. Thus, the current for read operation is expressed as

$$I_{DD} \cong [m\, i_{DC}\, \Delta_t + C_{PT}\, V_{INT}]f + I_{DCP} \tag{8.7}$$

To reduce active power, the three issues of static current, voltage, and capacitance are vital similar to DRAMs. However, the static current charge, $m\, i_{DC}\, \Delta t$, should be reduced more intensively because it dominates the total active current, which differs between SRAMs and DRAMs. Obviously, the S/N issue is not so serious a problem in SRAMs as in DRAMs because of the ratio operation.

Eventually, DRAMs and SRAMs have evolved to use similar circuit techniques, although emphasis on each of the three issues is different between both types of RAM, as described in the following sections. To clearly show the state-of-the-art RAM designs, the range of cell design parameters for the active current in DRAMs and SRAMs are compared in Table 4. Note the peak current, mi_P, which is a good measure of active power. A partial activation scheme of a multi-divided word line, as described later, reduces the DRAM current down to the SRAM level, although it is still in an experimental stage for DRAMs. In addition to the active current described above, the subthreshold DC current of a MOSFET will be a source of active power dissipation for future DRAMs and SRAMs

Table 4 Comparison of determinants of active cell current between DRAMs and SRAMs [1] (estimated based on data of ISSCC Digests. DRAMs are 1 Mb to 16 Mb and SRAMs are 256 Kb to 4 Mb).

	DRAM	SRAM
number of cells on a data-line pair	256 ~ 512	256 ~ 1024
C_D (pF)	0.2 ~ 0.3	1.0 ~ 2.0
V_S (V)	0.1 ~ 0.2	0.1 ~ 0.3
ΔV_D (V)	1.5 ~ 2.5	0.1 ~ 0.3
$i_p/i_{DC}(r)$ (μA)	20 ~ 50	5 ~ 50
Δt (ns)	10 ~ 20	5 ~ 50
m	2k ~ 8k (128 ~ 512)[a]	64 ~ 128 [a]
$mi_p/mi_{DC}(r)$ (mA)	100 ~ 160 (6 ~ 10)[a]	6 ~ 25 [a]

a. Partial activation of multi-divided word line

supplied with ultra-low voltages of less than 2 V. As V_T becomes small enough to no longer cut off the transistor, the subthreshold current increases exponentially with decreasing V_T. Temperature and V_T variations enhance the current, causing an active current increase and cell stability degradation. Moreover, the subthreshold current is proportional to total channel width of the FETs. Therefore, more attention should be paid to interactive circuit blocks such as the decoder, word driver, and memory cell. The subthreshold current generated from an overwhelmingly large number of inactive (non-selected) circuits would eventually be larger than the AC current from the small number of active (selected) circuits, dominating the total chip active current. All the SRAM flip-flop cells could be DC substrate current sources [10].

8.3.2. Data Retention Power Sources

DRAM — In the data retention mode, a memory chip is not accessed from outside and the data are retained by the refresh operation. The refresh operation is performed by reading data of the m cells on a word line and restoring them for each of the n word lines in order. Note that n in the logical array in Figure 8.9 corresponds to the number of refresh cycle in catalog specification. A current given by Eq. (8.5) flows every time m cells are refreshed at the same time. The frequency f at which the refresh current flows is n/t_{REF}, where t_{REF} is the refresh time of the cells in the retention mode and increases with reducing junction temperature. Thus, from Eq. (8.5) the data retention current is given by

$$I_{DD} \cong [m\, C_D \Delta V_D + C_{PT} V_{INT}]\,(n/t_{REF}) + I_{DCP} \tag{8.8}$$

t_{REF} is much longer than t_{REFmax} which is guaranteed in the catalog specification and shown in Figure 8.17. This is because t_{REFmax} is for the active mode when operating the memory at the maximum frequency of around 10 MHz and where cell leakage current is maximized with highest junction temperature. On the other hand, t_{REF} is an extremely slow refresh frequency ($n/t_{REF} << 62$ KHz) where the current is minimized with lowest junction temperature. In any event, I_{DCP} becomes relatively large for other AC current components because of small n/t_{REF}. This implies the necessity of reducing both AC and DC components.

SRAM — In low power CMOS SRAMs, the static cell leakage current, mni_{hld}, is the major source of retention current because of the negligibly small I_{DCP}. The leakage current or retention current have been maintained at a small value by the following memory cell innovations. A high-resistance polysilicon load cell has been widely used for its high-density and low i_{hld} characteristics, although a full CMOS 6-T cell provides the smallest i_{hld}. For 4-Mb or higher density SRAMs, however, the polysilicon load cell starts to be replaced by a TFT-load cell since it faces difficulty in keeping low power retention characteristics. For ultra-low voltage operation involving scaled V_T, subthreshold currents from almost all circuits in DRAMs and SRAMs could drastically increase the retention current, again posing a serious concern.

8.4. Low Power DRAM Circuits

8.4.1. Active Power Reduction

The power has been gradually decreased in spite of the increase in memory capacity, as shown in Figure 8.1 and Figure 8.11 [1]. This is due to low-power circuits developed at each generation. For a given memory capacity chip, successive circuit advancements have produced a power reduction equivalent to 2 to 3 orders of magnitude over the last decade. Figure 8.12 shows the power dissipation of a 64-Mb DRAM, hypothetically designed with the NMOS circuit of the 64-Kb generation in 1980, compared with the CMOS circuit presented in 1990 [2]. Almost the same process and device technologies are assumed. The drastic reduction in power by about two orders of magnitude is due to many sophisticated circuits: partial activation of a multi-divided data-line and shared I/O which reduce mC_D in Eq. (8.5); CMOS NAND decoder which reduces $(n+m)C_{DE}$; external supply voltage (V_{DD}) reduction from 5 to 3.3 V, half-V_{DD} data line precharge and on-chip voltage- down conversion which reduce V_{DD}, ΔV_D, and V_{INT} in Eq. (8.3)

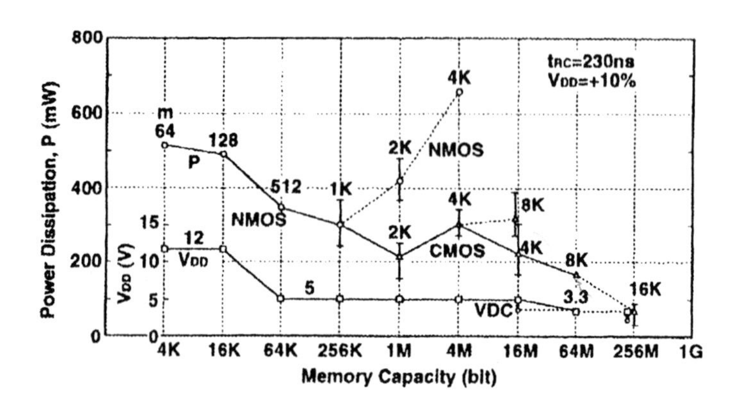

Figure 8.11 Trend in power dissipation of DRAM chips [1].

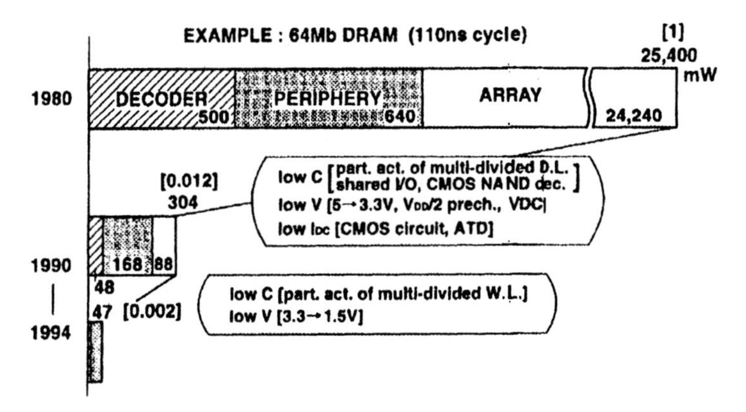

Figure 8.12 Low power circuit advancement of DRAMs over the last decade [1].

and Eq. (8.5); and CMOS drivers and pulse operation of column circuitry and amplifiers which reduce periphery static current, I_{DCP}. An exploratory 64-Mb DRAM chip, also hypothetically designed using the state-of-the-art technology, could further reduce operating power to about 1/10 that of the 1990 chip. Subthreshold current reduction to enable 1.5-V V_{DD} and partial activation of a multi-divided word line are responsible for this reduction. In this section, these key circuit technologies are reviewed. The details of CMOS circuits, although crucial to reduce the power of mega-bit DRAMs, are omitted because they are well known, [8].

8.4.1.1 Charging Capacitance Reduction

The charging capacitance of a data-line is reduced by partial activation of multi-divided data lines, which is now widely accepted in commercial 16-Mb chips. Partial activation of multi-divided word lines further reduces the capacitance, although this is still in the experimental stage. Thus, a combination of multi-division of the data-line and word-line minimizes the capacitance.

Partial Activation of Multi-divided Data Line — With increasing memory capacity, the ever-increasing number of memory cells connected to one data-line causes an increased C_D and thus increased power as well as a poor signal-to-noise ratio, as expressed by Eq. (8.5) and Eq. (8.6). A practical solution is to divide one data line into several sections and to activate only one section. In the early days, the number of divisions was increased by simply increasing the number of Y decoders by using additional chip area. This Y-decoder division was applied to various sense amplifier (SA) arrangements such as one SA at each division (nor-

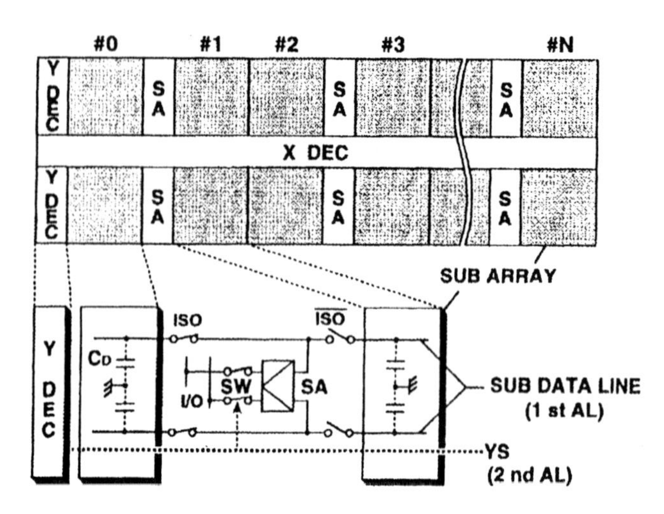

Figure 8.13 Multi-divided data-line architecture with shared SA, shared I/O and shared Y-decoder scheme.

mal arrangement) and a shared SA [9]. However, this division was almost exhausted at the 16-Mb generation with the combination of shared SA, shared I/O, and shared Y decoder, as shown in Figure 8.13 [11]. Shared I/O further divides a multi-divided data line into two, which are selected by the isolation switches, ISO and /ISO. The shared SA provides an almost doubled cell signal with halved C_D. A shared Y decoder can be made without any increase in area by using the second-level metal wiring for the column selection line (YS). The partial activation is performed by activating only one sense amplifier along the data line. Figure 8.14(a) shows trends in total data-line charging capacitance, C_{DT} ($=mC_D$). The C_{DT} has been minimized by sharing SA, I/O, and Y decoder, and by increasing the number of n, as described later. Figure 8.14(b) shows trends in total data-line dissipating charge, Q_{DT} ($=mC_{DD}V_D$). The Q_{DT} has been suppressed as much as possible with the help of ever-reducing operating voltage, as shown in Figure 8.11 and described later.

Partial Activation of Multi-divided Word Line — The concept behind partial activation of multi-divided word line is shown in Figure 8.15. Figure 8.15(a) represents the non-divided word line, a method which has been used in all the commercial chips since the 1-Mb generation. Each poly-Si word line is shunted with Al wiring once every 64 or 128 cells to reduce the large RC delay due to the considerable resistivity of poly Si. In this case, all the data lines are activated once a word line is activated. For the multi-division shown in Figure 8.15(b), one word

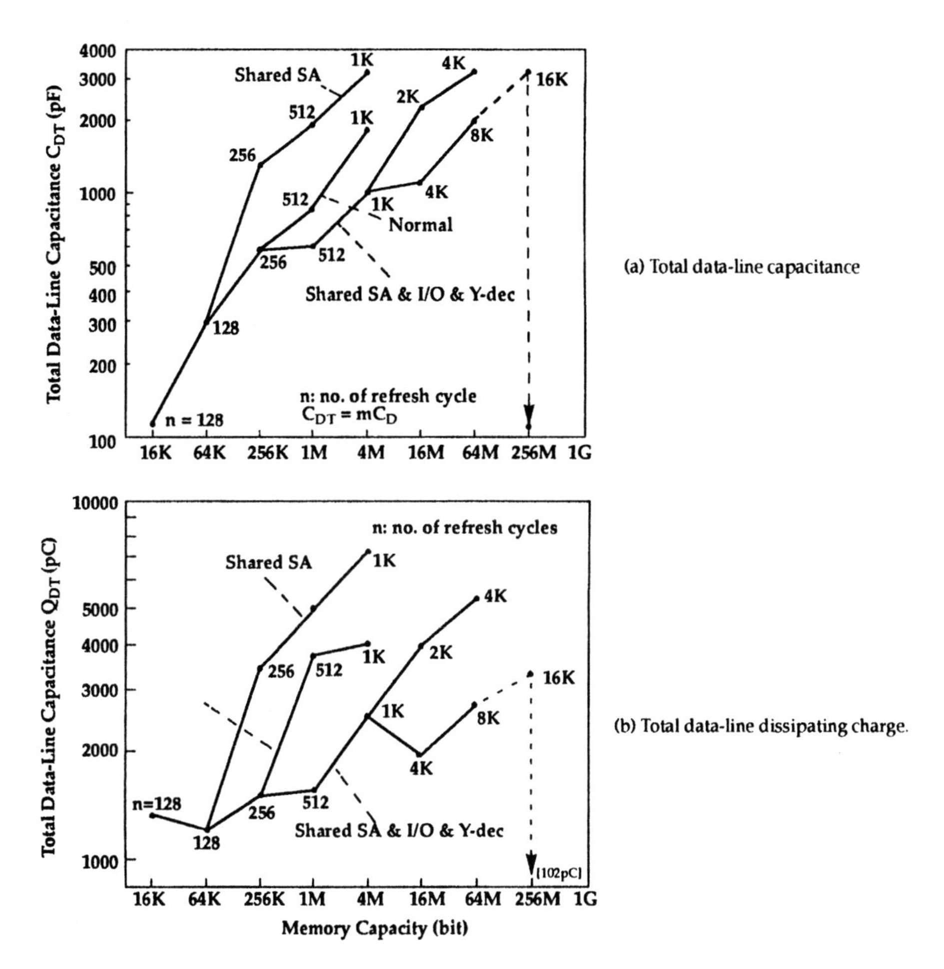

Figure 8.14 Trend in total charging capacitance and dissipating charge of data lines [1]. C_D of 200 fF and ΔV_D of 1 V are assumed for 256 Mb.

is divided into several parts by the sub-word-line (SWL) drivers. Any SWL is selected by the coincidence of the selected main word line (MWL) and the selected row line (RX). High-speed SWL selection is performed by the use of two-level Al wirings: Al_1 for MWL and Al_2 for RX. Only the signals from the cells on the selected SWL are amplified by the corresponding amplifiers, allowing for a partial activation of the word line. Thus, the multi- division method provides less data-line charging capacitance than the no-division scheme. Note that the number of cells connected to one SWL in this physical array corresponds to m in the logical array in Figure 8.9. Figure 8.16 shows the actual word-line structure

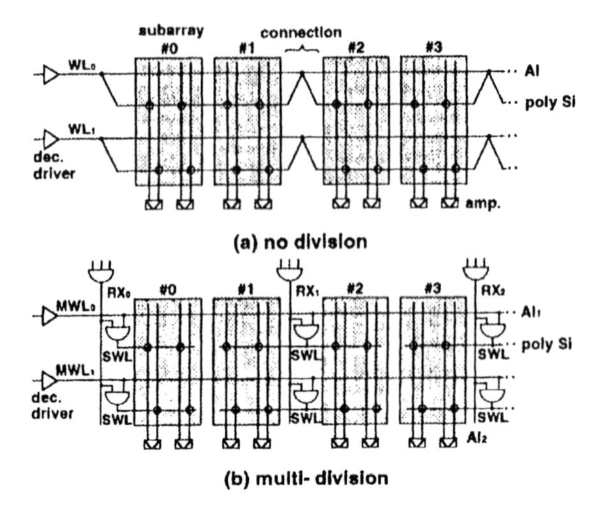

Figure 8.15 Concept behind partial activation of multi-divided word-line.

recently proposed in an experimental 256-Mb DRAM [12]. The MWL pitch can be relaxed to 1/4 with this configuration. SWL drivers are placed alternately to relax the tight layout pitch of the DRAM, although the architecture is similar to that of an SRAM, as described later. The architecture seemingly does not meet requirements for the traditional address multiplexing scheme since it increases the number of row address signals or introduces a speed penalty involved in additional selection. However, as far as power reduction is concerned, it has great potential. For example, the C_{DT} could be reduced down to about 100 pF at 256 Mb, as shown in Figure 8.14, with assuming a C_D of 200 fF, a ΔV_D of 1 V and the

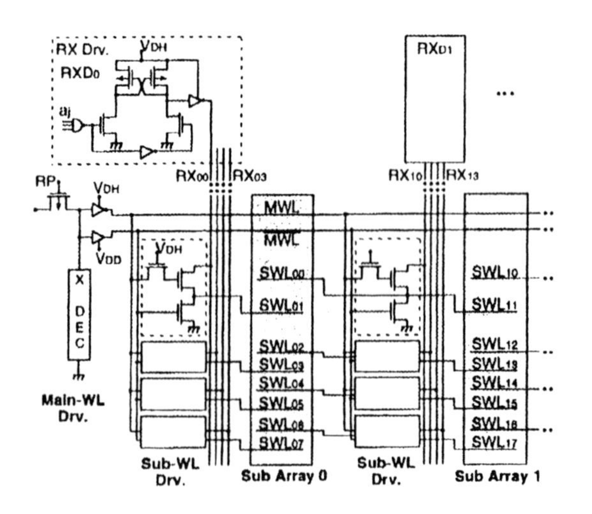

Figure 8.16 Hierachical word-line architecture [12].

number of activated data lines reduced by 1/32 as in reference [12]. Consequently, the architecture almost halves the chip power of the partial activation of the data line.

Refresh Time Increase — Even though reduction of C_D is achieved by the use of partial activations described above, reduction of m is never achieved without the help of increasing the maximum refresh time of the cell, t_{REFmax}. This stems from the necessity to preserve the refresh-busy rate [9], γ, expressed by

$$\gamma = t_{RCmin}/(t_{REFmax}/n) = (M/m)(t_{RCmin}/t_{REFmax}) \tag{8.9}$$

where t_{RCmin} and M are the minimum cycle time and memory capacity, respectively. It expresses what percentage of the time the memory is not accessible from outside. A smaller γ is preferable since it involves less conflict between refresh and normal operation. Hence, for a fixed M, it is necessary to maintain mt_{REFmax} to keep γ constant assuming a fixed t_{RCmin}. This implies a reduced m accompanied by an increased t_{REFmax}. Moreover, to quadruple M, mt_{REFmax} must be quadrupled. This has been achieved by almost doubling both m and t_{REFmax}, which results from compromising the power with cell leakage current [9]. As a result, $n(=M/m)$ has gradually increased with each successive generation with increased t_{REFmax} for commercial DRAMs [13], as shown in Figure 8.17. Alternative choices for n and m, as shown in Figs. 11 and 14, have been eventually rejected by this compromise. However, it seems difficult to keep the pace of doubling t_{REFmax} at each generation because it is determined by cell leakage current. One solution is the use of a new refreshing scheme [12] favorable to partial activation

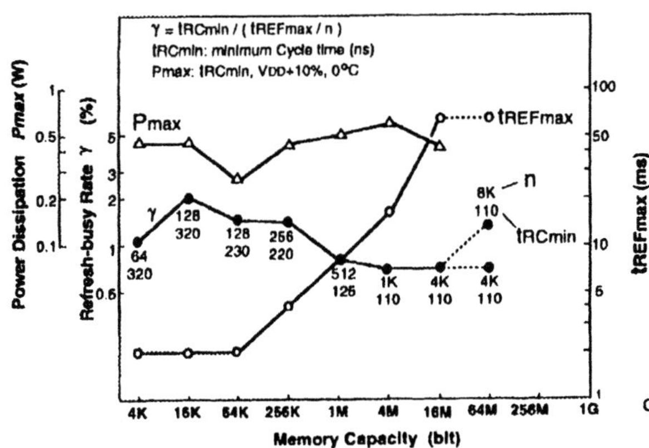

Figure 8.17 Trends in power dissipation (P_{max}), refresh-busy rate (γ) and maximum refresh time (t_{REFmax}) [13].

of a multi-divided word line. Note that the traditional scheme uses the same n, that is, the same m for both normal and refresh operations. The new scheme, however, uses a reduced m for normal operation which is a determinant of maximum power, while maintaining the same m as in the traditional scheme for refresh operation to preserve γ. The resulting power reduction allows an increased t_{REFmax} with a reduced junction temperature.

8.4.1.2 Operating Voltage Reduction

V_{DD} reductions from 12 to 5 V at the 64-Kb generation and then to 3.3 V at the 64-Mb generation, as shown in Figure 8.11, and half-V_{DD} data-line precharge which has been widely used since commercial 1-Mb DRAMs are well known as contributors to low power. On-chip voltage down converters [9] which are employed in commercial 16-Mb DRAMs to maintain a standard 5 V supply also contribute to low power. This low power operation necessitates an improvement in signal-to-noise-ratio, as described in section 8.3.

Half-V_{DD} Data Line Precharge — An excellent circuit for reducing the array operating current is half-V_{DD} data-line precharge [8][14]. Table 5 shows full-V_{DD} precharge with an NMOS sense amplifier (SA) which was popular until the 256-Kb generation and half-V_{DD} data-line precharge favorable to CMOS SA. In

Table 5 Half-V_{DD} Precharging scheme [8].

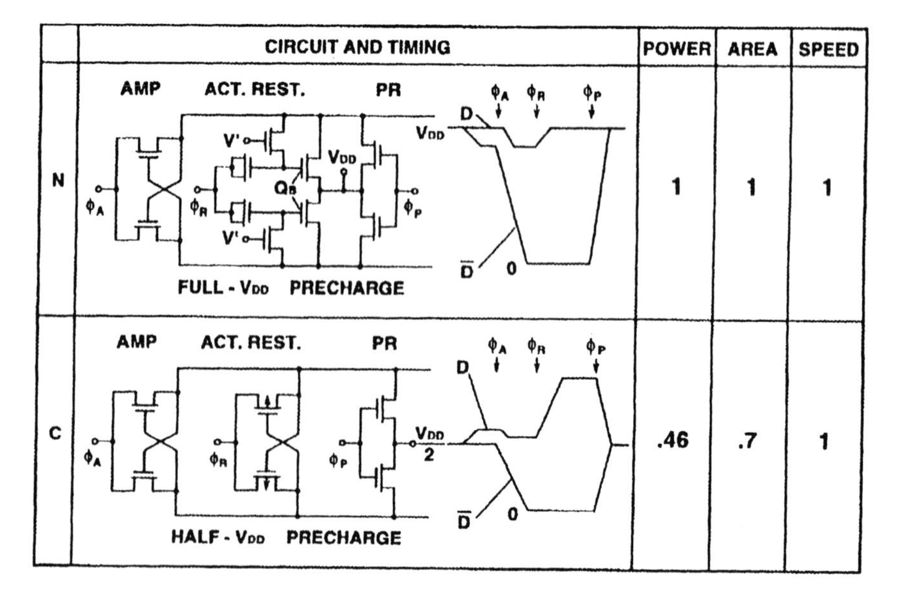

	CIRCUIT AND TIMING	POWER	AREA	SPEED
N	FULL - VDD PRECHARGE	1	1	1
C	HALF - VDD PRECHARGE	.46	.7	1

full-V_{DD} precharge, the voltage difference on the data lines is amplified by applying pulse ϕ_A and then the resultant degraded high level is restored to full-V_{DD} level by applying pulse ϕ_R to the active restore circuit. In contrast, the restore operation in half-V_{DD} precharge is simply achieved by applying pulse ϕ_R to the cross-coupled P-MOSFETs. In principle, half-V_{DD} precharge halves the data-line power of full-V_{DD} precharge with halved data-line voltage swing. A large spike current caused during restoring or precharging periods is also halved with less noise generation, allowing a quiet array.

On-chip Voltage-Down Converter (VDC) — For a fixed external supply V_{DD}, a voltage-down conversion scheme combined with scaled-down devices reduces chip power, as shown in Table 6. An internal voltage, V_{INT}, reduced by scaling factor (k >1), permits the use of scaled devices in core circuits while maintaining the same electric field. This eventually provides a low power (reduced by 1/k) with higher speed and smaller chip area. This is justified by the negligibly low VDC current (about 3% of total 16-Mb chip current) and the negligibly small VDC area (less than 1% of 16-Mb chip area). The breakdown voltage of devices in the VDC can be adjusted to a sufficiently high level through slight modifica-

Table 6 Impact of VDC approach on CMOS-chip performances under a fixed V_{DD}

approaches / performances	conventional approach (I_{DD}, V_{DD}; V_{INT}; chip; core circuits)	VDC approach (I_{DD}, V_{DD}; VDC; $\sim I_{DD}$, V_{INT}; chip; core circuits)
external voltage (V_{DD})	1	1
internal voltage (V_{INT})	1	1/k (k>1)
FET dimensions (L_i, W_i, t_{ox}, x_j)	1	1/k
electric field for FETs (E)	1	1
FET current (I_{DSi} *1)	1	1/k
power dissipation (I_{DD} *2 · V_{DD})	1	1/k
on resistance ($R_{oni} = V_{INT} / I_{DSi}$)	1	1
delay ($\tau_c = R_{oni} C_{Gi}$ *3)	1	1/k
circuit area	1	1/k

*1 $I_{DSi} = (W_i / L_i)(V_{INT} - V_T)^2 / t_{ox}$, *2 $I_{DD} = \Sigma I_{DSi}$, *3 $C_{Gi} \propto L_i W_i / t_{ox}$

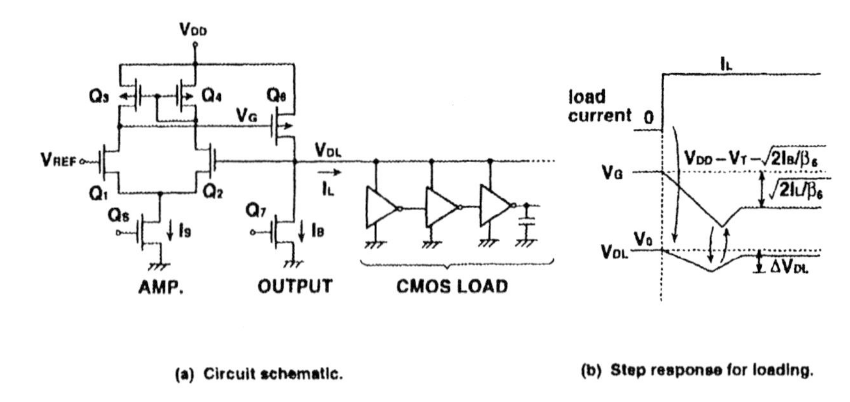

Figure 8.18 Voltage down converter [15].

tions involving device parameters. The keys to designing VDC are provision of a stable and accurate output voltage under rapidly changing load current and provision of on-chip burn-in capability. Figure 8.18 shows a schematic of a typical VDC [15] and the step response for the load current, I_L. The almost fixed output voltage, V_{DL}, is about 3.3 V for V_{DD}=5 V, 16-Mb DRAMs. For accuracy and load current driving capability, it consists of a current-mirror differential amplifier (Q_1-Q_4,Q_S) and common-source drive transistor (Q_6). As shown in Table 4, the array current for a DRAM is fairly large compared with that of a SRAM. The peak height is more than 100 mA with a peak width of around 20 ns. Thus, the gate width of Q_6 has to be more than 1000 μm. In order to minimize the output voltage drop ΔV_{DL}, the gate voltage of Q_6, V_G, has to respond quickly when the output goes low. An amplifier current, I_S, of 2 to 3 mA enables such a fast response time. Bias current source I_B is needed to clamp the output voltage when the load current becomes almost zero. To ensure enough loop stability with minimized area and operating current, phase compensation [15] is indispensable. The reference voltage, V_{REF}, must be accurate over wide variations of V_{DD}, process, and temperature for stable operation, because the voltage level determines the amount of cell signal charge as well as the speed performance. A band-gap V_{REF} generator and a CMOS V_{REF} generator [16] utilizing the threshold voltage difference have been proposed to meet the requirements. A burn-in operation with the application of a high stress voltage to devices is indispensable in VLSI production, both for reliability testing and for chip screening. For this purpose, the V_{REF} generator is designed to output a raised voltage when V_{DD} is higher than the value for normal operation [16]. Otherwise, the fixed voltage fails to apply a higher stress voltage.

Signal-to-Noise-Ratio Improvement — Low-voltage operation has to be accompanied by high S/N design [9] because it reduces the cell margin with reduced signal charge, as described previously. The operating margin of the cell shown in Figure 8.19 can be roughly expressed as

$$Q_S > Q_L + Q_C + Q_N$$
$$Q_S \text{ (signal charge)} = (V_{DD} - V_P)C_S$$
$$Q_L \text{ (leakage charge)} = i_L\, t_{REFmax}$$
$$Q_C \text{ (maximum charge collected at the cell node by an } \alpha \text{ particle hitting)}$$
$$Q_N \text{ (data-line noise charge)} = C_D\, v_N \qquad (8.10)$$

and where the available signal voltage for "1" or "0" is V_{DD}-V_P for reference voltage V_P (= $V_{DD}/2$), i_L is the cell leakage-current, and v_N is the noise caused by capacitive couplings on the data line from other conductors, the electrical imbalance between a pair of data lines and amplifier offset voltage. It is essential to maximize the signal charge, C_S (V_D-V_P), while minimizing the effective noise charges in order to achieve stable memory operation. The following are developments that have been made to meet these requirements.

1. **Signal Charge** — Word-line bootstrapping with a word voltage higher than V_{DD}+V_T is an well-known circuit for increasing Q_S with storing a full-V_{DD} level. The half-V_{DD} cell plate shown in Figure 8.20 is also an important concept. Until the 64-Kb generation, the V_{DD} plate, in which a gate-to source MOS capacitor was utilized as a cell capacitor, was standard (Figure 8.20(a)). The capacitor-plate (gate) voltage is V_{DD} while the storage-node (source) voltages are V_{DD} or 0 depending on the stored information. In the 256-Kb generation, however, a change was made to the V_{SS} (0V) plate, since the V_{DD} plate suffered from signal charge loss due to V_{DD} bounce at the plate. The n^+ layer can form capacitance

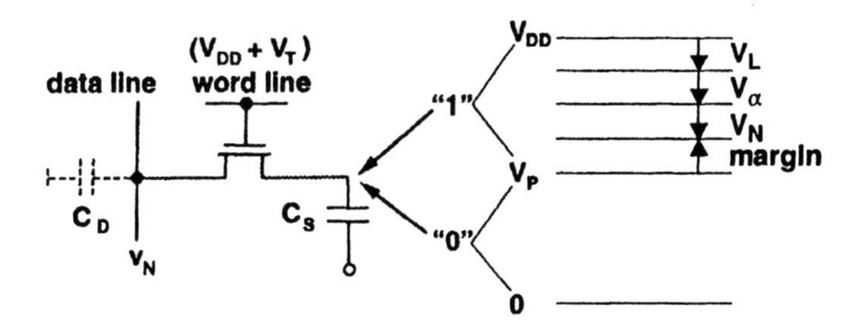

Figure 8.19 Memory cell margin [9].

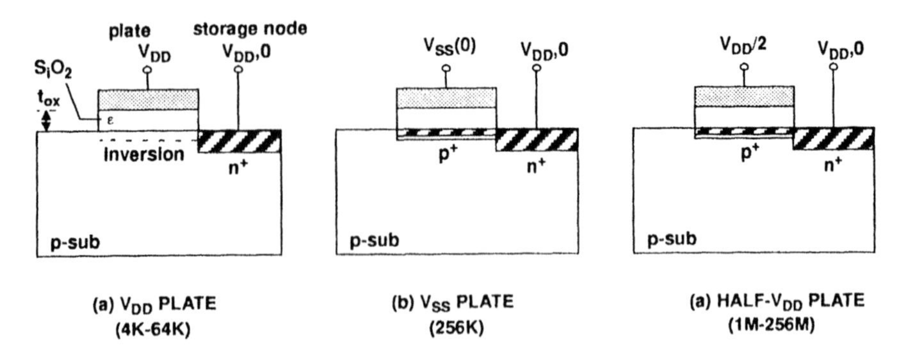

Figure 8.20 Voltage supplying schemes to the cell capacitor plate.

despite the V_{SS} plate. The p^+ layer increases the capacitance and works as a barrier against soft error. Note that the maximum voltage across the gate insulator is V_{DD} for both the V_{DD} plate and V_{SS} plate. For the 1-Mb generation and beyond, the half-V_{DD} plate has become standard, since it halves the insulator thickness for the same electric field, enabling doubled C_S and Q_S. In addition to thin insulator and new insulating films having a higher dielectric constant, vertical capacitors such as trench and stacked capacitors, as shown in Figure 8.21, are vital to increase the capacitor area. They have been used in commercial DRAMs of 4-Mb and beyond instead of a planar capacitor cell. Due to such innovations, signal charge reduction from the 64-K to 64-Mb generations is less than 1/5 while the cell area becomes 1/100, as shown in Figure 8.22. The resultant Q_S- decrease, however, more strongly requires noise reduction techniques.

2. **Leakage Charge** —Cell-node structures with less *p-n* junction area such as stacked capacitors are essential, as is suppressing the increase in junction temperature, T_j, as well as improving the quality of the fabrication process. This is because a longer t_{REFmax} is required with increases in memory capacity, as

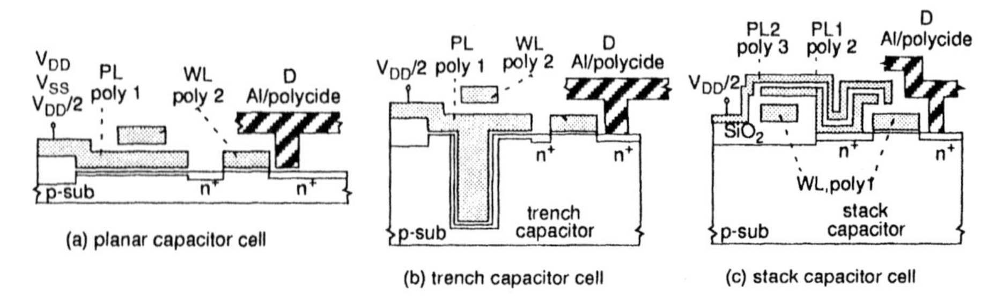

Figure 8.21 Trends in cell capacitor.
WL: word line; D: data lines; PL/PL 1/PL2: capacitor plate.

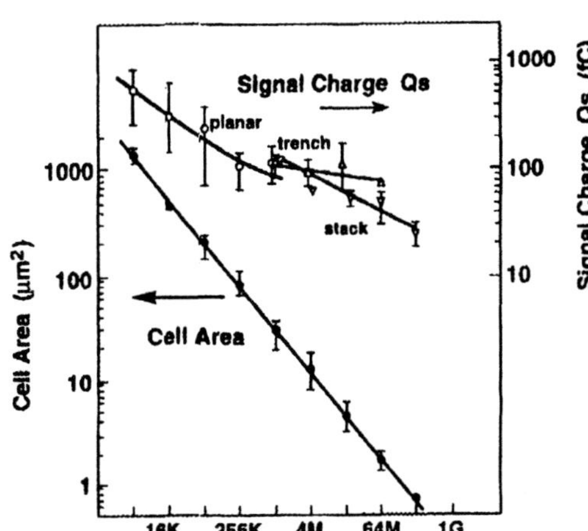

Figure 8.22 Trends in memory cell area and signal charge [1].

described above. Thus, low power dissipation and low thermal resistance packages are important in order to reduce T_j.

3. **Critical Charge** —The α-particle-induced soft efforts have been a major problem. Fortunately, however, it has been found that the critical charge Q_C decreases with reduced diagonal length of the depletion region in the memory cell. This suggests that a smaller Q_C is obtained by increasing cell density. Furthermore, less charge collection structure with reducing the volume of the depletion region is efficient for attaining even smaller Q_C. Thus, stacked capacitor cells and p^+ barrier are favourable. Chip coating and purification of materials are both indispensable to reduce the number of α-particles from package and materials in a chip.

4. **Data-Line Noise Charge** — C_D has been suppressed effectively through the multi- division of data lines, as shown before. In addition, many attempts have been made to reduce the differential noise v_N on the pair of data lines. Obviously, the noise increases in the number of its sources and in its magnitude with an increase in memory capacity. Furthermore, higher speed requirements at every successive generation have increased noise. Hence, the S/N ratio decreases with increases in memory capacity, considering that the actual cell signal v_S is almost constant, independent of the memory capacity. The S/N ratio of the given memory cell can be degraded by two kinds of internal noise. The first is generated in the memory array. It is higher in larger logical and physical array sizes and is further increased by the scaling down of cells. The other is coupled from peripheral circuits to the memory array. Large spike currents due to peripheral circuit operations, which increase with chip size, are the noise sources during the memory cell operation. These noises are closely related to data-line arrangements, as shown in Figure 8.23. They seriously degrade the cell margin of an open data-line arrangement, which was standard until 16-Kb generation. However, they can be effi-

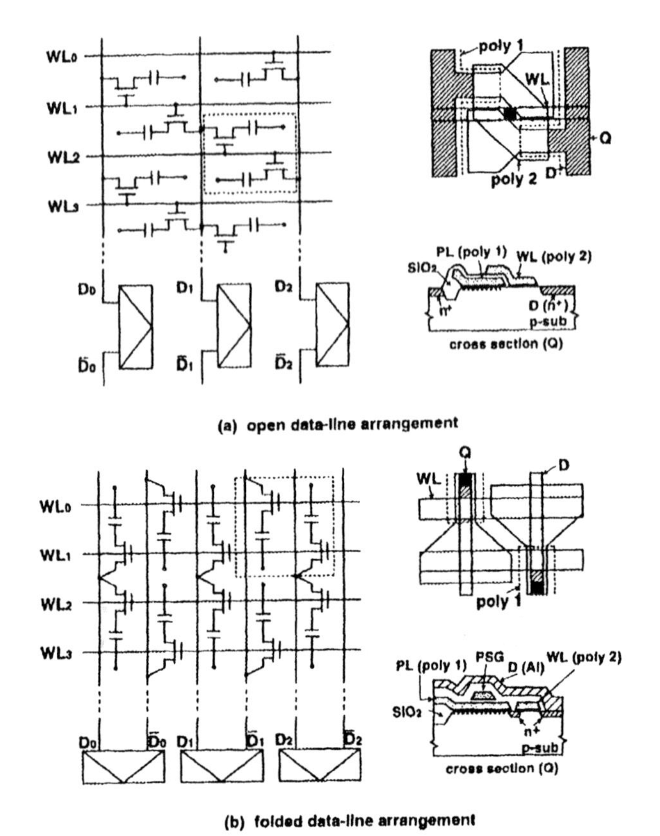

Figure 8.23 Data-line arrangements and their structures exemplified by planar capacitances [9]

ciently suppressed to an acceptable level by employing a folded data-line arrangement. The data-line interference noise as a serious limiting factor for 16-Mb and beyond, can be classified as being the one generated in the memory array. The noise becomes prominent with a decrease in data-line pitch. There are two approaches [1][9] to overcome this problem: a data-line transposition with a folded data-line arrangement, and a data-line shielded cell structure.

8.4.1.3 DC Current Reduction

Figure 8.24 shows column signal path circuitry [1] which is a main source of static current. It consists of a pair of data lines, a column switch, a pair of I/O lines, a load circuit for the I/O lines, and a differential amplifier ("main amp") for detecting the small signal voltage on the I/O lines. DC current flows from the I/O line load to data lines while the column switch is on. Also, the main amplifier consumes DC current for amplifying the signal, because it usually employs the conventional current mirror differential amplifier. The current reductions are

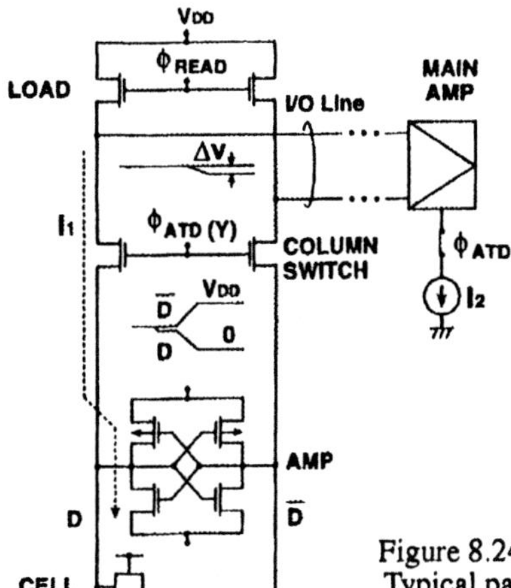

Figure 8.24 Pulse operation of column signal path circuitry [1]. Typical parameters for 16-Mb DRAMs are: $I_1/I_2 = 0.5 / 1$ mA, $\Delta V = 0.5$V, $C_{I/O} = 2$ pF.

essential especially for multi-data-bit chip because of the increased number of column circuitries. The DC currents are shut down with pulse-operation technique that will be described in detail in section 8.5. For example, in the static column mode where a DRAM operates as an SRAM for the column address signals, the column switch and the main amplifier is activated only when address-signal transition occurs. The ATD generates such control pulses, as described in section 8.5.

8.4.2. Data Retention Power Reduction

Reduction of both DC and AC current components in data-retention mode is a prime concern. Minimizing the power of on-chip voltage converters such as VDC, voltage-up (V_{DH}) converter, substrate back-bias (V_{BB}) generator, V_{REF} generator, and half-V_{DD} generator reduces the DC current component. Extending the refresh time and reducing refresh charge reduce the AC current component. They reduce I_{DCP}, $1/t_{REF}$, and $mC_{DD}V_D$ in Eq. (8.8), respectively.

8.4.2.1 Voltage Conversion Circuits

A precisely controlled V_{DH} level is important for eliminating V_T loss in memory cells and the periphery. However, a dynamic type booster which utilizes a capacitor to overdrive the output to the V_{DH} level suffers from uncontrollable output voltage for variation of V_{DD}, because V_{DH} is determined only by the capacitance ratio. A possible solution is a static word driver which directly operates from a V_{DH} DC power supply [1].

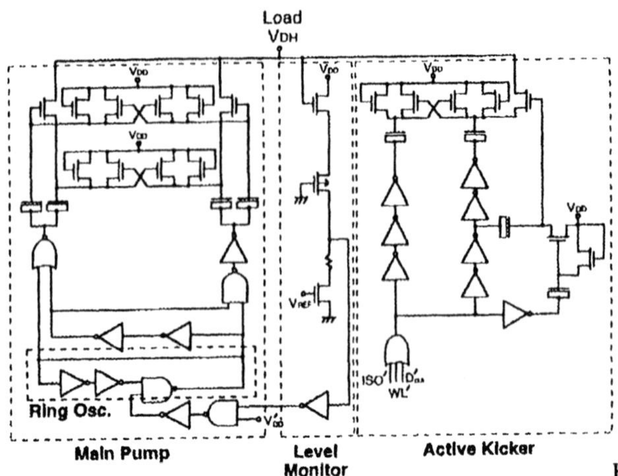

Figure 8.25 V_{DH} generator [1]

Figure 8.25 shows a V_{DH} generator. It features the use of two kinds of charge-pump circuits to provide charges for a pure capacitive output load: a main pump and an active kicker. The main pump compensates for a small charge loss due to the leakage current of the load. It is driven by the ring oscillator which is activated when V_{DH} is lower than the level determined by the level monitor. The active kicker operates synchronously with the load-circuit operation such as ISO driving (Figure 8.13), word-line driving, and output buffer driving. This circuit compensates for a large charge loss due to load circuit operations. Non operation of the main pump and extremely slow cycle operation of the active kicker in the data-retention mode provide a minimized retention current.

A V_{BB} generator is indispensable for stable operation of a DRAM, especially for the array. This provides a negative DC voltage of around -2 to -3 V to the P-substrate which is almost capacitive. Figure 8.26 shows the V_{BB} generator featuring two sets of charge pump circuits: a slow cycle ring oscillator 1 for supplying a small current during retention and stand-by modes and a fast cycle ring oscillator 2 for supplying a sufficiently larger current during the active cycle (/RAS on) or when the level monitor detects the V_{BB} level is high. Thus, it minimizes the retention current by shutting down the fast cycle circuit. A similar approach is useful for a VDC design. Another alternative is to stop the oscillation of the V_{BB} generator while the DRAM is not in an active cycle. This configuration is also useful for a VDC design.

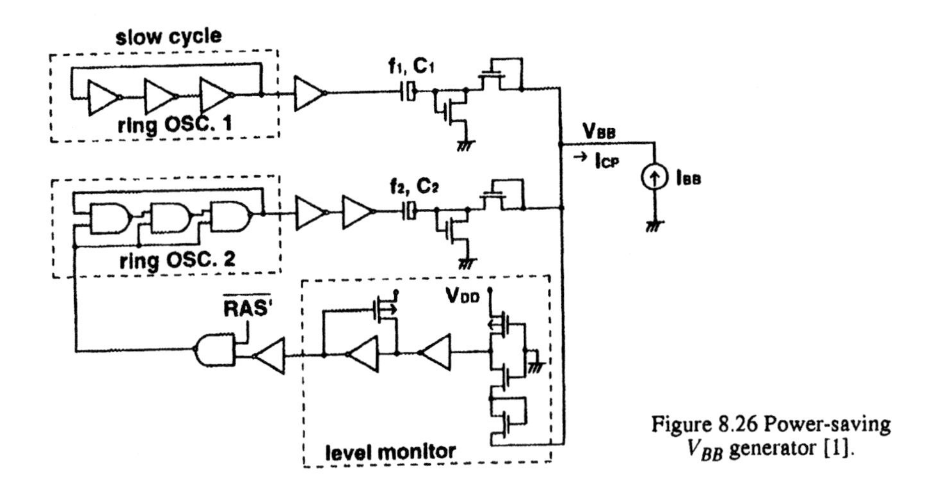

Figure 8.26 Power-saving V_{BB} generator [1].

With regard to a V_{REF} generator, a static V_{REF} generator has been proposed. However, a practical generator consumes more than 10 μA, because it needs a gain-trimmable amplifier circuit in order to get an accurate voltage level [16]. Gain trimming with poly silicon resistance makes it difficult to get high resistance values of over 1 MΩ without a large area and poor noise immunity against substrate bounce. Thus, an exploratory dynamic V_{REF} generator, as shown in Figure 8.27, has been proposed. It features dynamic operation like a sample and hold

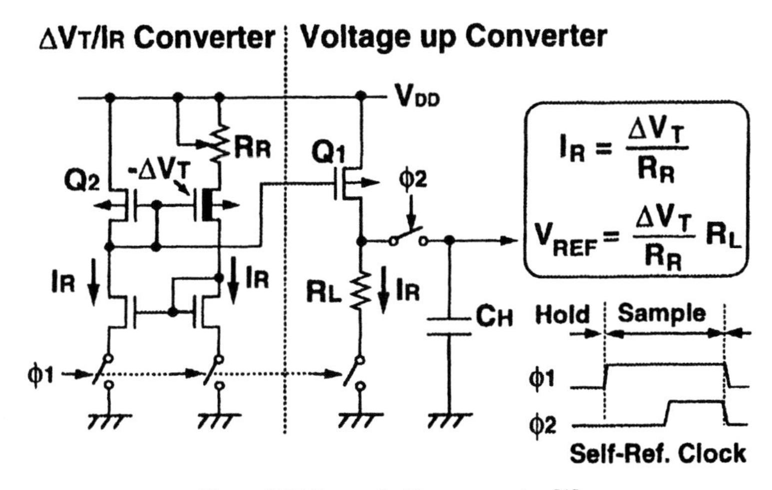

Figure 8.27 Dynamic V_{REF} generator [1].

technique. Reference current I_R, which is determined by the threshold-voltage difference ΔV_T and resistance R_R, is mirrored to the output node and converted to the voltage determined by the value of resistance R_L. The output voltage is determined only by the V_T difference and resistance ratio. Thus an accurate voltage can be obtained by trimming R_R, with well-known poly silicon fuses, even when ΔV_T fluctuates due to process variations. The current path is enabled while ϕ_1 is applied and the output voltage is sampled on the hold capacitance C_H while ϕ_2 is applied. These control pulses are generated from an on-chip self refresh circuit. For a pulse width ϕ_1 of 200 ns and a sampling interval of more than 100 ms, the total current is reduced to as low as 0.3 µA.

8.4.2.2 Refresh Time Extension

To extend the refresh time t_{REF} according to a reduced junction temperature in data-retention mode, a self-refresh control with an on-chip temperature detection circuit and the use of a cell-leakage monitor circuit on the chip [1] have been proposed.

8.4.2.3 Refresh Charge Reduction

One practical way to reduce the refresh charge is to reduce n (increase m) from that for active operation. This effectively reduces the operating frequency for the periphery while maintaining the array power constant. Another possible way is to reduce the voltage swing of data lines in the data-retention mode. The resultant reduced signal charge and increase in soft error rate are additional issues with this scheme. A charge recycle refresh scheme [1] has been proposed to reduce the data-line dissipating charge. In this scheme, the charges used in one array, conventionally poured out in every cycle, are transferred to another array and used, enabling the data-line charging current to be halved.

8.4.3. Perspective

8.4.3.1 Implication of V_T Scaling

As is described in section 8.1., V_T scaling is the major concern for achieving ultra-low-voltage and hence ultra-low-power VLSIs. To evaluate the subthreshold current caused by V_T scaling [17], the definition of MOSFET threshold voltage must be clarified. There are two kinds of V_T [18]: the extrapolated V_T and constant-current V_T, as shown in Figure 8.28. The extrapolated V_T is defined by extrapolation of the saturation current on the $\sqrt{I_{DS}}$-V_{GS} plane, neglecting the tail-

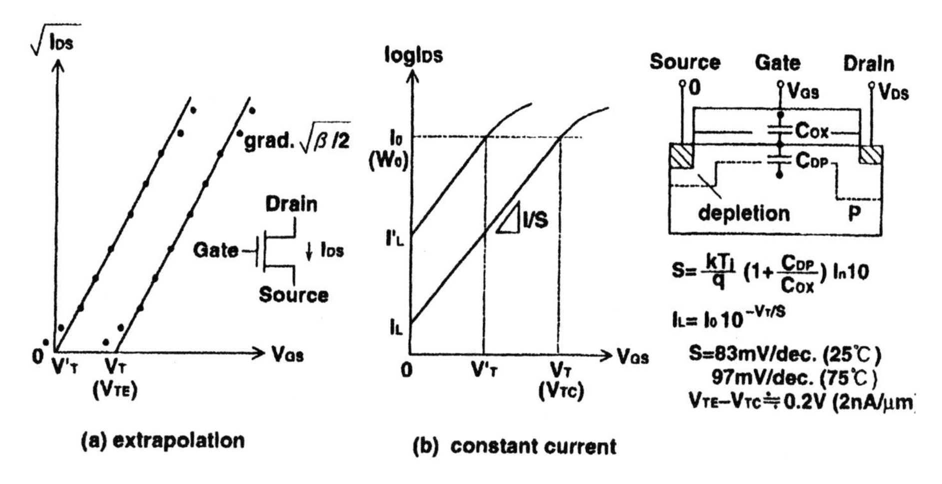

Figure 8.28 Definition of MOSFET threshold voltage.

ing current actually developed at approximately V_T. Our major concern is the subthreshold current which is developed at 0V V_{GS}. If V_T is high enough, the subthreshold current is 0. With decreasing V_T, however, a substantial subthreshold current starts to be developed, at a V_T higher than expected. This current is not expressed in this definition, although extrapolated V_T is familiar to circuit designers. Thus, constant current V_T is indispensable for evaluating the current. The V_T is defined as a VGS for a given current density on the log I_{DS}-V_{GS} plane. This constant current V_T (V_{TC}) is empirically estimated to be smaller than the extrapolated V_T (V_{TE}) by about 0.2V. The subthreshold swing, S, is expressed by FET parameters and junction temperature (T_j), as shown in the figure. S is about 83 mV/decade at 25°C while it is 97 mV/decade at 75°C for conventional MOS FETs. Hence, subthreshold current (I_L) exponentially increases with decreasing V_T.

The subthreshold current is a source of DC current for the basic CMOS circuit shown in Figure 8.29. For V_{DD} higher than approximately 2 V where the V_T (= V_{TE}) can be higher than around 0.4 V, either the NMOSFET or the PMOSFET in the inverter are completely cut off. For V_{DD} less than 2 V, however, a subthreshold current is developed. The V_T reduction, accompanied by lowering V_{DD} to maintain high speed, no longer cuts off the FETs. It causes an exponential increase in subthreshold current with decreasing V_T, and thus enhances the current with high temperature and V_T variation. This issue poses a serious concern regarding the power dissipation of memory chips as well as microprocessor chips,

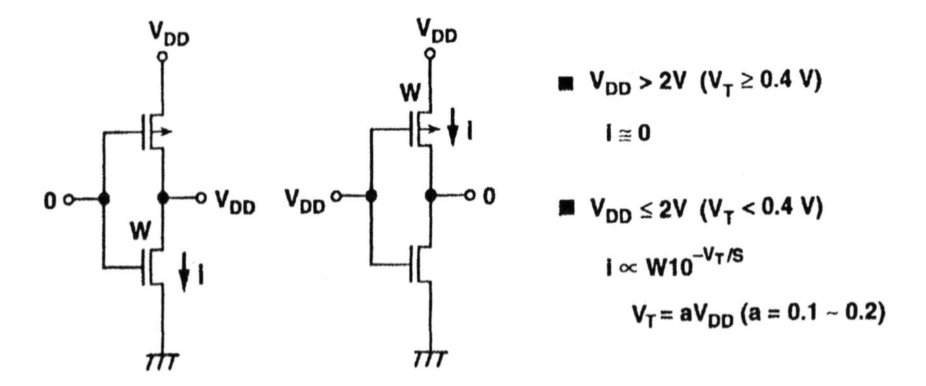

Figure 8.29 Subthreshold current increase in CMOS inverter with V_T scaling.

which comprise random circuits and iterative circuit blocks, as shown in Figure 8.30. Note that the subthreshold DC chip current, I_{DC}, from the overwhelmingly large number of inactive circuits, increases exponentially with the V_T reduction accompanied by lowering V_{DD}, and would increase not only the data-retention current, but also the active current. As a result, I_{DC} would exceed even the AC chip current, I_{AC}, which is the total charging current for capacitive loadings of the small number of active circuits, and would eventually dominate the active current of the chip, I_{ACT}, as shown in Figure 8.31 [22]. This is because, in addition to the huge difference in the number of circuits, AC capacitive current from each active circuit decreases with lowering V_{DD} while subthreshold current from each inactive circuit increases with lowering V_{DD}. Thus, our special concern is how to reduce the current from inactive circuits, especially from iterative circuits such as

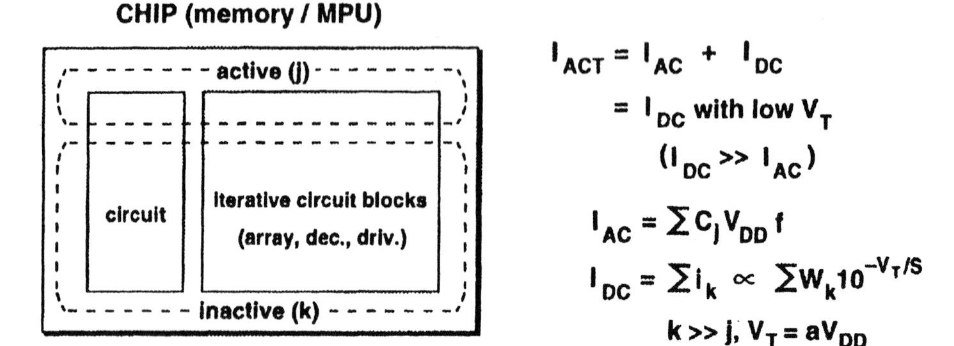

Figure 8.30 Subthreshold current increase in CMOS LSIs.

array, row and column decoders, and drivers. Attempts proposed so far to solve the problem are: source-gate back biasing and cutting off the leakage path with a high-V_T MOS FET while using a low-V_T MOS FET for the main signal path during the active period. These techniques reduce the DC current during the inactive period. This section gives details from a circuit design standpoint.

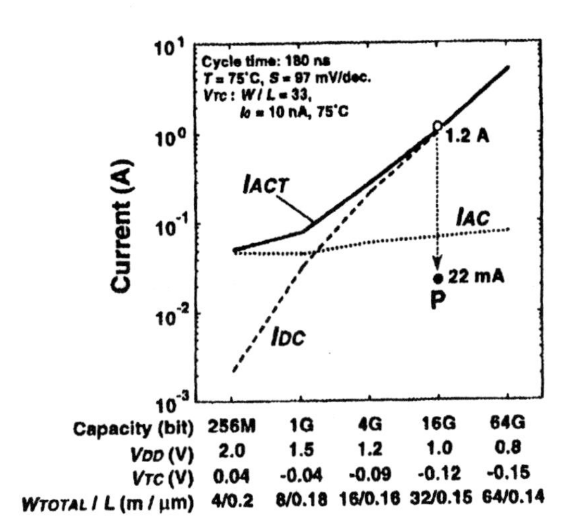

Figure 8.31 Trends in active current in future DRAM chips [22]. The threshold voltage is defined for a constant drain-source current of 10 nA.

8.4.3.2 Subthreshold Current Reduction Circuits

CMOS Basic Circuit — Figure 8.32(a) shows a schematic of the switched-source-impedance (SSI) CMOS circuit [19]. It features a switched

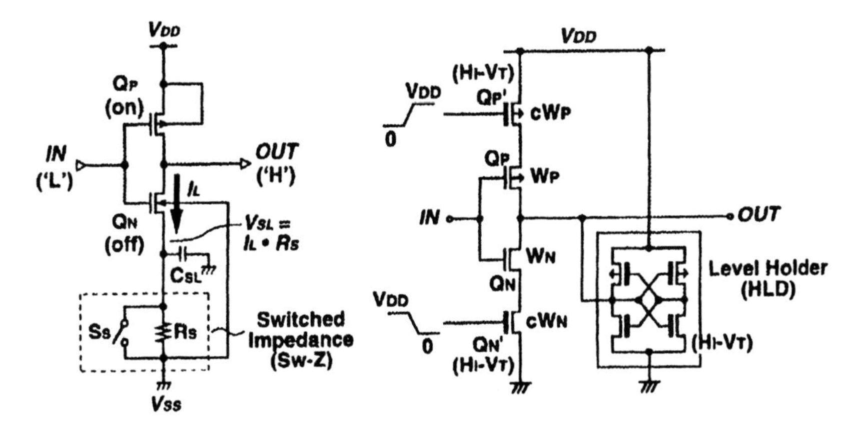

Figure 8.32 Subthreshold current reduction for basic CMOS circuit [19][20].

impedance (Sw-Z) at the source of the N-MOSFET Q_N, consisting of a resistor R_S for limiting the subthreshold current and a switch S_S for bypassing R_S. The switch keeps on during the active period so that the circuit works as a conventional CMOS circuit. For the inactive period when the switch is off, the subthreshold current raises the source voltage to V_{SL}, giving a gate-source back-biasing so that the current is reduced. This is just a negative feedback effect. Thus, the circuit features immunity to V_T fluctuations as a result of negative feedback. Resistor R_S may be implemented as a highly resistive poly-Si wire or as a MOSFET with a small channel width/length ratio. In fact, an intentional resistor is not necessarily needed. When the resistor is eliminated, R_S is regarded as the leakage resistance of switch SS. This scheme is also applicable to other logic gates as long as the input voltages are predictable. Another circuit for reducing the subthreshold current is to use a switched-power-supply inverter with a level holder [20], as shown in Figure 8.32(b). This is useful for some applications in which input voltage is not predictable as in the active mode of a DRAM, although there is a speed degradation due to Q_N' and Q_P', and high-V_T eventually restricts the lower limit of V_{DD}. The power supply of the CMOS circuit is controlled by FET switches Q_N' and Q_P'. The V_T's of all FETs except Q_N and Q_P are high enough, allowing negligible subthreshold current. As soon as the input level has been evaluated at high speed due to low-V_T of Q_N and Q_P and the resultant output is held in the holder, the switches are turned off. Thus, the output level is maintained without subthreshold current. The level holder can be laid out with minimized area since it plays only a role of level holding.

Iterative Circuit Blocks — One of the most efficient applications of the SSI scheme is to iterative circuit blocks. To reduce the subthreshold current in stand-by mode, it has been implemented in the decoded word driver of an experimental 256-Mb DRAM [21], as shown in Figure 8.33. A P-channel switching FET, Q_S, was inserted between the power-supply line V_{DH} and the driver-FET's (Q'_s) common-source terminal. The node PSL decreases in the standby mode by V_{SL} related to the subthreshold current (nI) in the P-MOSFETs in the driver circuits. V_{SL} is proportional to $\ln(nW_D/W_S)$. The channel width of the switching transistor W_S can be narrowed to an extent comparable to that of the driver FET W_D without speed degradation in active mode, since only one of n ($n = 256$, for example) driver FETs turns on. Then, the subthreshold current decreases exponentially with V_{SL}, and the descent of the node PSL stops within 200-300 mV below V_{DH}. This enables a high-speed recovery (2-3 ns) of the node PSL back to the V_{DH}

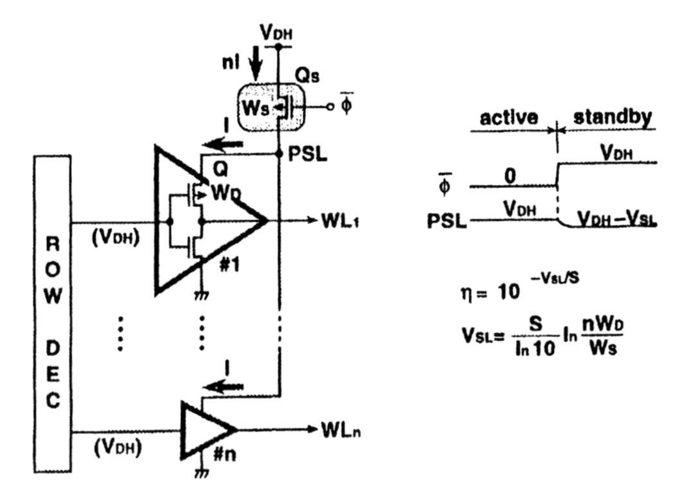

Figure 8.33 Subthreshold current reduction of DRAM word driver [22]. η is the current reduction ratio.

level in the transition from the standby mode to the active mode. The sub-threshold current in active mode is another concern for iterative circuit blocks, although that in stand-by mode is reduced by the circuit described above. After one selected word line is activated, all the drivers are sources of subthreshold current, eventually dominating the total active current. The issue is overcome by partial activation of multi-divided power-line [20], as shown in Table 7. One-dimensional (1D) power-line selection features a selective power-supply to part of the

Table 7 Partial activation of multi-divided power-line [22].

		Conventional	1-D Selection	2-D Selection
Configuration		VDD, m•n•I, W, 0, m•n Ckt., I, VDD, VDD	Activated, Inactivated, VDD, a•W, 0, VDD, a•I, VDD, n•I, I, n, m	VDD, 0, 0, VDD, b•W, a•W, (n/k)•I, a•I, n/k, VDD, b•I, k, m
Active Subthreshold Current	Ideal	m•n•I	n•I	(n/k)•I
	Actual	m•n•I	n•I + (m-1)•a•I	(n/k)•I + (m-1)•a•I + (k-1)•b•I

circuit block by dividing it into m sub-blocks each consisting of n circuits. The operation is performed by turning on a switch corresponding to a selected (activated) sub-block, while the others remain off. All the non-selected (in-activated) sub-blocks substantially have no subthreshold current since the same voltage relationship, as in stand-by mode in Figure 8.33, is established in each sub-block. This reduces the current to $n \cdot I$ with an m-fold reduction. For further reduction of the current, two-dimensional (2D) selection [22] has been reported. In this configuration, a circuit block having $m \cdot n$ circuits is divided into $m \cdot k$ sub-blocks, m in a row and k in a column. The subthreshold current is reduced, in inverse proportion to the number of sub-blocks, to $(n/k) \cdot I$ with an $m \cdot k$-fold reduction. In the conventional scheme, the active DC current is as high as 1.2 A for a hypothetically designed 16-Gb DRAM [20]. The dominant factor is the subthreshold current in the iterative circuit blocks such as word drivers, decoders, and SA driving circuits. These DC currents could be reduced with a 2-D selection scheme down to 22 mA, as shown by point P in Figure 8.31. When this scheme is combined with partial activation of multi-divided word line, an additional AC power and layout area due to 2-D control is negligible. This is because sub-array decode signal can be utilized for driving switches.

Memory Cell — The subthreshold current of a cell transistor would flow from the cell storage node to the data line while the data line is held at the low level. The current flow can be prevented by the use of a source-gate back-biasing scheme. Figure 8.34 is an example of the application to a 256-Mb DRAM [23]. It is called the boosted sense-ground scheme. In an active cycle, the CMOS latch-type amplifier (SA) and Q_1 are activated and the voltage of the sense ground

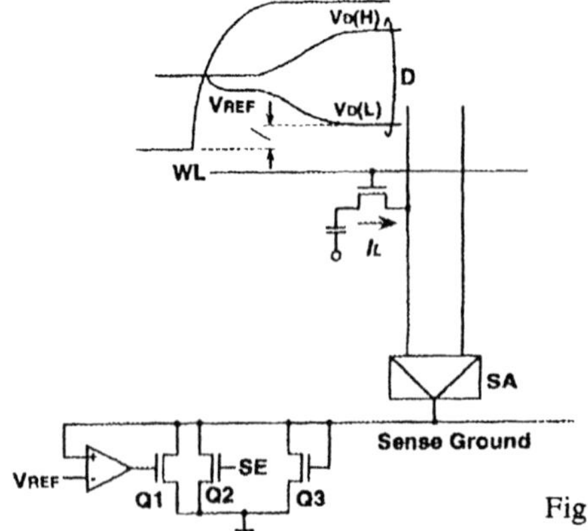

Figure 8.34 Boosted sense ground scheme [23].

becomes V_{REF}. Q_2 is turned on by using signal SE at the beginning of the sensing operation in order to accelerate data-line discharging. Both Q_1 and Q_2 are turned off during the stand-by period and Q_3 is used to clamp the voltage level of the sense ground to minimize the stand-by current.

8.5. Low Power SRAM Circuits

8.5.1. Active Power Reduction

SRAM chip development [24][25] has been driven by low power circuit applications, although the emphasis has recently been on more enhanced speed. Partial activation of a multi-divided word line and pulse operation of word-line circuitry are typical examples of low power circuits, which drastically reduce the DC current that dominates total active current with decreasing m and Δt in Eq. (8.7). Pulse operation of the column/sense circuitry is another example of low power techniques. Column/sense circuitry inevitably includes differential circuits, which unfortunately consumes much DC current to achieve high speed. Therefore, the pulse operation is essential to reducing I_{DCP} in Eq. (8.7). ATD (Address Transition Detection) plays an important role in the pulse operations for word-line and column/sense circuitry. For column/sense circuitry, lowering the operating voltage, while maintaining high-speed amplification capability for small signals, is also critical.

8.5.1.1 DC Current Reduction

Partial Activation of Multi-divided Word Lines — The multiple row decoder scheme and double-word line scheme [1], which divides a word line into sub word lines, greatly reduces the static current of SRAMs. A more sophisticated word-line division, called DWL (Divided Word Line), adopts a two-stage hierarchical row decoder structure [26], as shown in Figure 8.35. The DWL scheme requires two levels of metal layers; one for a main word line and the other for a data line. The number of sub-word lines connected to one main word line in the data-line direction is generally four (at most eight), compromising an area to a main row decoder with an area to a local row decoder. DWL features two-step decoding to select one word line, greatly reducing the capacitance of the address lines to a row decoder and the word-line RC delay. Figure 8.36 [26] shows that the column current (mi_{DC}) and word-line delay decrease as the word-line division number increases. The DWL scheme has been used in most high-density SRAMs of 1 Mb and greater. In a recent 16-Mb SRAM [4], a word line was divided into

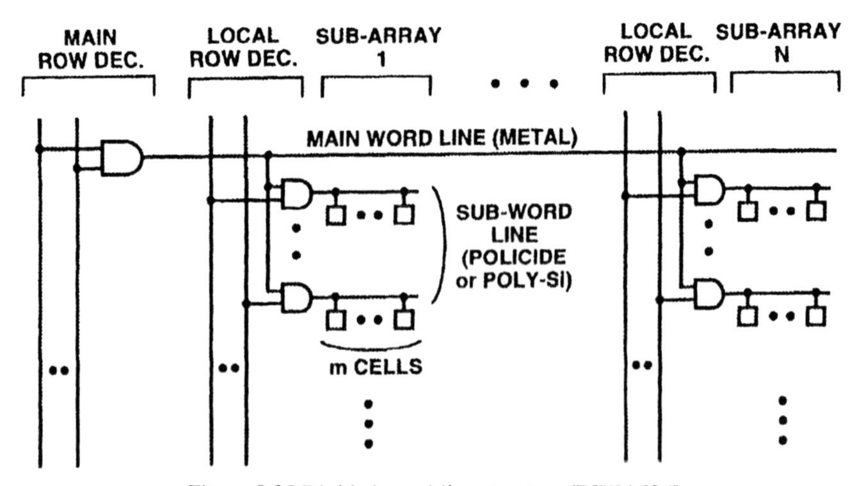

Figure 8.35 Divided word-line structure (DWL) [26].

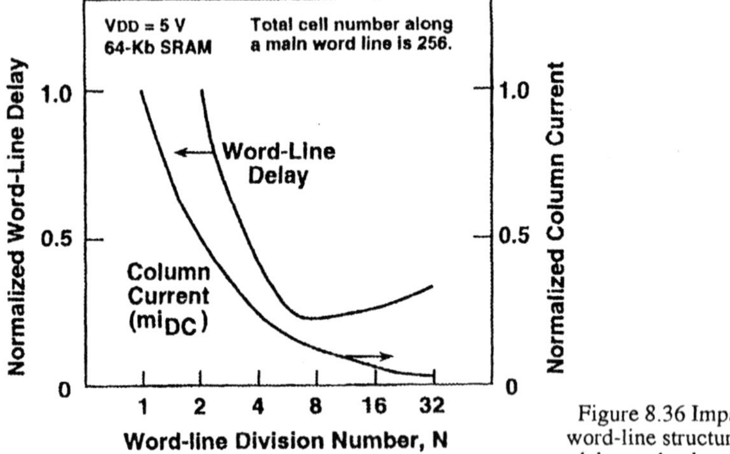

Figure 8.36 Impact of divided word-line structure on word-line delay and column current [26].

32 sub-word lines: N=32 and m=256. The cell current was reduced to one thirty-second of its original level with the DWL scheme. However, this scheme eventually results in a long main word line having a load capacitance that increases due to the increase in the number of local row decoders. This is because the word-line division number, N, must increase while keeping m (number of cells selected simultaneously) small. To overcome the problem, two approaches [1] have been proposed: a combination of a multiple row decoder and a DWL, and a three-stage hierarchical row decoder scheme. An experimental 4-Mb SRAM with

the three-stage hierarchical decoder reveals a reduction of total capacitance in the decoding pass by 30% and a reduction in delay by 20% compared with the DWL scheme.

Pulse Operation of Word-Line Circuitry — The duration of the active duty cycle can be shortened by pulsing the word-line for the minimum time required for reading and writing in a cell array, as shown in Figure 8.37 [27]. This reduces the power by the duty ratio of the pulse duration to the cycle time. The word-activating pulse, ϕ_X, is obtained by lengthening out the original ATD pulse, described shortly, enough to build up the data-line signal and latch the amplified signal by

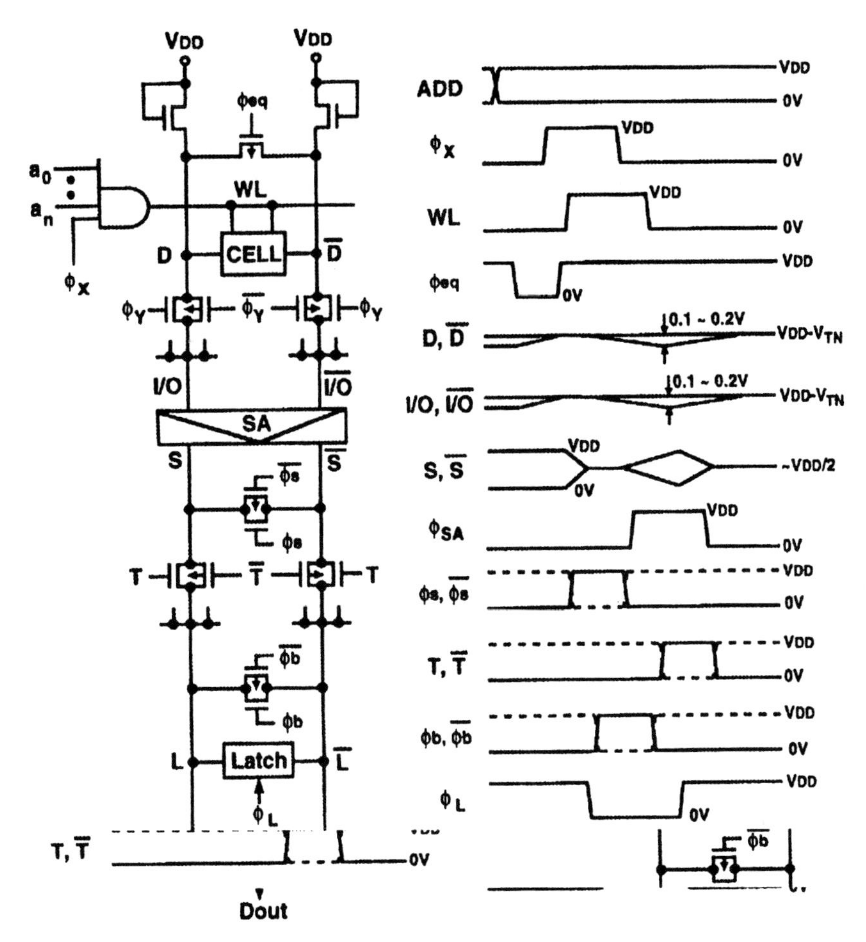

Figure 8.37 Simplified view of pulsed operation of word line, sense amplifier SA, and latch circuit [27].

ϕ_L. This scheme is usually employed with a pulsed sense amplifier and a latch circuit, as shown in Figure 8.37. One of data line signal pairs (D, /D) is selected by ϕy and /ϕy, and transmitted to I/O and /I/O, respectively. The I/O signals are amplified by a sense amplifier SA. The amplified S and /S signals from a sub-array are selected by signals T and /T and then are transmitted to the latch where the signals are latched to keep the data output valid after the word line and sense amplifier are inactivated. An on-chip pulse-generating scheme using ATD appeared in SRAMs first, then in DRAMs. An ATD circuit comprises delay circuits and an exclusive OR circuit, as shown in Figure 8.38 (a) (b) [1]. An ATD pulse, $\phi(a_i)$, is generated by detecting "L" to "H" or "H" to "L" transitions of any input address signal a_i, as shown in Figure 8.38 (c). All the ATD pulses generated from all the address input transitions are summed up to one pulse, ϕ_{ATD}, as shown in (d). This summation pulse is usually stretched out with a delay circuit and used to reduce power or speed up signal propagation.

An additional DC current in the write cycle, as shown in Figure 8.39 [1], is another concern. An accurate array current expression for the read cycle, exclusive of periphery AC and DC currents in Eq. (8.7), is given by

$$I_{DDA}(r) = [mi_{DC}(r)\Delta t + mC_D\Delta V_r]f \qquad (8.11)$$

and a corresponding current for the write cycle is given by

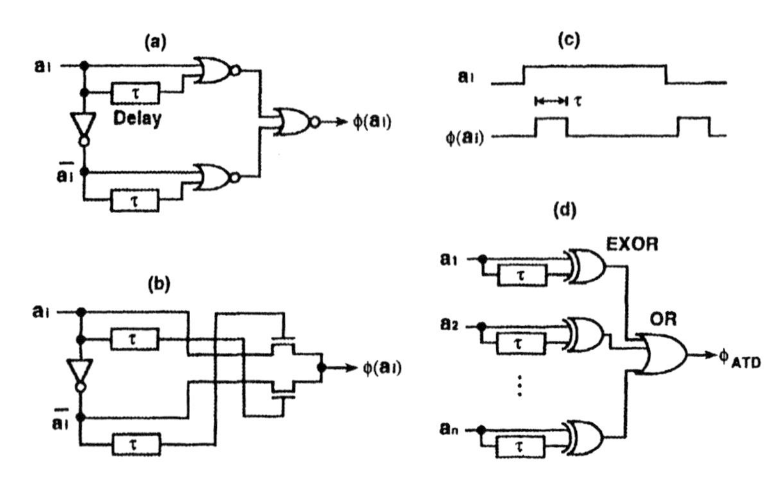

Figure 8.38 Address transition detection circuits [1].
(a) (b) ATD pulse generating circuits; (c) ATD pulse wave form; (d) Summation circuits of all ATD pulses generated from all address transitions.

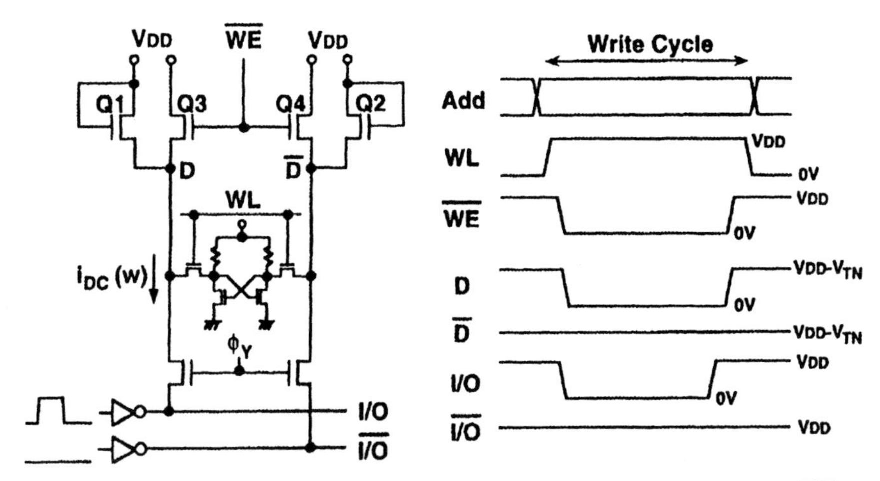

Figure 8.39 Simplified diagram of data line load control with write-enable signal $\overline{WE}$ (variable imepdance data-line load) [1].

$$I_{DDA}(w) = [(m\text{-}p)i_{DC}(r)\Delta t + pi_{DC}(w)\Delta t + pC_D\Delta V_w]f \qquad (8.12)$$

where $i_{DC}(r)$ and $i_{DC}(w)$ are data-line static currents in the read and write cycles; ΔV_r and ΔV_w are data-line voltage swings in the read and write cycles, respectively; p is the number of data which are simultaneously written into cells. In a typical 5V 4-Mb SRAM, $I_{DDA}(r)$ and $I_{DDA}(w)$ are 6.4 mA and 10.4 mA, respectively, assuming $i_{DC}(r) = 100\ \mu A$, $i_{DC}(w) = 1.0$ mA, $\Delta V_r = 0.2$ V, ΔV_w @ V_{DD}, C_D = 1 pF, $m = 128$, $p = 8$, $\Delta t = 30$ ns, and f = 10 MHz. Inherently large $pi_{DC}(w)$ and $pC_D\Delta V_w$ make $I_{DDA}(w)$ larger than $I_{DDA}(r)$, especially for multi-bit SRAMs which make p large. To reduce $I_{DDA}(w)$, both $i_{DC}(_w)$ and C_D must be decreased. To reduce $i_{DC}(w)$, the variable impedance load [1] makes all the data-line load impedances high in the write cycle by cutting off Q_3 and Q_4 with the write enable signal, /WE, as shown in Figure 8.39. In some SRAMs the loads are entirely cut off during the write cycle to stop any DC current. C_D is reduced by partially activating the data lines that are divided into two or more portions, just as in DRAMs.

Pulse Operation of Column/Sense Circuitry — A DC current of 1 mA to 5 mA flows in a sense amplifier on the I/O line. The power dissipation becomes the larger portion of the total chip power as the number of I/O lines increases to obtain higher data throughput for high-speed processors. Figure 8.40 shows a switching scheme of well-known current-mirror sense amplifiers [27]. Two amplifiers are serially connected to obtain a full supply-voltage swing output,

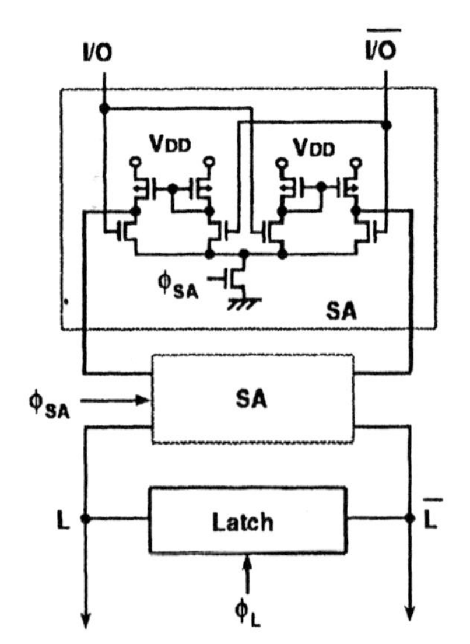

Figure 8.40 Power switching of current-mirror sense amplifier [27].

since one stage of the amplifier does not provide a gain enough for a full swing. A positive pulse, ϕ_{SA}, activates the sense amplifiers just long enough to amplify the small input signal; then the amplified output is latched. Hence, the switching scheme combined with a pulsed word line scheme reduces the power, especially at relatively low frequencies, as shown in Figure 8.41 [1]. Pulse operation of word-line circuitry and sense circuitry was also applied to fast SRAMs [1], achieving 24-mA power consumption at 40 MHz and 13-ns access time. Further current reduction is gained by sense amplifier current control which switches the

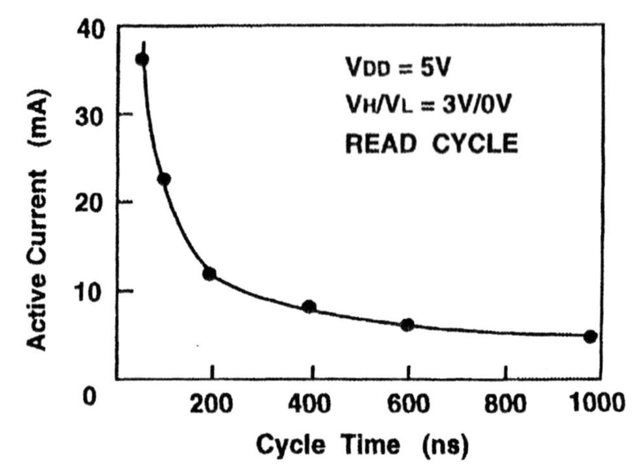

Figure 8.41 Measured active current in a 4-Mb CMOS SRAM with pulse word line and pulse sense amplifier control [1]. V_H and V_L are high and low input voltage levels, respectively.

current to the minimum level required for maintaining data from a high level necessary only during amplification. Moreover, a latch-type sense amplifier, such as a PMOS cross-coupled amplifier [28], as shown in Figure 8.42, greatly reduces the DC current after amplification and latching, since the amplifier provides a nearly full supply-voltage swing with positive feedback of outputs to PMOSFETs. As a result, the current in the PMOS cross-coupled sense amplifier is less than one-fifth of that in a current-mirror amplifier, while affording fast sensing speed. Equalizers EQL equilibrate the paired outputs of the amplifier before amplification, which requires much more accurate timings for stable operation.

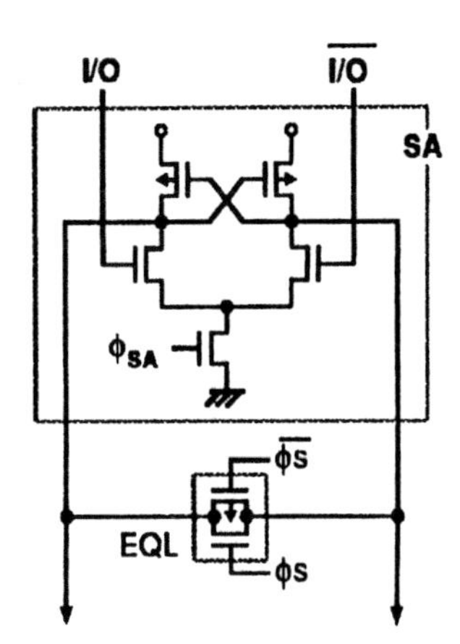

Figure 8.42 PMOS cross-coupled sense amplifier [28].

8.5.1.2 Operating Voltage Reduction

Low Voltage Sense Amplifier — A 5-V power supply and NMOS data-line loads, as shown in Figure 8.37, have been widely used for SRAMs. An NMOS V_T drop provides intermediate input voltages, essential to obtain large gains and fast sensing speeds, to sense amplifiers. For lower V_{DD} operation, however, if a resultant data-line signal voltage close to V_{DD} is amplified, PMOS data-line loads without the V_T drop are more suitable. This is made possible by level-shifting the data-line voltages. A resulting intermediate voltage allows it to use conventional voltage amplifiers for the succeeding stage. The use of NMOS source followers and scaled low-V_T NMOSs are good examples of level shifting [1].

A current sense amplifier combined with PMOS data-line loads [29], as shown in Figure 8.43, facilitates low voltage, high speed operation. A normally-on equalizer makes a small current difference between the paired data lines

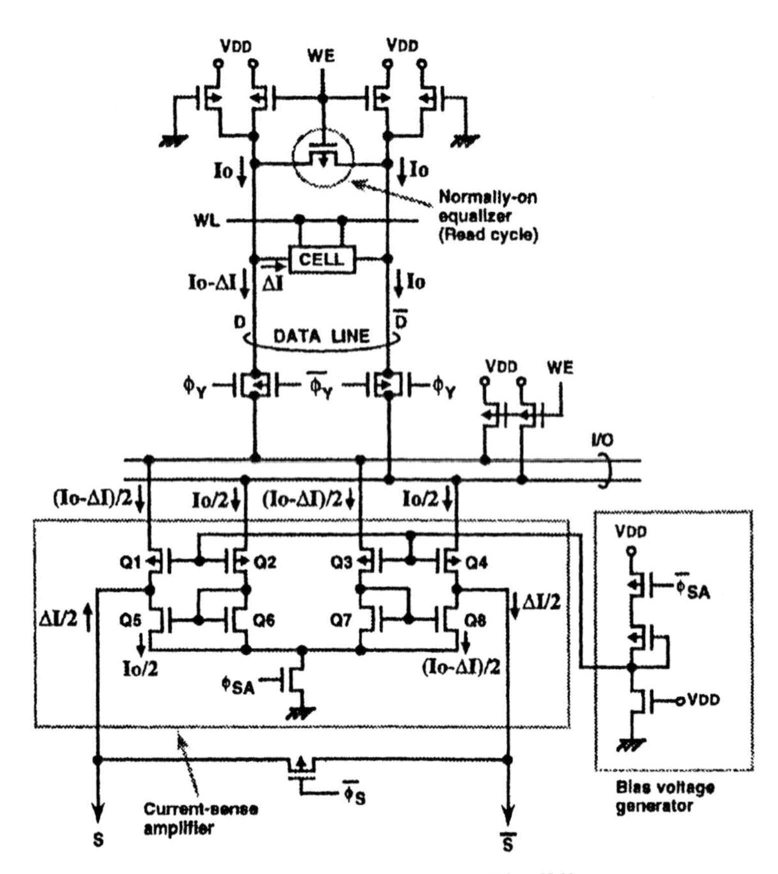

Figure 8.43 Current sense amplifier [29].

during a read cycle, depending on cell information. The current-mirror configuration of the amplifier circuit makes a current difference of ΔI/2 between Q1 and Q5 (Q4 and Q8), which eventually discharges and charges the outputs, S and /S. The bias voltage generator provides an intermediate voltage of 1 V to 1.5 V at 3.3-V V_{DD} to increase gain by operating Q1 - Q4 close to the saturation region. Current sensing provides the advantage of an extremely small data-line voltage swing of less than 30 mV, and eliminates the need for pulsed data-line equalization, which allows for fast sensing. Note that the required voltage swing in a conventional voltage amplifier is 100 mV to 300 mV, as shown in Table 4. For a fixed delay of 1.2 ns, the amplifier reduces the current consumption by about 2 μA compared with a conventional current-mirror voltage amplifier.

On-chip Voltage Down Conversion [1] — On-chip power supply conversion was first used for a 256-Kb SRAM to internally supply 3.3 V to the 0.7-μm devices with a 5-V V_{DD}. The SRAM has a power down mode to reduce standby chip current to 5 μA. A 4-Mb SRAM, which turns off one of two VDCs (voltage down converters) to provide a 50-μA standby current, automatically shuts down the two VDCs when the external supply voltage is reduced to 3.3 V to obtain a 1-μA data retention current. An experimental VDC which achieves sub-μA standby current has been reported. The VDC circuits in SRAMs are basically the same as DRAM VDCs, as shown in section 8.4.

8.5.1.3 Charging Capacitance Reduction

Charging capacitance reduction techniques, initially targeted at obtaining high speed in SRAMs, also contribute to power reduction, which is more of an issue in DRAMs. These techniques include data-line division, I/O line division, and pre-decoding scheme. Inserting a pre-decoding stage between an address buffer and a final decoder optimizes both speed and power, and has been used in most SRAMs.

8.5.2. Data Retention Power Reduction

On-chip voltage converters in commercial SRAMs have not been widely used, although there have been many attempts which differ from the DRAM approach. This is mainly due to an inherently wider voltage margin and a different operating principle of the SRAM cell. Thus, data retention current has been sufficiently reduced solely by the memory cell improvement, as described in section 8.3. Switching the power supply voltage to 2 V to 3 V, from 5 V at normal operation, further reduces the data retention current. However, this voltage switching approach is eventually restricted by the ultra-low voltage ability of SRAMs, described in the following section.

8.5.3. Perspective

Low voltage operation capability of a SRAM memory cell is a critical issue. A full CMOS memory cell with an inherently wide voltage margin can be operated at the lowest supply voltages, as exemplified by a 1-Mb SRAM with a 200-ns access time at 1 V [1]. For polysilicon-loads or polysilicon PMOS load cells, a two-step word voltage scheme and an array operational scheme with a raised DC voltage have been proposed for widening the voltage margin [1]. The raised DC voltage supplied to the source of PMOSFETs in a cell allows the storage node voltage of the cell to quickly rise in the write operation. However, the low-voltage SRAMs mentioned above are quite slow because of relatively high V_T of 0.5 V and more. Thus, the low V_T designs are indispensable for high-speed operation. However, this in turn increases the cell leakage currents tremendously [10] due to subthreshold current. Subthreshold current increases in the periphery could be overcome by the use of reduction circuits as described in section 8.4. In addition to a soft error increase due to a decreased signal charge, the V_T mismatch that continues to increase between paired FETs in a cell, and V_T variation, are also prime concerns [1]. Circuit techniques to suppress the mismatch in sense amplifiers with a 3 V power supply have been recently reported [1]. However, there are still challenges facing ultra-low voltage SRAMs.

References

[1] K. Itoh, K. Sasaki and Y. Nakagome, "Trends in Low-Power RAM Circuit Technologies," *Proc IEEE.*, April 1995

[2] Y. Nakagome et al., "An experimental 1.5-V 64-Mb DRAM," *IEEE J. Solid-State Circuits*, vol. 26, no. 3, pp. 465-472, April 1991.

[3] K. Satoh et al., "A 4Mb pseudo-SRAM operating at 2.6¿1V with 3uA data- retention current," *ISSCC Dig. Tech. Papers*, pp. 268-269, Feb. 1991.

[4] M. Matsumiya et al., "A 15ns 16Mb CMOS SRAM with Reduced Voltage Amplitude Data Bus," *ISSCC Dig. Tech. Papers*, pp. 214-215, Feb. 1992.

[5] HITACHI MEMORY CATALOGS, Aug. 1993 and Aug. 1994.

[6] M. Lamson et al., "Lead-On-Chip Technology for High Performance Packaging," *Proc. 43rd Electronic Components and Technology Conference*, pp.1045-1050, June, 1993.

[7] M. Taguchi, "High-Speed, Small-Amplitude I/O Interface Circuits for Memory Bus Application," *IEICE Trans. Electron.*, vol. E77-C, No.12, pp. 1944-1950, Dec. 1994.

[8] K. Kimura et al., "Power reduction techniques in megabit DRAM's," *IEEE J. Solid-State Circuits*, vol.SC-21, pp. 381-389, June 1986.

[9] K. Itoh, "Trends in megabit DRAM circuit design," *IEEE J. Solid-State Circuits*, Vol. 25, pp. 778-789, June 1990.

[10] M. Aoki and K. Itoh," Low-Voltage, Low-Power ULSI Circuit Techniques," *IEICE,Transactions on Electronics*, vol. E77-C, No. 8, pp. 1351-1360, August 1994.

[11] K. Itoh et al., "An experimental 1Mb DRAM with on-chip voltage limiter," *ISSCC Dig. Tech. Papers*, pp. 106-107, Feb. 1984 .

[12] T. Sugibayashi et al., "A 30ns 256Mb DRAM with multi-divided array structure," *ISSCC Dig.Tech. Papers*, pp. 50-51, Feb. 1993 .

[13] K. Itoh, "Reviews and prospects of deep sub-micron DRAM technology," *International Conf. Solid State Devices and Materials*, Yokohama, pp. 468-471, Aug. 1991.

[14] N.C.C. Lu and H. Chao, "Half-V_{DD} bit-line sensing scheme in CMOS DRAM," *IEEE*

J. Solid-State Circuits, vol.SC-19, pp. 451-454, Aug. 1984.

[15] H. Tanaka et al., "Stabilization of voltage limiter circuit for high-density DRAM's using pole-zero compensation," *IEICE Trans. Electron.*, vol.E75-C, no.11, pp. 1333-1343, Nov. 1992.

[16] M. Horiguchi et al., "Dual-regulator dual-decoding-trimmer DRAM voltage limiter for burn-in test," *IEEE J. Solid-State Circuits*, vol.26, no.11, pp. 1544-1549, Nov. 1991.

[17] S. M. Sze, "*Physics of Semiconductor Devices* (2nd ed.)," Wiley, New York, 1981.

[18] Y. P. Tsividis, "*Operation and Modeling of the MOS Transistor*," New York, McGraw-Hill, 1988.

[19] M. Horiguchi et al., "Switched-source-impedance CMOS circuit for low standby subthreshold current giga-scale LSI's," *IEEE J. Solid-State Circuits*, vol.28, pp. 1131-1135, Nov. 1993.

[20] T. Sakata et al., "Subthreshold-current reduction circuits for multi-gigabit DRAM's," *Symp. VLSI Circuits Dig. Tech. Papers*, pp. 45-46, May 1993 .

[21] T. Kawahara et al., "Subthreshold current reduction for decoded-driver by self biasing," *IEEE J. Solid-State Circuits*, vol.28, no.11, pp. 1136- 1144, Nov. 1993.

[22] T. Sakata et al., "Two-dimensional power-line selection scheme for low subthreshold-current multi-gigabit DRAMs," *ESSCIRC Dig. Tech. Papers*, pp. 33-36, Sept. 1993.

[23] M. Asakura et al., "A 34ns 256Mb DRAM with Boosted Sense-Ground Scheme," *ISSCC Dig. Tech. Papers*, pp. 140-141, Feb. 1994.

[24] M. Takada, et al., "Reviews and Prospects of SRAM Technology," *IEICE Transactions*, vol. E74, No. 4, pp. 827-838, April 1991 .

[25] K. Sasaki, "High-Speed, Low-Voltage Design for High-Performance Static RAMs," *Proc. of Tech. Papers, VLSI Technology, System, and Applications* , pp. 292-296, May 1993.

[26] M. Yoshimoto, et al., "A 64Kb CMOS RAM with Divided Word Line Structure," *ISSCC Dig. Tech. Papers*, pp. 58-59, Feb. 1983 .

[27] O. Minato, et al., "A 20ns 64K CMOS RAM," *ISSCC Dig. Tech. Papers*, pp. 222- 223, Feb. 1984.

[28] K. Sasaki, et al., "A 9-ns 1-Mbit CMOS RAM," *IEEE J. Solid-State Circuits*, vol. 24, pp. 1219-1225, Oct. 1989.

[29] K. Sasaki, et al., "A 7Mb 140mW CMOS SRAM with Current Sense Amplifier," *ISSCC Dig. Tech. Papers*, pp. 208-209, Feb. 1992.

[30] H. W. Johnson and M. Graham, "*High-Speed Digital Design*," Prentice-Hall, 1993.

PART III

Architecture and System Design Levels

9

Low-Power Microprocessor Design

Sonya Gary

The selection of low-power microprocessors was in the past limited to low-performance, application-specific products or microcontrollers. Recent industry trends toward portable computers and energy-conscious, or "green," PCs have spurred the development of microprocessors that can achieve a balance between power consumption and performance. This balance must satisfy the broad range of applications included in these new trends, such as notebook computers, personal digital assistants (PDA), and digital personal communicators.

Power and performance have generally been at odds in microprocessor design. The design focus has traditionally been on boosting performance with higher frequency operation and additional circuitry for expanded functionality or throughput. The increase in power dissipation accompanying these performance advances was previously a concern only with regard to heat dissipation. Until recently, this could be resolved easily with simple packaging techniques and heat sinks. Since battery operation was not a requirement, the energy source was considered limitless. Features might have been rejected on the basis of additional area if cost was a factor, but seldom were they rejected based on additional power consumed. Power was never a limiting factor in determining performance.

Designing for lower power, this perspective changes. Microprocessor performance is to some degree limited by the power constraints of the system for which it is designed. In PCs and workstations the constraints may include energy conservation requirements or a limited ability to provide adequate cooling. In portable products, these constraints are dictated by battery technology. For example, in the case of a typical notebook computer, to maintain a reasonable discharge time for the currently available notebook batteries, the power budget of the computer might be around 15 Watts. About one third of this, or less, might be allocated to the microprocessor. Industry-standard microprocessors designed for the desktops often dissipate well over this total budget number of 15 Watts, making them unsuitable alternatives for portable products. Consumer demands are quickly closing the performance gap between portables and desktops. Therefore, despite the power limitation, a notebook microprocessor's performance must still approach desktop standards for the notebook to be marketable. The challenge then is to maximize microprocessor performance for the given power constraints.

This challenge has motivated microprocessor designers to reevaluate their design methodologies with power as a concern in addition to performance. The development of low-power process technologies and the scaling of device feature sizes play an important role in enabling low-power microprocessors, but microprocessor designs must optimize trade-offs to meet the dual goals of high-performance and low-power system operation. Design considerations for low power can comprise all steps in the design hierarchy, including power management at the system level, architectural options to minimize power, and low-power implementation techniques. This chapter discusses possible design alternatives to satisfy the growing need for high-performance, low-power microprocessors. It concludes with a comparison of current low-power microprocessor implementations.

9.1. System Power Management Support

9.1.1. Low-Power Standby Modes

In low-power systems, power management can be implemented at the system level, with system logic or software controlling the power consumption of the processor and peripherals. Low-power microprocessor designs have included a variety of low-power "sleep" modes to support system power management. These modes usually reduce the processor clock frequency or gate clocks to prevent switching. This can be extremely effective, reducing standby power to a small fraction of the power consumed during normal operation.

If processor clock frequency is reduced, functionality will be maintained during the low-power mode and the processor can still service low-priority tasks that do not require full frequency performance. Processor clocks may only be stopped if the machine state is maintained statically since a fully dynamic design will lose the machine state unless it is clocked. In some designs, clocks are gated only in certain portions of the processor so that some functions are still active. An example of a function that may need to remain active is bus snooping for memory coherency. Clocks may also be gated at the global level so that no functions are active and the maximum power is saved. Stopping clocks globally in a static design offers the possibility of disabling any on-chip phase-locked loop (PLL) used for clock generation, and disabling the external input clock so that power is reduced to the small amount due to leakage current. Different modes may be offered to provide different levels of functionality and power savings. Figure 9.1 shows a comparison of different standby modes offered by three low-power microprocessors.

Low-power modes may be implemented with software or hardware control. Software control requires specific instructions or designated control register bits to signal the processor to enter a sleep mode when the processor will be idle. This code can be included in a low priority idle loop in the operating system code. Instruction execution returns to this loop when no other tasks are required. It exits this loop when higher priority activity, such as an operating system time slice counter expiration, generates a processor interrupt.

Hardware control requires manipulation of dedicated pins to signal that a sleep mode should be entered. Control may also be a combination of hardware and software if a specific externally generated interrupt corresponds to interrupt handler code that enters a sleep mode. This interrupt might be generated based on an external activity timer which monitors bus activity. In this case, the interrupt handler might check scheduler status to determine whether the processor is actually idle, or not generating bus activity because all necessary code and data are cached, and then enter a sleep mode accordingly.

9.1.2. Compatibility Constraints - x86 System Management Mode

Implementation choices for system power management support depend on the limitations of the architecture and the systems or software for which the pro-

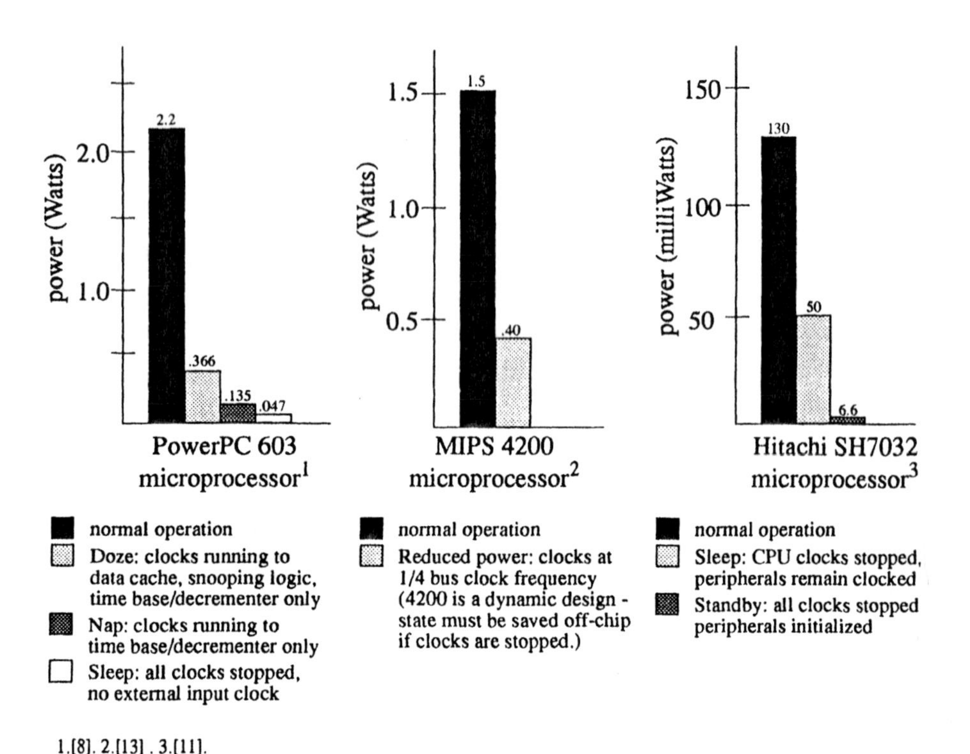

Figure 9.1: Power consumption of microprocessor standby modes

cessor is being designed. Most new architectures have introduced simple forms of software power management control that may be incorporated into an operating system. Older architectures dependent upon "shrink-wrapped" operating systems and applications that cannot directly support power management may require more complex control mechanisms. For example, many low-power x86 implementations include an architectural extension called System Management Mode (SMM) which maintains compatibility with the PC/AT environment by managing power outside of the operating system and applications code. SMM provides support for both processor and peripheral shutdown with a special non-maskable hardware interrupt, called system management interrupt (SMI), and new system management address space and instructions. These are necessary for compatibility since PC applications often do such things as make direct access to system resources, remap memory space and make private use of interrupt vectors [22].

SMM includes three low-power states - local standby, global standby, and suspend, which offer increasing levels of power savings. A software HALT instruction or an SMI enables SMM, which maps a special memory, system management RAM, into the main memory space. The current processor state is saved in system management RAM starting at a specific address. A user-written SMM handler is then executed to enter one of the low-power states. A resume instruction at the end of the handler restores the processor state. SMM can also be entered if there is an I/O instruction to a powered-down peripheral. In this case an I/O trap can initiate an SMI and the SMM handler can power-up the peripheral. The I/O instruction automatically gets reexecuted when the resume instruction is encountered [7].

Since SMM activity is independent of the operating system, the processor and peripherals are powered down based on a external timers expiring and not based on system inactivity as known by the operating system. This means that the processor and peripherals may still be powered up during some idle times while waiting for these timers to expire. Also, they may be powered down right before they need to be active, which may affect performance. For example, if access to a hard disk should take X milliseconds, but the disk drive has been powered down right before the disk access is requested, the access time will increase by the amount of time it takes to spin up the disk.

9.1.3. Supply Voltage For Standby

Since reducing the supply voltage reduces power but also reduces performance, it may be desirable to reduce the supply voltage below that required for minimum performance during times when performance is not an issue, such as during standby. If a voltage and frequency can be specified below which the processor will lose data, the system might reduce the voltage and clock frequency during times when the processor will be idle for a considerable length of time. It is also possible to reduce the voltage to 0 V during these times if the necessary processor state may be saved off-chip and then reloaded when activity resumes. This should only be done when the power and time required to save the processor state is insignificant compared to the power savings achieved. Future options may include selective control of the internal voltage to different portions of the processor, though the additional design complexity is currently prohibitive.

9.2. Architectural Trade-Offs For Power

Depending on the intended use of a microprocessor or microprocessor family, an architecture may be defined or altered to save power, sometimes with performance or area improvements, sometimes with trade-offs. Currently, the major obstacle to weighing architectural decisions with respect to power is the difficulty in estimating the power effects of these trade-offs, especially in the absence of an implementation. Even with an implementation, it is very difficult to gauge the effect of a specific design decision on power. Architectural power modeling tools are being developed, but decisions may still be approached based on general knowledge of the amount of circuit activity or additional die area that might be generated as a result of a decision. Given the function of a microprocessor, the optimal architecture for low power should use the least number of cycles, the least amount of I/O activity, and implement the least number of transistors to perform that function.

Limiting the power optimization of many designs are the predefined standards of a specific instruction set architecture, such as the Intel x86 architecture. These standards allow a microprocessor to be compatible with systems and software designed for this architecture, but require that it support a specific instruction set and a specific minimum feature set. Microprocessors designed for a specific instruction set architecture may still incorporate other microarchitectural decisions based on power concerns - examples are choices in pipeline staging, superscalar features, and cache memory. Newer microprocessor designs for products such as PDAs are less dependent on an existing software base since the final PDA products ship with most or all of the software installed. For these applications, designers may choose to define the architecture from scratch so that the processor can provide the best performance/power balance for its markets. Several architectural issues that may be addressed are discussed here including instruction set architecture, instruction execution pipelining, datapath design, caching and memory management.

9.2.1. Instruction Set Architecture

In creating a new architecture, the instruction set may be defined to reduce the amount of memory required to store a program encoding. Because a large portion of the energy required to perform an operation is the energy required to fetch the associated instructions, smaller program encodings, or those with greater code density, are generally more energy-efficient. Reducing the size of a program or critical code loop such that it may fit in an on-chip cache can cut the instruction fetch energy in half. Even reductions of less magnitude can significantly reduce instruction fetch energy. Reducing code size also reduces system power and cost by reducing the amount of memory required to contain it.

Many higher performance notebook processors are following performance-oriented design trends of recent years which preclude the use of some of power-saving instruction set techniques discussed here. Many of these trends need large opcodes, larger register files and fixed instruction length to maintain their performance gains. Some of these trends are reduced instruction sets (RISC), superscalar organization, which allows parallel instruction execution, load/store architectures, which manipulate values through registers instead of memory, and very long instruction word (VLIW) architectures.

9.2.1.1 Instruction Complexity: RISC or CISC

One way of improving code density is to create more complex instructions so that fewer instructions are required to code a function. Complex instruction set computer, or CISC, architectures were originally developed because memory speeds were so slow that designers created more complex instructions to more efficiently use the time between instruction fetches. CISC architectures have in recent years been rejected in favor of reduced instruction set computer, or RISC, architectures because the flexibility of a simpler instruction set can provide further performance improvements. Figure 9.2 shows a code size comparison of benchmarks compiled for a CISC architecture, the Motorola 68000 architecture, and a RISC architecture, the PowerPC™ architecture, jointly developed by IBM, Motorola and Apple [19].

Unfortunately, a complex instruction set also naturally complicates instruction decode, which can be an energy issue for larger instruction sets due to the amount of decode logic required. A small, complex instruction set may be an appropriate option for application specific designs that do not require a broad spectrum of instruction types. Alternatively, the addition of certain complex instructions to a RISC instruction set may provide greater code density while maintaining relative decode simplicity and higher performance. Examples of CISC-style instructions which improve code density and reduce memory traffic are load and store multiple operations and indivisible byte instructions for semaphore operations to memory.

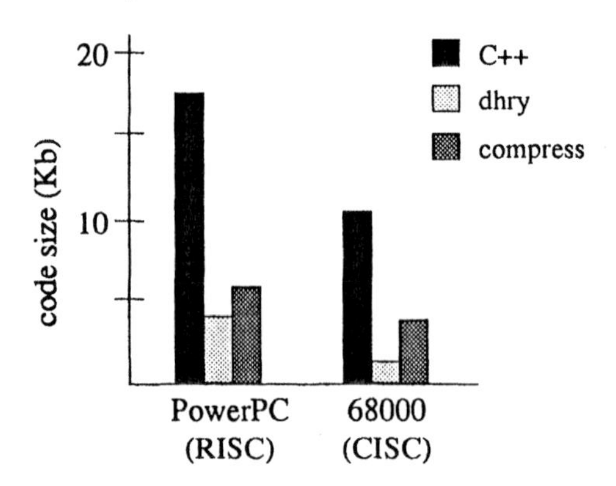

Figure 9.2: Code density comparison of RISC and CISC

9.2.1.2 Register Variables vs. Memory Variables

The use of registers to hold variables has many advantages. These include increased performance since registers are faster than memory, reduced memory traffic since operations are performed on registers, and increased code density in some cases since registers can be specified with fewer bits than memory in an instruction. Conversely, using variables directly in memory may provide better code density by reducing the number of instructions to perform an operation. This is particularly true for cases where the value being stored to memory is the result of a single operation. Support for a limited number of memory-memory or register-memory operations can be provided within a RISC instruction set for selected use in improving code density. Good choices might be memory-to-memory moves and logic instructions that modify memory directly.

9.2.1.3 Reduced Instruction Length

Another way to improve code density is to reduce the instruction length. Though more instructions may be required to perform an operation, fewer total instruction words are fetched and instruction decode is greatly simplified. Studies have shown that for the same basic RISC architecture, choosing a 16-bit, two address instruction set over the more traditional 32-bit, three address instruction set provides greater code density and increases instruction fetch energy efficiency by as much as 40% using instruction caching and buffering [3].

One example of a recent RISC architecture with a 16-bit fixed-length instruction set is the Hitachi SH architecture which is used for the SH7000 series integrated processors designed for PDAs. Like many processors for embedded applications, the SH7000 series products use a non-Harvard architecture with no instruction or data caches and a shared memory system. Since the SH architecture has a fixed 16-bit instruction length, two instructions may be fetched per cycle using a 32-bit bus, freeing the bus every other cycle for data transfers. Using a 16-bit bus, fewer memory chips are required to support the bus width, which can decrease system power.

The shorter instruction length of this architecture can significantly impact program size - Dhrystone compiled for SH is about 40% smaller than the version compiled for the SPARC or MIPS RISC architectures [11]. The shorter instruction length does impose some limitations for this architecture. SH processors are limited to 16 32-bit registers so that all operations can be encoded in the 16-bit format. Arithmetic instructions have only two operands, while RISC instructions typically have three. Also, conditional branch and load/store instructions are limited to 8-bit displacements. These limitations may, in a strict sense, reduce performance for some code sequences, but since the SH processors do not incorporate an instruction cache, minimizing the number of memory accesses is important for both power and performance.

9.2.1.4 Variable Instruction Length

In order to avoid the most common limitations of moving from a 32-bit instruction length to a 16-bit instruction length, some new RISC architectures have opted for variable length instructions. This technique has already been used in CISC architectures for increased code density to improve performance. In these new instruction set architectures shorter instruction lengths are used for the majority of instructions, while optional longer lengths are used to include commonly used formats that cannot be accommodated in the shorter length. Several current PDA architectures have done this including the NEC V800. The V800 architecture includes both 16-bit and 32-bit instructions, with the 32-bit length being used for loads and stores, long-displacement branches, and three-operand calculations.

With variable instruction length, program coding is simplified and the code density is still much improved. The V800 version of Dhrystone is approximately the same size as the SH version [11]. The 32-bit instructions do not provide a significant performance advantage for V800 since many of the long format instructions, the function of which can only be replicated on SH with multiple single cycle instructions, take two or more cycles to execute. This architecture does allow for 32 32-bit general purpose registers and this larger register file makes fewer memory references necessary.

The major disadvantage of a variable length instruction set is the complexity of sequential instruction decode, particularly if the instruction set is large and varied. This decode complexity may limit performance and require considerably more decode logic. For example, superscalar processors, i.e. those which may dispatch and execute multiple instructions in parallel, may have performance difficulties with parallel decode since the length of the instructions being decoded must first be sequentially determined.

9.2.1.5 Memory Addressing

Memory addressing modes can impact code density by affecting the size of instructions and by offering opportunities to code with fewer instructions. Depending on the type and number of addressing modes, many instruction bits may be consumed encoding these modes. Additional addressing modes also increase design complexity. Beyond the minimum that are necessary, it may be optimum to provide select addressing modes which have the greatest effect on reducing the number of instructions required. RISC architectures usually include addressing modes such as register, immediate, displacement, indexed and absolute. There are a variety of memory addressing modes not typically included in RISC architectures which allow a selection of functions to be performed using fewer instructions. Examples are postincrement and predecrement memory addressing options for loads and stores. These are useful for reducing the code necessary to step through arrays within a loop or to implement a stack.

9.2.1.6 Instruction Encoding and Addressing

Another alternative for creating a more energy-efficient instruction set is to group instruction encodings for common sequences so that these sequences minimize instruction bit switching. In fetching these instruction sequences, switching of the instruction bus is reduced. In moving these sequences through an execution pipeline, switching of internal instruction busses is reduced.

Support for Gray code instruction addressing provides another opportunity for power savings. Gray code is an alternative to binary code that encodes only one different bit in consecutive numbers. The locality of program execution and the knowledge of instruction addresses upon coding helps Gray code minimize the amount of bit switching for instruction address sequences. Experiments have shown Gray code addressing for typical RISC programs to reduce the switching activity of instruction address lines from the processor to cache or memory by 30-50% compared to normal binary code addressing [21].

9.2.2. Instruction Execution Pipeline

As with most other microarchitectural decisions, the choice of execution pipeline is one in which performance, area and power requirements must be considered. Pipelining is a key technique used to improve microprocessor throughput. Two parameters to be considered are the number of pipeline stages and the number of execution pipelines. Major high-performance trends in RISC architectures, superscalar and superpipelined design, have involved increasing both of these numbers. Performance advantages and power disadvantages of using these techniques are listed here.

Superscalar

- **performance:** increases throughput by providing multiple execution units so that parallel execution pipelines may be implemented
- **power:** multiple units increase design complexity and area requirements; instruction dispatch logic increases considerably due to extra data dependency checking requirements

Superpipelined

- **performance:** simplifies and increases the number of pipeline stages so that each pipeline stage may be performed faster and clock frequency may be increased
- **power:** increases design complexity and the number of necessary latches and clock elements; has the distinct power disadvantage of needing to increase frequency for performance

The standard RISC execution pipeline is five stages - instruction fetch, instruction decode, execute, memory access (if necessary), register writeback - with the execute stage for most instructions being a single cycle. Five stages allows enough pipelining to maintain throughput during stalls or cache misses, while maintaining a low average cycles per instruction. Microprocessors designed for low-power have typically chosen pipelines of five stages or less, with superscalar design used as an option only for those with higher performance requirements and less restrictive area and power requirements.

9.2.3. Datapath

The processor datapath includes execution units and the communication paths between them. These essentially define the functions of the processor, so in many cases, little can be spared at the architectural level that will reduce the energy required for a computation. Though area may be saved, associated performance losses can result in more cycles per computation.

General low-power logic and circuit implementation methods may be applied to a processor datapath to save power, but in some cases, datapath switching factors can be very low. Table 9.1 shows switching factors for the integer unit result bus of the PowerPC 601™ microprocessor based on functional simulation of SPEC92 code segments [2]. This data shows that the average datapath switching factor may be as low as 20%, which indicates that the focus of saving power in datapaths should be on heavily loaded signals and busses.

Instruction Type	Number of Instructions	Average Per Cycle			Switching Factor
		# of 1s	# of 0 to 1s	# of 1 to 0s	
ALL	27277	7.70	3.05	3.13	0.19
rotate/shift/logical	3175	3.09	1.98	4.63	0.21
add/subtract	4937	5.90	3.41	4.31	0.24
effective address calculation	15496	9.76	3.15	2.37	0.17
multiply/divide	1070	2.11	1.50	1.56	0.10
control register	192	2.12	1.44	2.26	0.12
compare	2349	7.10	3.81	4.11	0.25
branch	58	9.24	5.03	13.50	0.58

Total number of cycles: 40633
Total number of cycles with valid integer instruction execution: 27277 (67%)
Total number of cycles with valid data on integer result bus: 24870 (61%)

Table 9.1 PowerPC 601 Integer Result Bus Activity

9.2.3.1 Register Files

Along with being responsible for the performance of load/store architectures, register files save system power by reducing memory traffic. Operations may be performed on register values and results written to registers instead of memory. Several ways that power consumption resulting from the operation of the register file itself may be reduced are:

- Disable read ports when operands are sourced from other locations.
- Block redundant register writes during stalls.
- Write results to the register file simultaneous with operand forwarding.
- Only read operands from the register file (no operand forwarding) to eliminate extra bussing. (Note: This reduces performance for both superscalar implementations using register renaming and superpipelined implementations due to additional stalls.)
- Use tristate busses with logic to hold the last values read and written - this minimizes switching on these busses which are usually heavily loaded.

9.2.4. Cache Effects

One method of significantly decreasing memory traffic is to include instruction and/or data caches. Though caches can consume considerable power and die area themselves, they in turn can improve system power and performance by reducing I/O and memory activity and by making instructions and data available more quickly.

9.2.4.1 To Cache Or Not To Cache

In most cases, the choice to include caching is a performance one. Processors designed for consumer computer markets cannot achieve acceptable levels of performance without caches. Cache size becomes particularly important for processors trying to achieve desktop performance for software-emulated code sequences. Operating systems and applications that must be emulated in software run less efficiently than those that are native code. Architectures dependent upon emulation for a software base must have an instruction cache large enough to hold a significant measure of emulator code. This helps to maintain competitive performance with those already having native software support.

For embedded application processors, power and die size, or cost, are weighed more heavily in the cache decision. For some specific applications, a cache cannot buy much performance improvement unless it is large. Often, external memory of sufficient speed can be relatively inexpensive and may be a simpler alternative. Another popular option is to provide on-chip RAM and ROM with the CPU core; this can improve performance and decrease I/O power by keeping frequently used code or data on chip. Still, some processors for battery-powered embedded applications have found caches to be a good choice to reduce memory traffic power. Also, new embedded processor architectures designed for good code density can more easily achieve better cache performance. These architectures can choose a smaller cache to provide improved power and performance characteristics. For example, the Hobbit CPU, based on AT&T's CRISP architecture, includes a 3Kb, 3-way set associative on-chip instruction cache. The hit rate of this cache is roughly equivalent to that of an 8Kb, direct-mapped cache in a typical RISC architecture. This is due to the code density provided by the variable length CRISP instruction set. The Hobbit decoder converts the variable length instructions into 192-bit control words and a decoded instruction cache holds 32 of these values. This balances the performance limitations of decoding variable length instructions. It also limits the number of cache accesses, which further reduces power dissipation[15].

9.2.4.2 Cache Organization

The issue of cache organization is generally considered in terms of performance and, in most cases, choices that increase performance by decreasing miss rate also increase the power consumption of the cache. To decrease system power, this increase in cache power can be traded for the power saved due to the considerable decrease in memory traffic associated with the lower miss rate. Here, we briefly discuss the relationship between power and several major cache organization parameters.

Cache size. Miss rate decreases dramatically with increasing cache size up to the point when the cache is large enough to capture the majority of the code and data used in a relatively long execution period. Power due to memory traffic decreases in the same fashion, but cache power continues to increase at the same rate with cache size. Bit-line power essentially doubles when the cache size is doubled. Eventually, cache power dissipation dominates and the increased cache size contributes to performance, but not power savings. Figure 9.3a illustrates the relationship between cache size and total fetch energy.

Block size. Miss rate decreases with increasing block size until the smaller number of blocks begins to cause more conflict misses. As block size increases, the energy required to service a miss increases due to the increased memory access. It is possible to hide some of the time required to perform the larger memory access with critical-word-first forwarding and servicing hits during misses, but the energy remains the same. Eventually, this energy cost may outweigh the energy savings of a smaller miss ratio, since some of the instruction or data brought in on a miss will not be used due to context switches. Figure 9.3b illustrates the relationship between block size and total fetch energy.

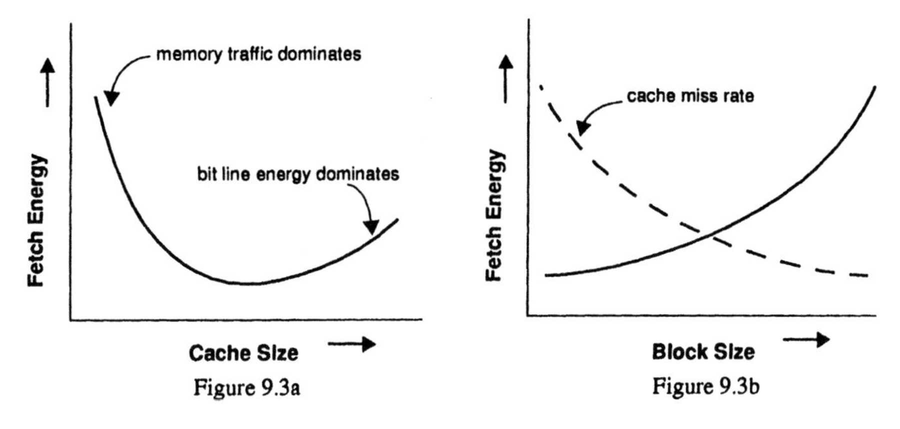

Figure 9.3a: Cache size vs. fetch energy
Figure 9.3b: Cache block size vs. fetch energy

Set associativity. Miss rate decreases with more degrees of associativity, though this effect diminishes with increasing cache size. Cache power does increase with more degrees of associativity since an n-way set-associative cache increases sense amp power by n times. Bit-line power is much greater than sense amp power, however, so cache size affects power more than degrees of associativity. Direct-mapped caches have the lowest access power and faster access times, but miss rates are much higher, increasing system power.

Unified vs. split. The power consumption difference is negligible between a unified cache and split instruction and data caches using the same total array size. Arbitration for unified cache access generally requires very little extra logic. The major difference in system power lies in the miss rates, which are heavily dependent on the percentage of program memory references to each cache.

Write Policies. There are two standard choices for write policy - write through, where data is written both to the cache and external memory, and write-back, where data is written only to the cache. With write-through, read misses never generate copyback writes to external memory. With write-back, writes to external memory require a cache block copyback, which consumes more power than a single-byte write, the most common type. But copybacks only need to occur when a cache block is replaced or to maintain external memory coherency, both of which are required very infrequently. Also, multiple writes to the same block may only require one block write to external memory - think about how much power this might save in executing a code loop containing stores to the same block! Though write-through is easier to implement, write-back is a more energy-efficient policy since it generally uses far less memory bandwidth. It is also a better performance solution since writes to the cache are much faster than writes to external memory.

Using either write policy, there is also the choice of whether or not to allocate a cache block on a write miss, known as write-allocate. The alternative is to write directly to memory on a write miss. Write-allocate is the better choice for power since it also reduces memory access requirements. If there are subsequent writes (or reads) to that block, they may be performed directly in the cache.

9.2.4.3 Buffering Techniques

Simple buffering techniques can create an extended multilevel cache structure, further exploiting the locality characteristics that make caching effective for power and performance. One example is buffering the most recently accessed cache block so that another cache access is not necessary until it is outside of this block. This is particularly effective for instructions since instruction accesses are usually sequential and no writes have to be handled. It is also more effective if the cache is powered down when idle. Figure 9.4 shows a plot of simulated switching energy for a 32-bit instruction set architecture using different cache sizes, a 256b block size, direct mapped placement, with and without block buffering [3]. With larger cache sizes, block buffering yields energy savings of as much as 50%.

A small cache which stores recently replaced cache blocks, also known as a victim cache, is another alternative, but it must be the size of several blocks to produce a noticeable improvement in overall hit rate. This makes its cost in power and area a questionable sacrifice for the decrease in cache access power.

9.2.4.4 Sub-Array Access

One solution for decreasing cache access power is to subdivide the data array based on access size, and then access only the minimum number of subar-

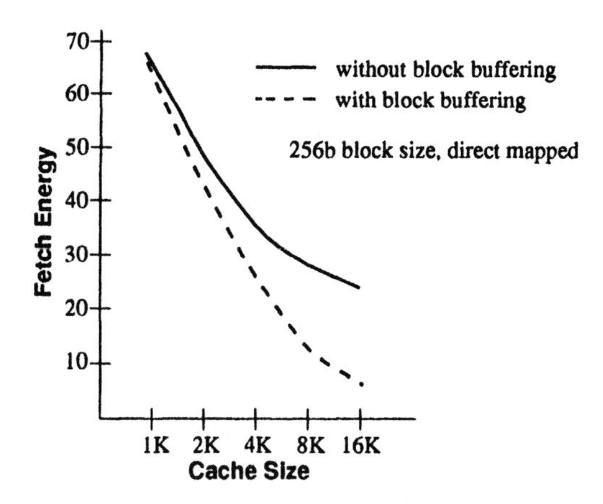

Figure 9.4: Cache size vs. fetch energy for block buffering

rays for a particular operation. For example, an 8Kb, 2-way set associative cache with an 8-word block size can be subdivided into 8 1Kb subarrays, each subarray containing one word of the cache line for both sets. For a single word access, only one subarray is used; for a doubleword access, two subarrays are used. Bit lines and output drivers of subarrays not being accessed can be held in a constant state of precharge and the bit line precharge signals can be turned off only when a subarray is being accessed [9]. A doubleword access in this case would require only 1/4 of the energy needed to access the whole cache line.

This technique is very effective and can be implemented with minimal logic cost. There is a possible small performance penalty in gating the precharge signals if this is the critical path. Subarrays may be created on any appropriate access boundary, such as byte, word or doubleword. It is also possible to create subarrays on block boundaries, and turn on only the necessary subarray based on address bits.

9.2.4.5 Sectoring

Cache sectoring has been used in the past to reduce tag array size and control logic. In reducing the size of the tags it also reduces tag power and memory traffic. Cache blocks are divided into sectors, each of which is still associated with the same tag. Accesses are performed on sectors instead of blocks. This generally increases the miss rate since only a sector of data is brought in at a time. This feature is then a trade-off between the potential increase in memory access power, which depends on the resulting miss rate and memory access size, and the reduced tag power and area.

9.2.4.6 Cache Locking

Some caches use a locking capability to prevent crucial instructions or data from being pushed out to memory. If a cache block is loaded then locked, it cannot be replaced. This feature may be used to decrease instruction fetch traffic or data traffic during specific code sequences to reduce power and improve performance. Caches may be designed to lock by block or by way, or the cache may be locked as a whole. This technique requires software support to initiate the lock.

9.2.4.7 Array Design

Crucial in reducing the power dissipation of a cache is the array design. Following are a few methods that have effectively been used to reduce cache and tag array power.

6T vs. 4T. Low-power cache designs typically use a six-transistor (6T) SRAM cell instead of a four-transistor (4T) cell, primarily due to the DC current always drawn by a 4T cell. A 6T cell is designed as a full-CMOS cross-coupled inverter circuit, each inverter using one n and one p transistor. A 4T cell replaces the p transistors with resistors, so that one of the resistors is always drawing current whether the cell is holding a zero or one value.

Pulsed word lines. Pulsed word lines can be used to eliminated unnecessary bit line discharge once the sense amps have been strobed. In some cases the bit line discharge can be limited to a certain voltage to save power at lower frequencies, where one of each pair of bit lines would discharge completely.

Single-ended bit line reads. With some sacrifice in access time, a single-ended bit line can be used for reads. Dual-ended bit lines are usually used for speed but single-ended bit lines can reduce power by cutting the chance of bit line discharge in half for unbiased data.

9.2.5. Memory Management

Operating systems that run on notebooks and PDAs most often require memory management support. Many of the techniques used to minimize cache power may also be used to reduce power for arrays dedicated to the use of virtual memory such as translation lookaside buffers (TLB), segment registers, and block address translation. This includes buffering and subarray techniques along with the design of the arrays themselves.

One of the few processor families to incorporate a memory management structure specifically designed for low-power systems is the ARM family, used in Apple's Newton PDAs. The ARM610 includes a virtual cache. Instead of requiring a TLB and cache flush on each context switch, as would be done in a traditional operating system, each process is assigned its own virtual address space within the operating system so that a flush on every context switch is not necessary. Permission changes in the memory management page tables also force TLB flushes. Apple uses a technique called domain memory management which allows permissions to be globally changed on a domain by writing to a single internal 32-bit register so that no TLB flushes are required [6]. No data indicating power savings due to this particular technique has yet been published.

9.3. Choosing the Supply Voltage

9.3.1. CV^2f - Why Voltage Is So Important

The total power for each node in CMOS digital circuits may be represented by the following equation:

$$P = \left(C_L V_{dd}^2 f_p\right) + (I_{SC} V_{dd}) + (I_{leakage} V_{dd}) \tag{9.1}$$

Here, the first term represents the power dissipated in charging and discharging a capacitive load, C_L. V_{dd} is the supply voltage and f_p is the switching frequency. The second term is the power dissipated due to the short circuit current, I_{SC}, generated when both NMOS and PMOS transistors are simultaneously active as during gate output switching. The last term is due to leakage current, $I_{leakage}$, which varies with fabrication technology [25]. Dominating this power equation is the first term, which is due to logic switching. Since this component is proportional to V_{dd} squared, any decrease in the supply voltage will provide a squared power savings. Thus, lowering the voltage is the most effective means of power reduction.

For current technologies with feature sizes below 1.0 micron, if the feature size is not adjusted, performance decreases linearly with decreasing supply voltage down to very low voltages. Despite this decrease in circuit speed, smaller process geometries are offering a greater relative increase in circuit speed due to the fact that gate delay is inversely proportional to the square of the effective channel length of the devices. The net effect is that circuit performance improves as CMOS technologies shrink into the sub-half micron region, despite V_{dd} reductions. Another consideration is that smaller geometries include thinner gate oxides which impose voltage limitations. As geometries decrease, the voltage must be lowered to support these new technologies. In effect, we cannot continue to push the performance of process technologies without lowering the supply voltage.

9.3.2. Moving From 5 V to 3.3 V

The goal of maximizing performance in microprocessors has kept designs using 5 V processes to maximize switching speed. Another factor in the reluctance to decrease the supply voltage is the availability of compatible external components. Most are designed to operate at the 5 V TTL standard. External level-shifting logic to interface to these components can consume valuable board space or increase system cost. Alternatively, protective circuitry on-chip to handle input voltages above the chosen supply, for example, 3.3 V I/O buffers that are 5 V tolerant, also add cost and decrease performance [24].

These concerns are now being reconsidered in light of the low power requirements of portable systems. Since microprocessor power may be a large fraction of the power budget, and since lowering the voltage can provide a quadratic decrease in power consumption, moving the voltage below 5 V can make a significant impact. The trend has been to move from a 5 V to 3.3 V supply voltage, which can offer a 50-60% power savings. New low voltage industry standards are being designated, such as LVTTL, which is the TTL standard with a new V_{cc} of 3.3 V. At a reduced voltage of 3.3 V, advanced CMOS and BiCMOS processes still offer a substantial improvement in circuit speed.

9.3.3. Moving Below 3.3 V

For very low power applications, it may be desirable to design a microprocessor with an optimum supply voltage for low power. This voltage may be lower than the 3.3 V standard, depending on the given process technology. As the supply voltage is reduced to a value closer to the threshold voltage, device current decreases with the square of the difference between the supply voltage and threshold voltage. Therefore, the supply voltage cannot be chosen too close to the threshold voltage without a significant speed loss. This would push a microprocessor out of the acceptable frequency range for many low-power markets. Also, since products such as PDAs, digital personal communicators, or even notebooks, are frequently placed in "sleep" modes during idle periods, leakage current must be considered. As the threshold voltage is reduced, leakage current increases, however, as the supply voltage is reduced, leakage current decreases. The power saved by choosing a very low supply voltage must be weighed against possible performance losses and increased leakage current affected by the choice of both the supply and threshold voltages.

In order to avoid the significant loss in performance of extreme reductions in the supply voltage, one technique being considered is scaling the voltage to the point just outside velocity saturation for the specific technology. When the electric field from the drain to source is great enough to accelerate carriers to their terminal velocity, the device is in velocity saturation. Before this point, the drive current is a quadratic function of the voltage. Beyond this point, the drive current increases linearly with voltage. Because electric fields increase as a process is shrunk, the supply voltage at which velocity saturation occurs will be lower for smaller process geometries.

Designing for a worst case supply voltage equal to the value at velocity saturation allows a power reduction to the point at which any further reduction in the supply voltage will produce a much greater loss in performance. It is also the point of the lowest power-delay product and therefore the most efficient design point in terms of power and performance [5]. Behavior in velocity saturation is part of the reason why the move from 5 V to 3.3 V decreases power dissipation without an unreasonable loss in circuit speed - the power decreases quadratically while the performance decreases linearly. Also, since the move to a lower voltage for a particular processor design is typically coupled with a smaller process technology, circuit speed can increase considerably despite the decreased supply voltage.

9.3.4. Supporting Mixed Voltage Systems

It will be some time before all component designs are standardized to a lower supply voltage. In the meantime, support for mixed voltage systems is important in the move toward lower power. There are several ways a microprocessor can support both 5 V and 3.3 V systems. One method is to provide a 3.3 V CPU with separate groups of power pins to control the voltage for the bus and DRAM interfaces, as on the Intel 486SL. This allows each to run at either 3.3 V or 5 V. Early systems designed with this microprocessor might operate only the CPU at 3.3 V, while later systems might include 3.3 V DRAM and peripherals, and eventually a 3.3 V bus standard. Another option is to operate internally at 3.3 V and provide on-chip level shifters to drive 5 V peripherals.

Some microprocessors have been designed with both 5 V and 3.3 V process technologies to provide a design easy to retrofit into current systems with the former, and a low-power design with the latter. The MIPS 4400 is an example of this. Others have been designed with 5 V processes but have been specified by the manufacturer with a 3.3 V to 5 V operating range, such as the Hitachi SH7034. A lower maximum frequency and a lower maximum wattage are specified for 3.3 V, appropriate for lower performance applications. Another option, used on the PowerPC 603™ microprocessor, is to design with a 3.3 V process but maintain I/Os that are 5 V TTL tolerant. The addition of features to support mixed voltage systems allows system designers to choose a low-power microprocessor with the flexibility of choosing other components based on cost, availability and power savings.

9.4. Low-Power Clocking

Power dissipation increases linearly with frequency, as shown in Equation 9.1. Clocks typically switch with the highest frequency on a microprocessor, and, since clocks must be distributed across a microprocessor, clock switching has a considerable impact on power. Clock power can be as much as twice the logic power for static logic and three times the logic power for dynamic logic [18]. To minimize clock power, a microprocessor should be operated at the minimum frequency necessary to attain the required level of performance. Along with this, clock generation and distribution may be designed to further reduce clock and system power demands.

9.4.1. Clock Generation

To generate a global clock, many low-power microprocessors incorporate a PLL (phase-locked loop) to maintain proper edge alignment between the external input clock and the processor's internal clock. If a PLL is not used, clock latency is an issue. This is commonly resolved by reducing routing resistance with increased wire width, which increases routing capacitance and therefore power. The global clock is driven across a processor and is used to create local clocks fed to processor circuitry. Regenerating or buffering clocks locally controls global clock loading, reducing power. All clocks, including scan clocks for LSSD (level sensitive scan design test protocol), may be generated locally from a single global clock to minimize power due to clock routing.

Microprocessor designs often employ a single type of clock regenerator or buffer throughout the design, adding various "dummy" loads to balance loading and control clock skew. Designing with power in mind, a better approach is to provide a family of clock regenerators with different output drive strengths that drive predefined values of clock line capacitance. With this selection of clock regenerators, an appropriate regenerator may be chosen to drive a group of latches based on the drive strength required for that particular load. In this way, clock skew may be controlled without any unnecessary power dissipation incurred in driving "dummy" loads.

9.4.2. Adjusting Clock Frequency for Power

In light of increasing processor frequency to achieve higher performance, many clock circuits, including PLLs, are also designed to be configurable to provide an internal processor clock that is faster than the external input clock or bus clock. With this feature, a system may take advantage of the performance of operating the processor at a higher frequency without the power or transmission line penalties associated with higher frequency operation of the external bus. The most popular option has been to double the frequency of the external input clock, though many others have been pursued, including generating multiples of the external clock or fractional ratios such as 3:2. Configurations have also included halving or quartering the external input clock frequency to reduce power during times when maximum performance is not a requirement. Controls to change the clock configuration may be done in software, possibly by setting appropriate bits in a control register, or it may be done in hardware, by configuring pins dedicated to this purpose.

One disadvantage of using a PLL is the inability to quickly change internal clock speed. Each time the processor frequency is changed, time must be allocated for the PLL to relock to the new frequency (10-100 microseconds). Depending on the PLL design, some processors may require that the processor clocks be stopped before the input frequency is changed to guard against frequency overshoot. If the frequency must be changed dynamically, a bypass option might be provided whereby the external input clock is fed directly into the global clock driver, bypassing the PLL. The system can then directly control processor frequency to manage power without the PLL relock penalty or any required processor support. The drawback in bypassing the PLL is that clock latency is introduced.

9.4.3. Clock Distribution

The most important factor considered in clock distribution is the difference in clock arrival times across the chip, or clock skew, which can reduce the effective clock period. There are several methods used for clock distribution that minimize skew, but not all methods have the same effect on power dissipation. Distributed clock buffers dissipate more active power than a passive clock distribution network. Though it may provide greater power savings, passive distribution may negatively affect performance by producing slower rise and fall times at the local clock regenerator inputs. Grid distribution networks can produce good clock skew results but may also produce very high routing capacitance, which increases power. One method that is effective in both minimizing skew and controlling power is an H-tree clock distribution network. This configuration is illustrated in Figure 9.5.

This network uses a recursive H structure to distribute the clock signal in such a way that the interconnections carrying the clock signal to the functional subblocks are equal in length. The clock signals are thus delayed equally and are

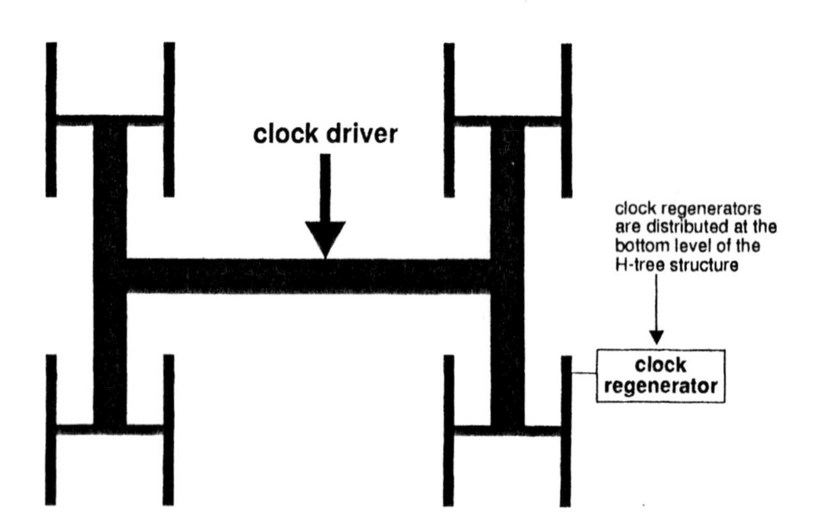

Figure 9.5: H-tree clock distribution network

synchronous in arriving at the subblocks. The H structure is duplicated until the subblocks are small enough to produce tolerable skew within the subblocks [1]. At each point of a new H in the tree, the resistance is halved and the capacitance is doubled. Therefore, a larger line width is used for the main H structure to minimize resistance, while a narrower line width is used at branching points along the tree to minimize capacitance. Routing resistance and capacitance can be carefully controlled to minimize both clock skew and power dissipation. The H-tree distribution method is very effective for processor designs that are not so large that the necessary depth of the H-tree produces excessive routing capacitance in a manner similar to a grid network.

9.4.4. Clock Gating

On any given clock cycle, some circuits in a microprocessor will be idle. Only a subset of the functions implemented will be active depending on the type of instructions being executed. For example, in a superscalar design incorporating multiple execution units, not all units are always busy, even if code is efficiently scheduled. Figure 9.6 shows estimated idle times of the execution units on the superscalar PowerPC 603 microprocessor. These idle times were estimated for the SPEC92™ benchmark suite using a trace-driven architectural simulator [10].

In a static design, it is possible to save substantial power by disabling clocks to different areas of the processor when these areas are idle. This saves both clock power and power consumed by logic resolution when registered values

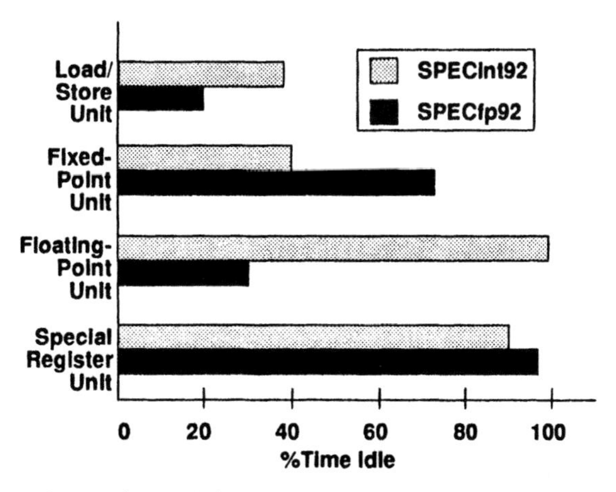

Figure 9.6: Execution unit idle time for the PowerPC 603 microprocessor

change. This technique may be approached at any level of the design hierarchy, from a small logic block up to the entire chip.

Clocks may be disabled at their respective regenerators or with higher granularity at the latches which they feed. Signals may be defined to dynamically do this cycle-by-cycle or to do this under state machine control. Clocks may be easily gated along functional boundaries. An example of this is a superscalar design where clocks are gated to different execution units based on the instructions being dispatched and executed each cycle. The PowerPC 603 microprocessor is a low-power design with this type of implementation [10]. This approach is very beneficial in cases when this processor is executing integer code since clocks are disabled to the floating-point unit, which is a large circuit block with a large number of clock regenerators and significant clock loading. Also, floating-point code does not always contain a large percentage of integer operations, so while executing this code, clocks to the fixed-point unit are disabled during a large percentage of the execution time. Figure 9.7 shows the power saved using this type of clock gating, sometimes known as "dynamic power management," for the PowerPC 603 microprocessor [8].

It is possible to implement dynamic power management using a central control logic block or distributed logic. The distributed approach can be more efficient since logic to control clock gating can often be composed of the same logic that determines instruction execution, possibly even the same gates if the

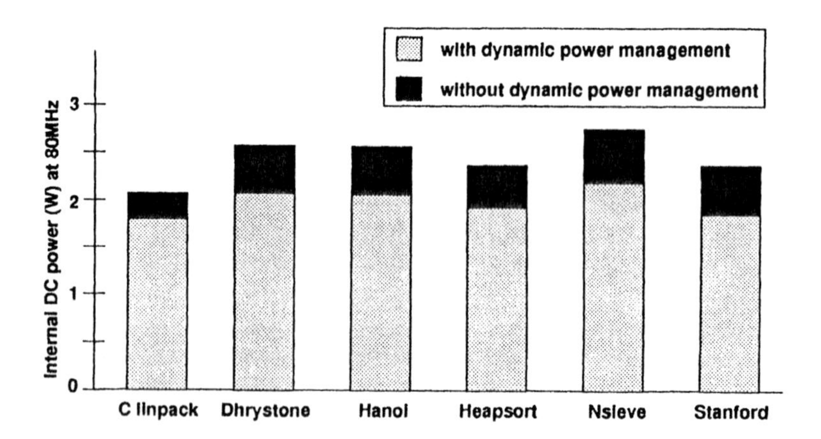

Figure 9.7: Power decrease for the PowerPC 603 microprocessor using dynamic power management

extra loading is not prohibitive. In this way, dynamic power management adds little area or complexity to the design. Clock gating may also be used to implement low-power modes used during processor idle times. (See Section 9.1, System Power Management Support.)

9.5. Implementation Options for Low Power

A variety of logic, circuit and floorplanning design techniques have been pursued for low power in CMOS digital design, and many of these can be used to substantially affect microprocessor power. Several possible options and their suitability for microprocessor design are briefly discussed here.

9.5.1. Static vs. Dynamic Logic

Static and dynamic implementations have both been chosen for low-power microprocessor designs and each has advantages and disadvantages for power consumption. Most recent implementations have been static, but the MIPS 4200 and the Fujitsu FR20 are current examples of dynamic low-power microprocessor designs. Static logic generally averages less switching activity. The major reason is that dynamic logic requires nodes to be precharged every cycle so a great deal of switching is generated by nodes that are precharged then discharged. Also, driving the precharge signal can consume considerable power.

Static logic does not have the precharge requirement, but it can generate spurious transitions or "glitches" during a cycle as the function resolves. Static logic can be designed to minimize glitching, but dynamic logic never generates glitches since each node can transition at most once each cycle. Static logic is also susceptible to short circuit current unless n and p transistors are designed for equal rise and fall times. Depending on the function to be implemented, it may be possible to implement that function with dynamic logic having fewer intermediate nodes that switch than with a static implementation.

To minimize power, a function must be evaluated to determine which implementation generates less switching and less capacitive loading on switching nodes. For example, Table 9.2 shows PowerPC instruction fetch switching factors of dynamic and static implementations for SPEC92 trace samples [2]. The trace samples were one million instructions and the results were based on a serial scan of the trace instructions. The dynamic case assumes a precharge to one, but results would be comparable to the static cases if the precharge was to zero.

Dynamic CMOS = total zero bits / total bits Static CMOS k bits = state change history, instructions fetched through k-bit wide register				
	Dynamic CMOS	Static CMOS 32 bits	Static CMOS 64 bits	Static CMOS 128 bits
SPECint92	0.69	0.36	0.37	0.37
SPECfp92	0.63	0.36	0.33	0.33

Table 9.2 PowerPC Instruction Fetch Switching Factors

One significant advantage of using static logic is the ability to stop clocks in order to reduce power without losing state. This can be very useful for power reduction during standby modes of portable applications. Unless static elements are included, dynamic logic requires that state be saved off to stop the clocks and then it must be restored when the clocks are restarted; this requires both time and energy.

If dynamic logic is used, precharging to $(V_{dd}-V_t)$ instead of V_{dd} may be appealing since the limited voltage swing can reduce both power consumption and delay. However, due to the body effect of the NMOS transistor used to bring down the precharge value, output values less than $(V_{dd}-V_t)$ may be produced. This can increase power due to short-circuit current since downstream static logic does not fully turn off. Therefore, precharging to $(V_{dd}-V_t)$ should only be used in cases where the wire load on the precharged node is large such that the power saved by reducing the voltage swing is greater than that lost due to short-circuit current in downstream logic.

9.5.2. Logic Synthesis

Since low-power microprocessors can be complex, large designs (>1 million transistors) and custom design is very time-consuming, synthesis is an important tool for control logic implementation. Logic synthesis parameters may be adjusted to optimize for low power. Synthesis may be biased to minimize spurious transitions and to choose gates that meet timing constraints yet dissipate less dynamic or static power. To choose gates based on power, gate libraries must be characterized for static and dynamic power dissipation and gates designed specifically for low-power should be available for non-critical paths.

9.5.3. Parallelism

One approach to maintaining throughput with a decreased supply voltage and corresponding lower frequency is logic parallelism. This involves duplicating hardware assuming that the parallel logic can provide the same throughput as the original logic while operating at greatly reduced frequency and voltage. This technique works well for functions that are easily parallelized, such as image processing, but it can be difficult to attain the same performance using this type of parallelism in a general purpose microprocessor. Though superscalar designs employ execution units that allow parallel instruction execution, the serial and dependent nature of most program code does not allow parallelism to achieve the performance gain necessary to outweigh the performance lost with lower frequency operation. Implementations that are highly parallelized also come at a large area sacrifice intolerable for some low-power applications where board space is limited.

9.5.4. Gate Sizing

Gates along a path are commonly sized according to timing constraints first and then possibly adjusted for area. Area reductions can reduce power by decreasing the loading capacitance. To optimize for low power, all non-critical paths should be resized to match the delay of the most critical path. In a microprocessor design, this path is usually the worst case pipeline stage delay. Since this path determines the maximum operating frequency, no performance is gained with other paths being faster. Down-sizing paths in this way can considerably reduce both area and loading capacitance.

9.5.5. Minimizing Switching

Minimizing the spurious switching that occurs during logic resolution can be accomplished in a variety of ways. In some cases logic may be restructured to reduce average switching activity. In other cases, latches may be strategically used to hold logic outputs until the final state of the logic has been reached or the change in latch inputs is relevant, at which time these latches may be opened. This prevents downstream logic from switching unnecessarily. One example of this use of latches in a superscalar microprocessor design is in latching instruction inputs to a particular execution unit. The instruction latch opened only if the instruction is actually dispatched. A more power-consuming alternative would be to latch the instruction every cycle and only record whether the instruction is valid based on whether it was actually dispatched to that unit.

Some circuits, such as incrementers, may be implemented in several ways, some of which switch less frequently than others. In the case of an incrementer, Gray code and Johnson counters are available, and though they do not produce a normal binary sequence, the maximum number of bits that will transition in a cycle is one. If the incrementer value is used only internal to the design, logic that uses the value may be adjusted to accommodate the non-binary count. An example of where a counter like this might be used in a microprocessor is for a pointer into an internal memory location.

9.5.6. Interconnect

Since microprocessors are often large designs, interconnect capacitance can be a large percentage of the loading capacitance. Power related to wires may account for over 40% of the total chip power [10],[18]. Chip floorplanning is most often driven by frequency performance but it can also be driven by power. Floorplanning can be adjusted to minimize the length of wires that switch often or are heavily loaded, such as register writeback busses. Strategically reducing the lengths of certain wires can optimize interconnect capacitance for low power.

9.6. Power and Performance: Comparing Microprocessors

It is difficult to compare the effects of different microprocessor features on power versus performance using real-world examples since each combines such a variety of features. Many microprocessors for very low power or embedded applications use a Dhrystone MIPS per Watt figure to convey this information, while notebook processors more often report SPECint and SPECfp results per Watt. Tables 9.3 and 9.4 show the vital statistics, including power and performance, of a variety of current low power processors for the embedded and notebook markets.

	Fujitsu FR20 [1]	Hitachi SH7032[2]	NEC V810[2]	AT&T Hobbit 92010[3]	ARM 700[4]	Motorola 68349[5]
Vdd	3.3-5 V	3.3-5 V	2.2-5 V	3.3-5 V	3-5 V	3.3 V or 5 V
frequency (MHz)	25 @5 V 16 @3.3 V	20 @5 V 12.5@3.3 V	25 @5 V 16 @3.3 V 10 @2.2V	30 @5 V 20 @3.3 V	33 @5 V 20 @3V	25 @5 V 16 @3.3 V
die area	70 mm2	92 mm2	53 mm2	92 mm2	46 mm2	100.3 mm2
technology	0.8um/ 2 metal	0.8um/ 2 metal	0.8um/ 2 metal	0.9um/ 2 metal	0.8um/ 2 metal	0.8um/ 2 metal
cache	1K Instr. 1K Data	8K RAM	1K Instr.	3K Instr.	8K Unified	1K Instr. or 2K SRAM
math on chip	none	MAC	FPU	none	none	none
instruction length	16 bit+ variable	16 bit fixed	16 or 32 bit variable	2, 6, 10 byte variable	32 bit	16-bit+ variable
ISA	RISC+	RISC	RISC	RISC/ CISC	RISC	CISC
Dhry MIPS*	28 @25MHz	16 @20MHz	18 @25MHz	20 @30MHz	30 @33MHz	10 @25MHz
typical power* (mW) at max freq.	600 @5 V	500 @5 V 130 @3V	500 @5 V 100 @3.3 V 40 @2.2V	900 @5 V 250 @3.3 V	550 @5 V 120 @3V	960 @5 V 300 @3.3 V

Table 9.3 Very Low Power Microprocessors (Embedded Applications, PDAs, etc.)

*Test conditions vary; source is vendor data.
1. [23]. 2.[11]. 3.[15]. 4. [14]. 5.[20].

	PowerPC 603 μP[1]	MIPS 4200[2]	Sun micro-SPARC[3]	QED Orion[4]	Intel 486DX4[5]	Intel P54C[5]
Vdd	3.3 V	3.3 V	5 V	3.3 V	3.3 V	3.3 V
frequency (MHz)	80@3.3 V	80@3.3 V	50@5 V	100@3.3V	100@3.3V	100@3.3V
die area	85 mm2	76 mm2	225 mm2	77 mm2	77 mm2	163 mm2
technology	0.5um/ 4 metal	0.6um/ 3 metal	0.8um/ 2 metal	0.6um/ 3 metal	0.6um/ 4 metal	0.6um/ 4 metal
cache	8K Instr. 8K Data	16K Instr. 8K Data	4K Instr. 2K Data	16K Instr. 16K Data	16K Unified	8K Instr. 8K Data
instruction length	32 bit fixed	32 bit fixed	32 bit fixed	32 bit fixed	2-14 byte variable	2-14 byte variable
ISA	RISC	RISC	RISC	RISC	CISC	CISC
SPEC92int*	75	55	22.8	68	51	100
SPEC92fp*	85	30	18.4	60	27	80
typical power* (W) at max freq.	2.2@3.3 V	1.5@3.3 V	3.3@3.3 V	4.0@3.3 V	4.3@3.3 V	4.0@3.3 V

Table 9.4 Notebook Microprocessor

*Test conditions vary; source is vendor data.
1. [10]. 2. [27]. 3.[4]. 4. [17]. 5. [12]

9.7. Summary

New low-power applications have brought about the development of microprocessors that combine features which enhance performance and reduce power. To truly optimize performance and power, all aspects of a microprocessor design must be addressed including operating voltage and frequency, clocking, instruction set, microarchitecture, logic and circuit design, floorplanning and process technology. Beyond the techniques discussed in this chapter to address these parameters, a variety of new and formerly abandoned methods are being examined for microprocessor design. Among these are asynchronous design, on-chip variance of frequency and supply voltage, double-edge triggered latching, application-specific instruction set architectures, and highly integrated design incorporating multiple system components on a single chip. With the demand for portable computer products continually increasing, the market for low-power microprocessors is progressively providing vast opportunities for innovative development in this field.

PowerPC, PowerPC 603 and PowerPC 601 are trademarks of International Business Machines Corporation.

REFERENCES

[1] Bakoglu, H. B., *Circuits, Interconnections and Packaging for VLSI*, Reading, MA: Addison-Wesley, 1990.

[2] Bunda, J. and M. Vaden of IBM Corp., Private communication.

[3] Bunda, J., W. C. Athas, and D. Fussell, "RISC Instruction Set Design for Low Power," Proceedings of the Hawaii International Conference on System Sciences, Wailea, Hawaii, January 1995.

[4] Case, B. "SPARC Hits Low End With TI's microSPARC," *Microprocessor Report*, Vol. 6, No. 14 (October 28), 1992, pp. 11-14.

[5] Chandrakasan A., S. Sheng, and R. Brodersen, "Low Power CMOS Digital Design," *IEEE Journal of Solid State Circuits*, Vol. 27, No.4, Apr. 1992, pp. 473-484.

[6] Culbert, M., "Low Power Hardware for a High Performance PDA," *Proceedings of COMPCON*, February 1994, pp. 144-147.

[7] Gallant, J., "Low Power uPs simplify design of portable computers," *EDN*, Vol. 38, No. 9, April 29, 1993, pp. 39-44.

[8] Gary, S., et. al, "PowerPC 603™, A Microprocessor for Portable Computers," *IEEE Design & Test of Computers*, Vol. 11, No. 4, Winter 1994, pp. 14-23.

[9] Gary, S., et al., "The PowerPC™ 603 Microprocessor: A Low-Power Design for Portable Applications," *Proceedings of COMPCON*, February 1994, pp. 307-315.

[10] Gerosa, G., et. al., "A 2.2W, 80 MHz, Superscalar RISC Microprocessor," *IEEE Journal of Solid State Circuits*, Vol. 29, No. 12, December 1994, pp.1440-1454.

[11] Gwennap, L., "Hitachi, NEC Processors Take Aim at PDAs," *Microprocessor Report*, Vol. 7, No. 8 (June 21), 1993, pp. 9-13.

[12] Gwennap, L., "Intel Extends 486, Pentium Families," *Microprocessor Report*, Vol. 8, No. 3 (March 7), 1994, pp. 1, 6-11.

[13] Gwennap, L., "MIPS Reaches New Lows With R4200 Design," *Microprocessor Report*, Vol. 7, No. 7 (May 31), 1993, pp. 1,6-9.

[14] Gwennap, L., "ARM7 Cuts Power, Increases Performance," *Microprocessor Report*, Vol. 7, No. 15 (November 15), 1993, pp. 12-13.

[15] Gwennap, L., "Hobbit Enables Personal Communicators," *Microprocessor Report*, Vol. 6, No. 14 (October 28), 1992, pp. 15-21.

[16] Hennessy, J. and D. Patterson, *Computer Architecture: A Quantitative Approach*, San Mateo, CA: Morgan Kaufmann Publishers, 1990.

[17] Killian, E., "Orion: One Year Later," *Proceedings of the Sixth Annual Microprocessor Forum*, October 18-20, 1993, pp. 24.1-10: MicroDesign Resources, 874 Gravenstein Hwy. South, Sebastapol, CA 95472.

[18] Liu, D., and C. Svensson, "Power Consumption Estimation in CMOS VLSI Chips," *IEEE Journal of Solid State Circuits*, Vol. 29, No.6, June 1994, pp. 663-670.

[19] Moat, K. and B. Lucas, "Benchmark Analysis of an Energy Efficient CPU Instruction Set," presented at the 2nd Annual Motorola War on Current Drain Symposium, September 22-23, 1994.

[20] Slater, M. "Motorola Extends 68300 Line for CD-I, PDAs," *Microprocessor Report*, Vol. 7, No. 8 (June 21), 1993, pp. 14-17.

[21] Su, C., C. Tsui, and A. M. Despain, "Low-Power Architecture Design and Compilation Techniques for High-Performance Processors," *Proceedings of COMPCON*, February 1994, pp. 489-498.

[22] Thorson, M., "System Management Mode Explained," *Microprocessor Report*, Vol. 6, No. 8 (June17), 1992, pp. 14-16, 21.

[23] Turley, J., "Fujitsu Starts New Embedded CPU Family," *Microprocessor Report*, Vol. 8, No. 13 (October 3), 1994, pp. 20-22, 25.

[24] Voldman, S. and G. Gerosa, "Mixed-Voltage Interface ESD Protection Circuits For Advanced Microprocessors In Shallow Trench and LOCOS Isolation CMOS Technologies," to be published in *Proceedings of IEDM '94*, December 1994.

[25] Weste, N. and K. Eshragian, *Principles of CMOS VLSI Design: A Systems Perspective*. Reading, MA: Addison-Wesley, 1988.

[26] Yeung, N., B. Zivkov, and G. Ezer, "Unified Datapath: An Innovative Approach to the Design of a Low-Cost, Low-Power, High-Performance Microprocessor," *Proceedings of COMPCON*, February 1994, pp. 32-37.

[27] Zivkov, B., B. Ferguson, and M Gupta, "M4200, A High-Performance MIPS Microprocessor for Portables," *Proceedings of COMPCON*, February 1994, pp. 18-25.

10

Portable Video-on-Demand in Wireless Communication

Teresa H. Meng, Benjamin M. Gordon, and Ely K. Tsern

Our present ability to work with video has been confined to a wired environment, requiring both the video encoder and decoder to be physically connected to a power supply and a wired communication link. This chapter describes an integrated approach to the design of a portable video-on-demand system capable of delivering high-quality image and video data in a wireless communication environment. The discussion will focus on both the algorithm and circuit design techniques developed for implementing a low-power video decompression system at power levels that are two orders of magnitude below existing solutions. This low-power video compression system not only provides a compression efficiency similar to industry standards, but also maintains a high degree of error tolerance to guard against transmission errors often encountered in wireless communication. The required power reduction can best be attained through reformulating compression algorithms for energy conservation. We developed an intra-frame compression algorithm that requires minimal computation energy in its hardware implementations. Examples of algorithmic trade-offs are the development of a vector quantization scheme that allows on-chip computation to eliminate off-chip memory accesses, the use of channel-optimized data representations to avoid the error control hardware that would otherwise be necessary, and the coding of internal data representations to further reduce the energy consumed in data exchanges. The architectural and circuit design techniques used include the selection of a filter bank structure that minimizes the energy consumed in the datapath, the data shuffle strategy that results in reduced internal memory size, and the design of digital and analog circuits optimized for low supply voltages. Our hardware prototype is a video decoding chip set for decompressing full-motion video transmitted through a wireless link at less than 10 mW, which is incorporated into a hand-held portable communication device with a color display. We will describe the design trade-offs of the prototype for low-power purposes, and quantify the system's performance in both compression efficiency and power dissipation.

10.1. Introduction

Video communication has rapidly become an integrated part of information exchange. A growing number of computer systems are incorporating multi-media capabilities for displaying and manipulating video data. This interest in multi-media combined with the great popularity of portable computers and portable phones provides the impetus for creating a portable video-on-demand system. Video-on-demand requires a bandwidth far greater than broadcast video, as each user can subscribe to different video programs at any time in any place. Because of the large bandwidth requirement, for both storage and transmission, data compression must be employed, requiring real-time video decoding in the portable unit. The key design consideration for portability is reduction of power consumption to allow for extended battery life.

Besides the constraint on power, the compressed video signals delivered by wireless communication are often corrupted in transmission. As wireless channel characteristics usually cannot be predicted beforehand, the received video quality may be degraded. To design a portable video-on-demand system that is sufficiently resistant to transmission errors, the compression algorithm not only need to deliver good compression performance, but also to maintain a high degree of error tolerance. To meet these goals, the design of a portable video-on-demand system must satisfy the following three criteria: low power consumption, high compression efficiency, and channel-error resiliency.

Of the three criteria, minimal power consumption is the guiding principle for both algorithm development and system trade-offs evaluation. Our video decoder operates at a power level that is more than two orders of magnitude below that of comparable decoders in similar technology. This tremendous saving in power consumption was attained through both algorithm reformulation and architecture innovations specifically targeted for energy conservation. How to reduce the overall power dissipation in real-time video compression/decompression is the main focus of this chapter.

As our application is the delivery of video-on-demand, the quality of decompressed video must not fall below the level of industry compression standards [1][2][3]. The selection of compression algorithms, however, is largely directed by their hardware implementations. Our goal is to design a compression algorithm with a performance meeting that of industry standards, while at the same time requiring only minimal power in both encoding and decoding operations.

In addition to the provision of high compression efficiency, error resiliency is another important factor in assessing system performance of a wireless portable application. Error resiliency as an aspect of video compression has not received much attention until recently [4][5][6], when wireless transmission of compressed video became regarded as a desirable feature. This criterion dramatically influences the design considerations in both algorithm development and hardware implementation. To combat channel error, most systems use error-correcting codes to protect data from being damaged in transmission and storage. One example is the Reed-Solomon code used on the data stored in compact discs [7]. On the other hand, error-correcting codes not only limit the bandwidth available for compressed data, but also result in greater hardware complexity, complicating the decoder design. Our approach to error resiliency, however, is to embed a high degree of error tolerance in the compression algorithm itself. We therefore guarantee consistent video quality under all error conditions, without the use of error-correcting codes.

The development of our prototype portable decoder required fabricating two custom chips and a low-voltage D/A converter. This chapter will discuss the architectural trade-offs and circuit techniques used in designing the prototype decoder. It will also offer an analysis of the effects of these trade-offs and techniques on power consumption. Implementation goals include low power, high throughput, display resolution-independent decoding, and good compression performance. The reduction in power consumption is achieved by using power-efficient compression algorithms and architectural mapping, low-power circuit design techniques, and a reduced supply voltage. An additional goal in developing this prototype is to establish an efficient design path from conception to transistor-level testing, creating cell libraries and identifying needed tools.

This chapter is organized as follows. Section 2 offers a brief overview of the standard video compression algorithms and their application to video compression in wireless communication. Section 3 details the compression algorithms developed for our portable video system and their performance in comparison with industry standards. Section 4 describes the error-tolerant capability of our algorithm in comparison with other algorithms under various channel error conditions. Section 5 discusses low-power circuit design techniques in general, while Section 6 discusses the low-power architectural strategies employed in designing our prototype decoder. In both Sections 5 and 6, the amount of power savings achieved in each step will be quantified and substantiated by measured data. Finally, in Section 7 we offer our conclusions and outlook for the future.

10.2. Video Compression for Portable Applications

10.2.1. Standard Compression Algorithms

The image and video compression standards (JPEG, MPEG, and H.261) have adopted the 2-dimensional discrete-cosine transform (2-D DCT) [8][9] as the basic information compaction engine. The 2-D DCT is performed on blocks of pixels of an image to obtain transform coefficients, each coefficient roughly corresponding to a spacial frequency. As most natural images have information concentrated at low frequencies, the 2-D DCT can be thought of as an energy compactor, compacting pixel energy in an block to those coefficients corresponding to low spacial frequencies. The concentration of energy into the lowest frequencies allows a more efficient, compressed representation.

Each transform coefficient is quantized to reduce the total number of bits necessary to represent the whole image. The standard image compression procedure recommended by JPEG (Joint Photographic Expert Group) uses a uniform quantizer with a different quantization step size for each of the 64 transform coefficients [1]. As most visual information is concentrated in low spacial frequencies, compression is achieved by quantizing the coefficients of low frequencies finely and the coefficients of high frequencies very coarsely. This strategy is designed to match the human visual system which is less sensitive to quantization error at higher frequencies. The values of these quantization step sizes were determined experimentally, under fixed viewing conditions with a constant display width and viewer distance. These conditions may not hold in general, but the quantization matrix seems to work quite well in most cases. To adjust for different compression ratios, one may uniformly scale the quantization step size for each transform coefficient by a constant factor.

As high-frequency coefficients are usually quantized by a fairly large step size, many of the coefficients will be quantized to zero. Instead of transmitting long sequences of zeros, zero run-length encoding is used to encode the length of repeated zeros, reducing transmission bandwidth. After zero run-length encoding, the zero lengths and non-zero coefficients are Huffman encoded, a procedure that assigns shorter binary representations to more frequently occurring symbols [10].

Both zero run-length encoding and Huffman encoding are lossless compression procedures, meaning that they do not decrease the information content, or entropy, of the data being compressed. Losslessly compressed data can be perfectly reconstructed bit by bit at the decoder side, if there is no bit error introduced in transmission. Figure 10.1 illustrates the standard JPEG compression procedure that uses lossless entropy encoding of zero lengths and Huffman codes to further reduce the bit representations of quantized DCT coefficients.

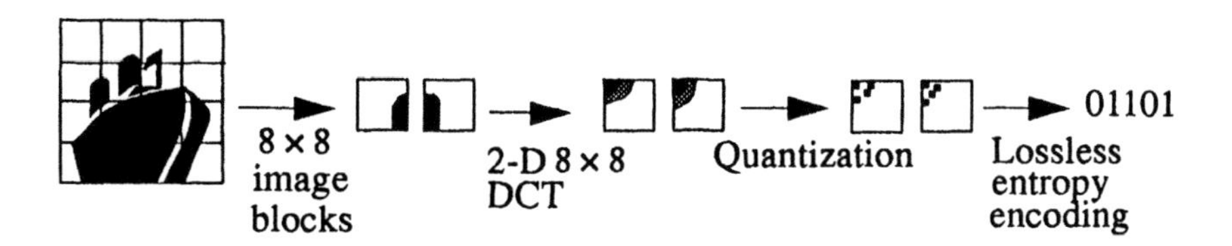

Figure 10.1 The JPEG compression procedure.

Although JPEG is a reasonably efficient compression procedure, the use of the 2-D DCT often introduces visual blocking artifacts in the decompressed image at high compression ratios. This corresponds to image content mismatches at the block boundaries, due to coarse quantization. These blocking artifacts are a general characteristic of block-based transform coding. We will return to this point in the next section.

10.2.2. Variable-Rate Codes and Error Propagation

One major feature of JPEG compression is that its compressed bit rate is not constant, but a function of image content given a fixed quantization matrix and the coding of zero lengths. Variable-rate codes usually provide better compression capability than fixed-rate codes because of the use of entropy encoding. Variable-rate codes, however, result in greater hardware complexity, requiring additional buffering and synchronization control in decoding. Variable-rate codes are also highly susceptible to bit errors, as a single bit error can cause error propagation and produce totally erroneous decoded results until the next synchronization point.

Although a single bit error in a fixed-rate encoded sequence can also cause some deviation of decoded pixel values, at least the error will not corrupt the information on pixel positions. As wireless channels are much noisier than wired channels, and their channel characteristics depend highly on the positions of and the interference pattern between the transmitter and receiver at any point in time, a fixed-rate code that does not cause error propagation would be desirable for wireless transmission. The objective is to design a fixed-rate code that delivers similar or better compression performance than a variable-rate code such as JPEG.

10.2.3. Error-Correcting Codes vs. Error-Resilient Compression

Because our portable video-on-demand system is to be embedded in a wireless communication environment, the first step in designing a suitable compression algorithm is to analyze the wireless channel characteristics. Wireless channels, depending on the channel modulation techniques used, display a wide range of error patterns. Experiments with mobile receivers and transmitters show that received signal power may experience fades exceeding 10 dB approximately 20% of the time, and 15 dB (measuring limit) up to 10% of the time [11], causing bursty transmission bit errors in the received data stream. Bursty bit errors are much more difficult to correct than random bit errors. To "randomize" burstiness, the transmitted data stream can be interleaved over a certain period of time so that if consecutive bit errors occur (as is typical with bursty bit errors) in the received data stream, the error pattern appearing in the decoded data stream will resemble that of random bit errors.

To reduce random bit errors, error-correcting codes are usually used. With the knowledge of a target channel bit error rate (BER) and the channel's signal-to-noise ratio (SNR), error-correcting codes can be very effective in correcting random bit errors. In wireless communication, however, the BER and the channel's instantaneous SNR vary widely over time and space, making a fixed-point design, as in the selection of an appropriate error-correcting code, a sub-optimal solution at best.

For wireless and mobile channels that experience very low BERs most of the time but with intermittent severe channel degradation occasionally, error-correcting codes are not as effective as one might think. On the one hand, under low distortion conditions (low BERs), when almost perfect transmission is available, error-correcting codes incur unnecessary bandwidth overhead, generating redundant bits that could be used for representing compressed video data to improve video quality. This bandwidth overhead can be substantial, from 10% to 50% in recent mobile transceiver designs [12]. On the other hand, under severe distortion conditions in which the BER exceeds the designed capacity, error-correcting codes may actually introduce more decoded errors than received ones. On top of all this, error-correcting codes require additional hardware at the decoder unit, making the design of our portable decoder a more difficult task.

Our approach to error tolerance is to design error-resilient compression algorithms so that if channel distortion does occur, its effect will be a gradual degradation of video quality, and the best possible quality will be maintained at all BERs. In this way, we do not pay a high premium in bandwidth overhead when protection against error is not needed (under low channel distortion), and still deliver reasonable-quality video when error-correcting codes would have failed (under severe channel distortion). Furthermore, an overriding assumption automatically made in using error-correcting codes is exact binary reproduction of the transmitted bits. Because our goal is to transmit compressed video data -- and human vision is fairly fault-tolerant -- exact binary reproduction at the decoder is not necessary. As long as the effects of error are localized and do not cause catastrophic loss of image sequences, a robust error-resilient compression algorithm without resorting to error-correcting codes for data protection can be the solution, and our best compromise among the goals of consistent video quality, minimum transmission bandwidth, and low-power implementation. The development of such an algorithm will be the focus of the next section.

To this point, we have discussed compression techniques for single frames, while the standard NTSC video displays 30 frames per second. The standard video compression recommended by MPEG (Motion Pictures Experts Group) uses a DCT-based procedure similar to JPEG for intra-frame compression, with inter-frame compression added in to exploit temporal dependency between frames [2]. Temporal dependency is extracted by an operation called block motion estimation, which estimates the motion vector of a particular block of pixels on an inter-frame with respect to a previous intra-frame. The motion vector represents a predicted block copied from the previous intra-frame. The pixel difference between this predicted block and the actual block on the inter-frame is calculated and transmitted along with the block motion vector.

At the decoder side, the block motion vector is used to copy a block of pixels from a frame buffer that stores the previous infra-frame, a procedure referred to as motion compensation. The pixel difference is then added onto the predicted block to generate the decoded block. If we compare the hardware complexity of video-rate JPEG and MPEG, the MPEG encoding is much more complicated than JPEG because it involves motion estimation, a procedure that was once implemented by 256 concurrent processing elements in a real-time MPEG encoder chip [18]. With regard to decoding, however, motion compensation in MPEG involves only a sequence of memory reads from the frame buffer. In most MPEG decoders, therefore, the calculation of the inverse 2-D DCT constitutes the major computation component, as is the case of video-rate JPEG decoders at 30 frames per second.

Our prototype portable decoder is an intra-frame decoder that does not use motion compensation. Hence the comparisons of our design with others in terms of compression performance and power efficiency will be made with video-rate JPEG decoder chips, one example of which was the C-Cube JPED decoder [15]. To exploit temporal dependency, our compression algorithm can be easily extended to 3-dimensional subband decomposition, with the third dimension in time to eliminate temporal redundancy [19]. In this work, however, we will focus on infra-frame based compression techniques and their low-power implementations.

10.3. Subband Decomposition and Pyramid Vector Quantization

In this section, we first give a brief overview of subband decomposition and pyramid vector quantization (PVQ). A more complete description of these two topics can be found in [16] and [13]. We then describe our subband/PVQ compression algorithm, which performs we well as the standard JPEG compression in terms of image quality and compression efficiency, while incurring much less hardware complexity and exhibiting a high degree of error resiliency.

10.3.1. Subband Decomposition

Many video compression algorithms are available with most performing a linear transformation, either a DCT or a filtering decomposition, on the image data. We chose subband decomposition over DCT for several reasons. First, subband decomposition delivers higher quality images than DCT by producing less noticeable visual artifacts (e.g. blockiness) and having higher resolution in low spatial frequencies. This is an important consideration in terms of compression performance. Second, subband decomposition is more amenable to rate control through selective transmission of different filter bands. Finally, and most importantly, subband decomposition allows for design parameters to be modified to achieve a low-power implementation.

As shown in Figure 10.2, subband decomposition divides each video frame into several subbands by passing the frame through a series of 2-D low-pass and high-pass filters, where denotes a down- sampling operation by a factor of two and denotes an up-sampling operation by a factor of two. Each level of subband decomposition divides the image into four subbands. We can hierarchically decompose the image by further subdividing each subband. We refer to the pixel values in each subband as subband coefficients.

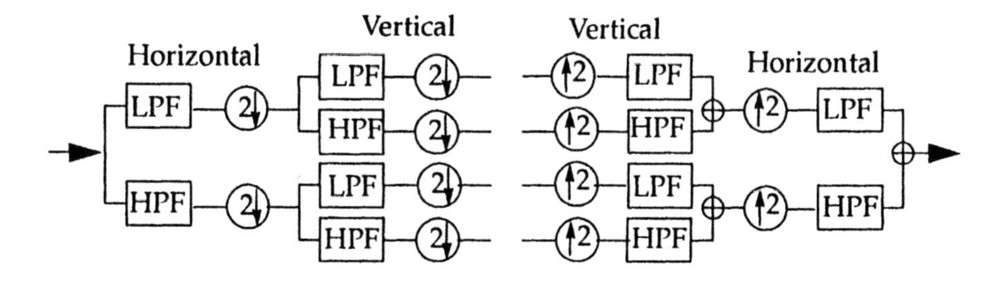

Figure 10.2 One level of 2-D subband decomposition.

The filtering process decorrelates the image information and compacts the image energy into the lower-frequency subbands, a natural result of the fact that most image information lies in low spatial frequencies. The information in high spatial frequencies, such as edges and small details, lies in the higher- frequency subbands, which have less energy (pixel variance). We achieve compression by quantizing each of the subbands at a different bit rate according to the amount of energy and the relative visual importance in each subband.

We apply four levels of subband decomposition in our compression scheme, using LL to represent the filtered output of the low-pass vertical and low-pass horizontal filter, LH the output of the low-pass vertical and high-pass horizontal filter, HL the output of the high-pass vertical and low-pass horizontal filter, and HH the output of the high-pass vertical and high-pass horizontal filter. Because each subband has a reduced frequency content, it is down-sampled by a factor of two in each dimension. The LL subband can be recursively filtered creating four levels of 13 subbands, as shown in Figure 10.3. For decoding, the four lowest-level subbands are each up-sampled and filtered, then summed together to form a reconstructed LL subband. This process is repeated through all four levels until the final image is formed.

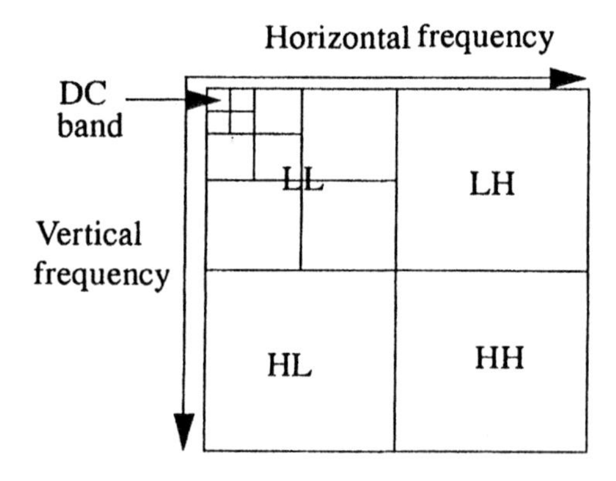

Figure 10.3 Four levels of subband decomposition.

Another pattern of decomposition is to filter the LL, LH, HL bands into 31 subbands, as shown in Figure 10.4, where a bit rate (bits/pixel) is assigned to each subband at a compression ratio of 12:1 for encoding image *Man*. This bit allocation was selected by using the marginal bit allocation strategy, i.e. recursively allocating bits to those frequency bands which best minimize the overall distortion per bit [21][22].The non-integer bit rate assigned to each subband is a result of using a vector quantization technique, as will be discussed in the next subsection.

The choice of low-pass and high-pass filter coefficients is determined by their ability to accurately reproduce the original image and cancel frequency aliasing between subbands. Examples are quadrature mirror filters (QMF) which have near-perfect reconstruction [20] and wavelet filters with perfect reconstruction in magnitude, though with a non-linear phase response [16][17].

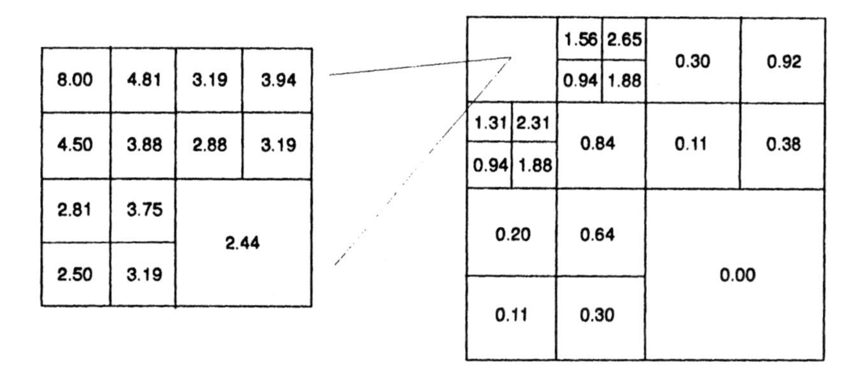

Figure 10.4 Subband bit allocation for encoding image *Man* at a 12:1 compression ratio.

10.3.2. Pyramid Vector Quantization

Vector quantization is a technique that groups a number of samples together to form a vector, and then represents each vector by an index defined in a previously established codebook. Vector quantization differs from scalar quantization in that a vector of samples is quantized together instead of individually. Pyramid vector quantization (PVQ) is a vector quantization technique that groups data into vectors and scales them onto a multi-dimensional pyramid surface. Fisher first introduced PVQ as a fast method of encoding Laplacian-distributed data and later applied this technique to encoding high-frequency DCT coefficients, which resemble the Laplacian distribution [13][24][26]. An example of a 3-dimensional pyramid surface is shown in Figure 10.5, where L is the vector dimension, which specifies the number of samples per vector, and K is an integer defined as the pyramid radius. In Figure 10.5, a pyramid surface with $L = 3$ and $K = 4$ is illustrated.

The lattice points on the pyramid surface, defined by all the points whose absolute norm equals the pyramid radius K, form the PVQ codebook. Each lattice point is assigned an index to represent the vector on the pyramid surface. The pyramid radius K and the vector dimension L determine the codebook size and therefore the coding rate (bits/pixel).

Pyramid surface defined by:

$$\sum_{i=1}^{L} |x_i| = K$$

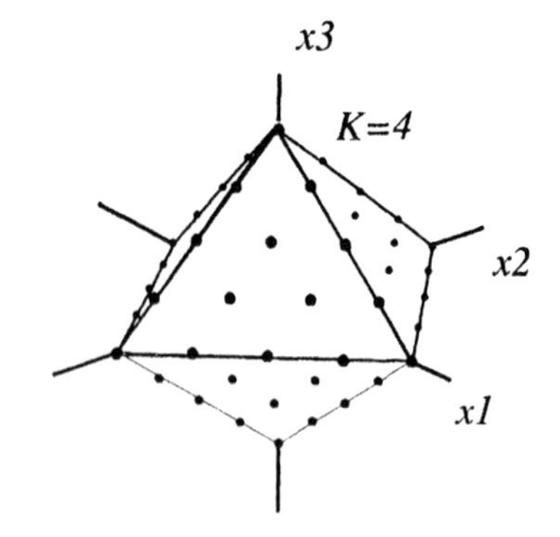

The signal space is represented by $x1$, $x2$, and $x3$, the values of three data samples.

Figure 10.5 A 3-D pyramid with the pyramid radius K equal to 4.

The basic quantization steps of PVQ are: (1) calculate the vector radius, r, defined as the absolute norm of the vector to be coded, (2) project the vector to be coded onto the pyramid surface of radius K by scaling each vector element by K/r, and (3) quantize the scaled vector to the nearest lattice point (identify the index of the nearest lattice point) on the pyramid surface. Both the lattice index and the vector radius are then transmitted. This version of PVQ is generally referred to as product PVQ [13].

Because the pyramid structure is regular, encoding and decoding lattice indices can be calculated with a simple formula, which has a computational complexity linearly proportional to the vector dimension and coding rate. Thus, PVQ eliminates the need for codebook storage and training found necessary in LBG-based VQ designs [25], effectively allowing for very large codebooks unrestricted by the size of physical memory. As will be discussed in Sections 10.5 and 10.6, memory operations consumes far more power than arithmetic computation, often by a factor of 10 or more for on-chip memory accesses, not to mention the power needed for off-chip memory accesses. PVQ therefore allows for reduction of power-consuming memory accesses in favor of arithmetic computation, offering low-power encoding and decoding at video rates.

The selection of vector dimensions in PVQ greatly impacts both the error-tolerance capability and hardware implementation of a PVQ code. Given a fixed coding rate, while larger vector dimensions achieve better compression performance because of a higher probability of statistical convergence, using a smaller vector dimension significantly improves error resiliency by localizing the effect of possible bit errors. This error-related design trade-off will be further discussed in Section 10.4.

One major benefit of using small vector dimensions is significant reduction in hardware, a result of smaller indices and correspondingly smaller datapath widths and memory. To locate the nearest lattice point to a scaled vector on the pyramid surface, the PVQ encoding and decoding algorithm requires table look-ups of pre-computed combinatorial calculations and index offset values. The size of total memory required for these tables is roughly proportional to the vector dimension, the pyramid radius, and the coding rate expressed in bits per pixel [5]. With a larger vector dimension, a larger radius K is required to maintain the same coding rate, which significantly increases the size of memory needed to store the tables. For example, for a given coding rate, to decode PVQ-encoded vectors of dimension 16 will require a memory size roughly eight times larger than that required for decoding vectors of dimension 4. However, to code subband coefficients in high-frequency subbands where the allocated bit rates are relatively low, we can afford larger dimensions for better compression performance. Our PVQ coder uses a vector dimension of one (PCM) on the lowest-frequency subband, up to a vector dimension of 32 on the highest-frequency subband.

10.3.3. The Subband/PVQ Compression Algorithm

This subsection describes the overall subband/PVQ compression algorithm. Our coding technique is entirely intraframe, i.e., each video frame is coded independently. The basic encoding scheme consists of the following three steps:

1. Each frame is filtered by four levels of subband decomposition.
2. Given its high entropy and non-Laplacian distribution, the coefficients in the lowest subband are scalar-quantized (PVQ of dimension one) to 8 bits.
3. For the other subbands, each subband uses a vector dimension between 4 and 32, depending on the visual importance of the subband (the higher the frequency, the larger the vector dimension). Each vector is then quantized using PVQ based on an error-resilient indexing scheme (discussed in Section 10.4). Each vector radius is quantized using a non-uniform logarithmic compandor, which matches the radius' probability distribution function.

10.3.4. Compression Performance

To demonstrate the compression performance of this fixed-rate subband/PVQ algorithm, we applied it to the USC database images *Lena, Lake, Couple*, and *Mandrill*. The subband filter used was taken from [20] and the subband bit allocation at various rates were obtained using marginal bit allocation [18][19], one example of which was shown in Figure 10.4. Our subband/PVQ compression exceeds previously reported PVQ results [24] for similar images [14]. As shown in Figure 10.6, it outperforms JPEG at all bit rates of interest (JPEG with scaled default quantization matrices and custom-generated Huffman tables), without requiring the use of variable-rate entropy codes. Image quality was measured by peak signal-to-noise ratio (PSNR), defined to be the maximum signal power over the squared compression error. Though the PSNR performance may not be a distortion metric faithfully representing human visual perception, our study on image quality assessment based on psycho-vision also suggested that the subband/PVQ algorithm delivers images of better quality than JPEG at the same compression ratio [23].

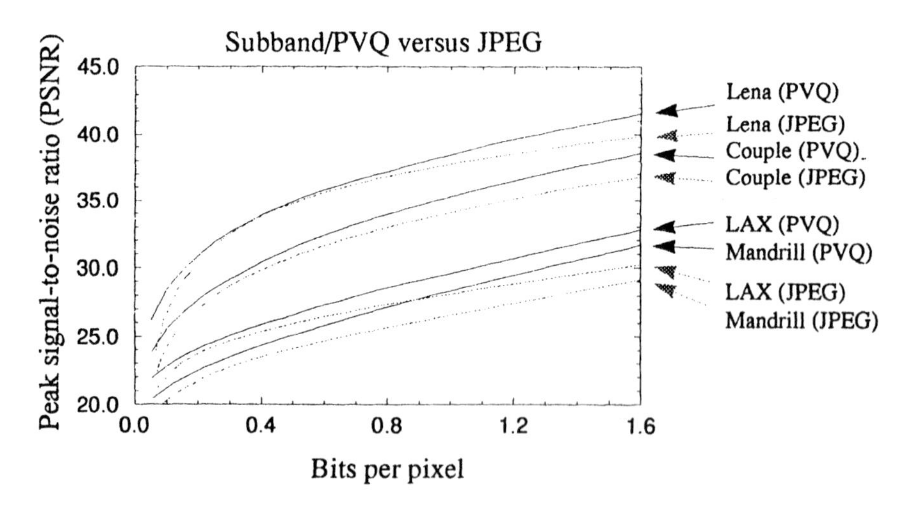

Figure 10.6 Compression performance in peak signal-to-noise ratio (PSNR) of subband/PVQ versus JPEG on USC database images *Lena, Couple, LAX*, and *Mandrill*.

10.4. Error-Resilient Compression

The effects of channel error on video quality can be grouped into two categories: the effects of error propagation and the effects of random bit errors on individual codes. As discussed in Section 10.2, bursty bit errors can often be modeled by random bit errors when data is interleaved for transmission. Because our subband/PVQ algorithm guarantees a constant compressed bit rate as determined by the subband bit allocation, random bit errors do not corrupt pixel position information, and the error effects will not propagate over vector boundaries. Of the two types of error effects, therefore, we need only consider the second type, namely, the effects of random bit errors on individual vectors.

For PVQ, random bit errors modify transmitted lattice indices and cause the decoded lattice points to deviate from their intended positions. If the index enumeration procedure, which assigns an index (a binary bit pattern) to represent each lattice point, is not well designed, a single bit error on an index may cause each vector element (one element representing a subband coefficient) of the decoded vector to deviate from its original value, sometimes to its negative value. In this section, we describe an index enumeration scheme that minimizes the effects of random bit errors on decoded subband coefficients [6].

10.4.1. Error-Resilient PVQ Index Enumeration

In our indexing scheme, each vector is described by four characteristics: (1) the number of non-zero vector elements, (2) the sign pattern, which describes the signs of the non-zero vector elements, (3) the vector pattern, which describes the positions of the non-zero vector elements, and (4) the vector shape, which describes the magnitudes of the non-zero vector elements. This method codes each characteristic separately, and then combines them all in a product code as described in [6]. Separating these characteristics in a product code limits the effects of bit errors.

To enumerate the sign pattern, we notice that there is a one-to-one correspondence between the number of sign bits and the number of non-zero vector elements. Let s be the number of non-zero elements, L be the vector dimension, and K be the pyramid radius. The number of ways to assign signs to s non-zero elements is 2^s. Thus, the index for the sign pattern can be concatenated together into a string that is s bits long.

To form an index for a vector, we first enumerate s, then the vector pattern, multiply that by the number of total vector shapes, and then add the enumerated vector shape. Finally, we append the sign bits to the least significant positions of the index. The decoding algorithm requires at most $(LK+5)$ memory fetches, $(LK+8)$ compares, $(LK+5)$ subtracts, and one divide.

To code the vector pattern, we notice that the total number of possible patterns of s non-zero elements equals $P(L,s) = \binom{L}{s}$, which describes selecting s positions from a vector of length L. The total number of possible vector shapes, given s non-zero vector elements whose magnitudes must sum to K, is $S(s, K) = \binom{K-1}{s-1}$.

The total number of points on a pyramid surface is obtained by taking the product of all possible sign combinations, patterns, and shapes and summing the product over all possible values of s:

$$X(L, K) = \sum_{s=1}^{m} 2^s P(L, s) S(s, K) = \sum_{s=1}^{m} 2^s \binom{L}{s} \binom{K-1}{s-1} \qquad (10.1)$$

where m denotes $min(L,K)$.

The enumeration is channel-error resilient because the sign bits form a binary product code -- a single bit error in the sign bits only causes a sign change in a single pixel without affecting other pixels. Furthermore, the pattern information rests in only a few of the most significant bits of the product code, while the shape information composes the majority of the index bits. Thus, single bit errors that occur in most bits only affect the shape (pixel values) without changing its sign or pattern; only bit errors that occur in the few most significant bits would corrupt both the pattern and the shape.

A slightly more robust enumeration can be used, which concatenates the pattern and shape bits, instead of multiplying them together. A single bit error can then affect only one characteristic, either pattern, shape or sign. This method also has the advantage of eliminating the need for a division in decoding, but can add an overhead of at most bit/pixel to the overall bit rate. This product code was used in our portable decoder design, for the purposes of reduced hardware and improved error resiliency.

10.4.2. Error-Resiliency Performance

We now evaluate our compression scheme based on its error-resiliency properties. To provide a relative measure of performance, we compare our fixed-rate, intra-frame technique to variable-rate JPEG, which is an intra-frame, DCT-based algorithm using zero run-length and Huffman codes to achieve a high degree of compression. To make the comparison fair, we applied error-correcting codes to JPEG-encoded data and used JPEG's resynchronization intervals of 6, as would be necessary if JPEG compression was used in wireless transmission. To test the effectiveness of different error-correcting codes in improving the image quality at various BERs, we selected three commonly used error-correcting codes: a (127, 120) Hamming code, which incurs a bandwidth overhead of 6% to the overall bit rate and is capable of correcting one bit error in 127 consecutive bits; a (73, 45) Weldon difference-set block code with a threshold decoding scheme, which offers good protection (Hamming distance of 10) and incurs a bandwidth overhead of 60% to the overall bit rate [27]; and a rate-1/2 (2,1,6) NASA planetary convolutional code, which offers a high degree of error correction but requires Viterbi decoding at the decoder and doubles the overall bit rate [28].

For subband/PVQ, we apply a simple 3-bit repetition code on the most significant bit (MSB) and the next MSB of each pixel in the lowest LL subband. Because the 4-level subband decomposition effectively compacts most image energy into the lowest LL subband, the size of which is only 1/256 of the total image, this simple DC protection scheme incurs a bandwidth overhead of merely 0.016 bit/pixel to the overall bit rate. In the following simulation the subband/PVQ algorithm actually uses a slightly lower bit rate than JPEG, even with the added DC protection. The simulation was conducted using image *Mandrill* compressed at 12:1 or 0.66 bit/pixel, a rate at which both JPEG and the subband/PVQ algorithm give good compression performance.

Figure 10.7 graphically shows the performance of our algorithm under noisy channel conditions for the *Mandrill* image. Under most error conditions, our subband/PVQ algorithm performs slightly better than JPEG with or without protection, as expected from the facts that our algorithm outperforms JPEG even without channel error (as shown in Figure 10.6), and that the subband/PVQ algorithm does not suffer from the bandwidth overhead incurred by the use of error-correcting codes. On the JPEG with error correction curves, we note that error-correcting codes do an excellent job of eliminating bit errors, but start to fail at BERs around 10^{-2}. Once the BER increases past this point, catastrophic errors occur, causing severe block loss and rapid drop in PSNR performance. Our fixed-rate code, however, maintains a gradual decline in quality, even under severe BER conditions. This gradual degradation in performance makes the subband/PVQ scheme well suited for situations where image quality must be maintained under deep channel fades and severe data loss, a characteristic of wireless transmission.

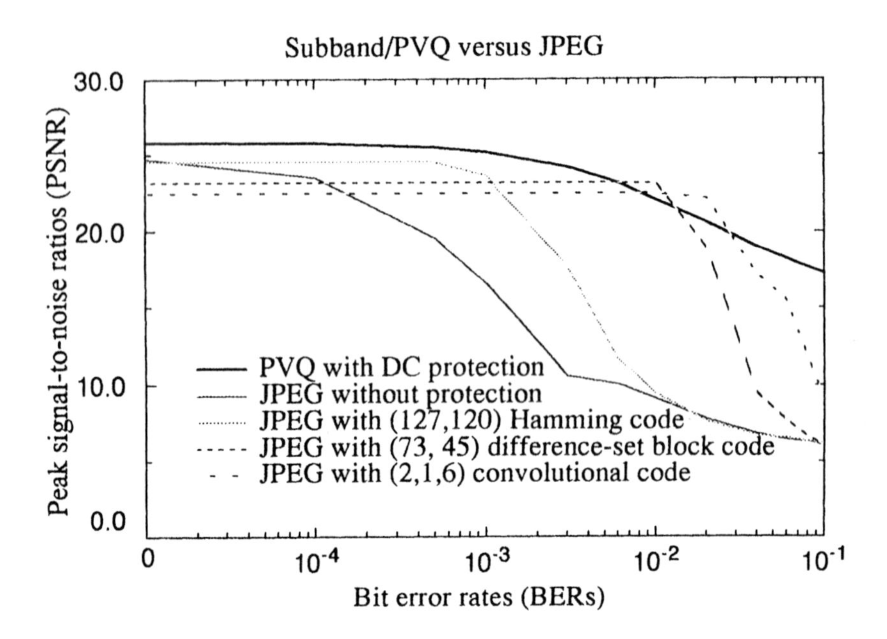

Figure 10.7 Peak signal-to-noise ratios (PSNR) versus bit error rates (BERs) for subband/PVQ and JPEG with error-correcting codes.

To show the effects of bit errors on decompressed images, we used image *Mandrill* compressed at 0.66 bit/pixel as an example. We corrupted the transmitted data at BERs of 10^{-3} and 2×10^{-2}. In order to limit error propagation in decoding the JPEG-compressed image (as a result of using variable-rate codes), we used the (73,45) block code mentioned above, reducing the effective bandwidth for the image data by approximately 40%. Figure 10.8 shows decompressed images under such channel conditions. At the BER of 10^{-3}, no noticeable artifact caused by channel error is present in the JPEG-compressed image (Figure 10.8 (c)), indicating that the error-correcting code is quite effective in correcting random bit errors at this rate. The JPEG-compressed image, however, displays lower quality than the subband/PVQ-compressed image (Figure 10.8 (a)), because 40% of the bandwidth had been used to provide for error correction. At the BER of 2×10^{-2}, the subband/PVQ-compressed image (Figure 10.8 (b)) is still recognizable with reasonable quality, though bit errors cause ringing artifacts, locally distorting the image. In the JPEG-compressed image (Figure 10.8 (d)), bit errors become very noticeable when error propagation occurs, causing total block loss.

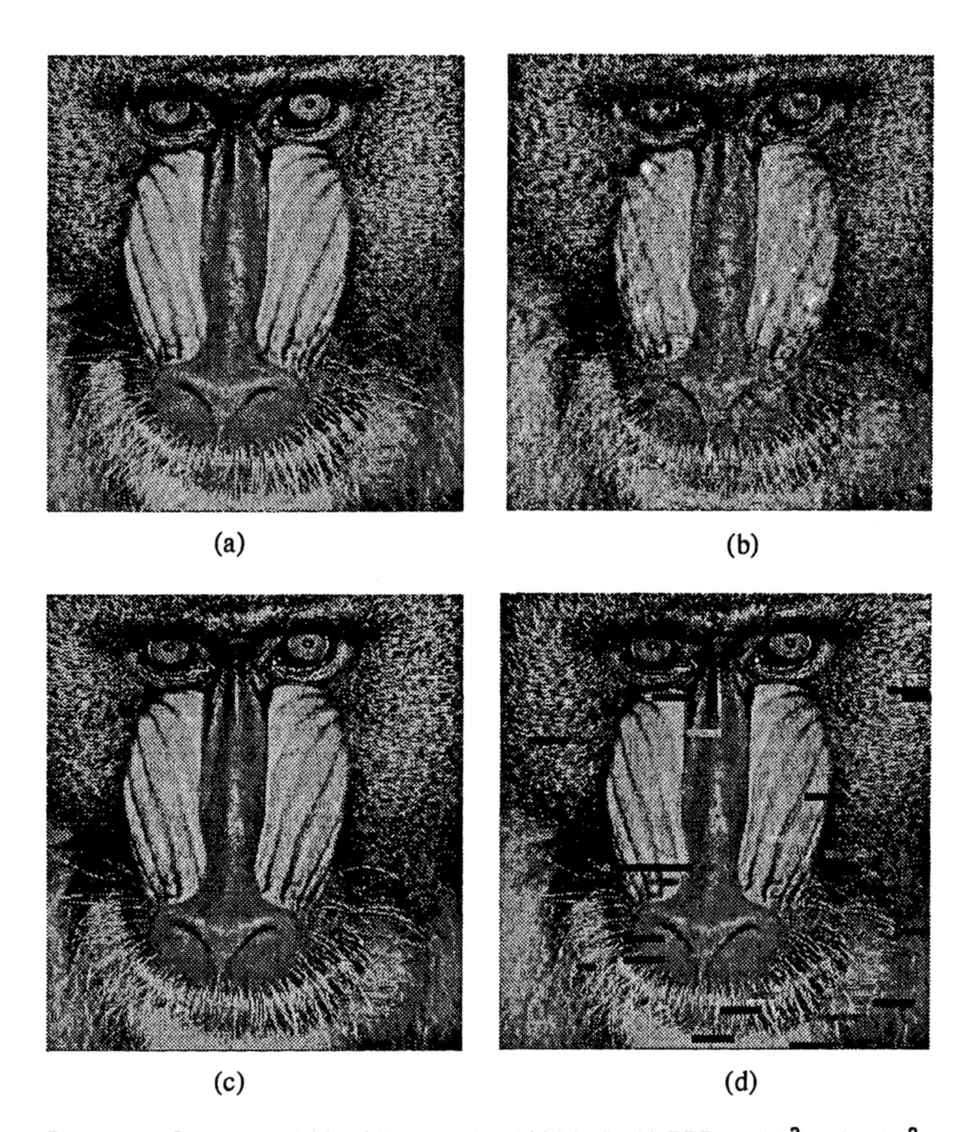

Figure 10.8 Compressed *Mandrill* at 12:1 (0.66 bit/pixel) with BERs at 10^{-3} and 2×10^{-2}: (a) and (b) were encoded using the subband/PVQ algorithm, while (c) and (d) were encoded using JPEG with the (73, 45) block code.

As for the quality of compressed *video*, when the BER approaches 10^{-3}, the random loss of blocks in JPEG-compressed video sequences becomes evident, causing continuous flickering that makes it difficult to view the video. In subband/PVQ-compressed video, increasing bit errors cause increased blurriness and wavering artifacts, but most of the video details remain distinguishable for BERs past 10^{-2}.

We attribute the error-tolerance capability of our algorithm to the following factors:

1. The fixed-rate nature of our algorithm prevents error propagation and eliminates the need for resynchronization.
2. Subband decomposition compacts most image information to the LL subband, where PCM is used to encode each pixel, limiting the effects of bit errors to a small region.
3. Error-resilient index enumeration helps improve image quality by up to 3 dB compared to random indexing under high BER conditions [6].

In summary, we demonstrate an entirely fixed-rate compression scheme based on subband/PVQ that both achieves the compression performance of the variable-rate JPEG standard and exhibits resiliency to channel error. In the next two sections, we will discuss the low-power design strategies used in implementing a prototype portable decoder based on our subband/PVQ algorithm, to be operated at a power level two orders of magnitude below existing JPEG decoders in comparable technology.

10.5. Low-Power Circuit Design Techniques

Compared with the C-Cube JPEG decoder [15], implemented in 1.2 μm COMS technology and dissipating approximately 1 W while decoding 30 frames of video per second, our subband/PVQ decoder is more than 100 times more power efficient, not accounting the power dissipated in accessing off-chip memory necessary in the JPEG decoding operation. Within this factor of 100, a factor of 10 can be easily obtained by voltage scaling of the power supply. Reduced supply voltage, however, increases circuit delay. This increase in delay needs to be compensated for by duplicating hardware, or chip area, to maintain the same real-time throughput. For example, reducing the supply voltage from 5 V to 1.5 V increases the circuit delay by more than a factor of 7 in standard CMOS design [29]. To maintain the same throughput, or real-time performance, the hardware must be duplicated 7 times, increasing the total chip area by at least the same amount. The fact that our video-rate decoder can be implemented with less than 700,000 transistors (including 300,000 transistors for on-chip memory) operating at a power-supply voltage less than 1.5 V, without requiring any off-chip memory support, indicates the efficiency of our decoding procedure. How we designed the decoder module to achieve minimal power consumption will be the focus of this and the next sections.

10.5.1. Low-Power Design Guidelines

The dynamic power consumption of a CMOS gate with a capacitive load C_{load} is given by

$$P_{dynamic} = p_f C_{load} V^2 f, \tag{10.2}$$

where f is the frequency of operation, p_f the activity factor, and V the voltage swing [30]. Reducing power consumption amounts to the reduction of one or more of these factors. Analysis in [29] indicated that for energy-efficient design, we should seek to minimize the power-delay product of the circuit, or the energy consumed per operation.

A more complete discussion on low-power CMOS circuit design is given in [31]. This subsection only briefly describes the guidelines used in designing our low-power circuit library. CMOS circuits operated at extremely low supply voltages allow a very small noise margin. A safe design is therefore of ultimate importance. The static CMOS logic style was chosen for its reliability and relatively good noise immunity. The transistors were sized in the ratio of 2:1 for PMOS and NMOS, multiplied by the number of transistors in series. A library of standard cells has been designed with these specifications in mind.

10.5.1.1. Scaling factor in a buffer chain

Well-known analysis of a multistage buffer chain indicates that the optimal scaling factor of the inverters in a buffer chain should be e (the exponential constant) [28]. This number is usually too small for practical use, however, as many stages of inverter buffers would be needed to drive a large capacitive load such as clock lines, consuming more power than necessary. Through simulations we have found that a scaling factor of 5 to 6 is optimal for low-power buffer chains operated at 1 V to 1.5 V, where the total delay would be comparable to that of a buffer chain with a scaling factor of e while consuming only two-thirds of its power.

Table 10.1 shows the energy consumed by a buffer chain at a supply voltage of 1.5 V with different scaling factors. The energy consumption of a buffer chain with a scaling factor of 6 remains a constant fraction of the energy consumed by buffer chains with a scaling factor of 4 or less over a wide range of loads, and its delay relative to the buffer chains of smaller scaling factors only increases slightly for large loads (over 5 pF). This is attributed to the fact that if we can save one or more stages in a buffer chain by increasing the scaling factor by a reasonable margin, the net result is a lower-power design with a comparable delay.

C_{load} (pF)	Scaling factor = *e*			Scaling factor = *4*		
	# of buffer stages	Energy/switch (pJ)	Avg. Delay (ns)	# of buffer stages	Energy/switch (pJ)	Avg. Delay (ns)
0.5	4	1.15	4.44	3	0.90	4.07
5	6	11.69	7.27	4	8.99	6.24
20	7	47.09	8.62	5	36.22	7.96
50	8	116.87	9.88	6	90.57	9.69

C_{load} (pF)	Scaling factor = *6*						
	# of bufferstages	Energy/switch (pJ)	Avg. Delay (ns)	Normalized to factor *e*		Normalized to factor *4*	
				Energy	Delay	Energy	Delay
0.5	2	0.75	3.83	0.66×	0.86×	0.84×	0.94×
5	3	7.66	6.24	0.66×	0.86×	0.85×	1.00×
20	4	30.67	8.63	0.66×	1.00×	0.85×	1.08×
50	5	76.64	10.11	0.66×	1.02×	0.85×	1.04×

Table 10.1 Comparison of buffer chains with different scaling factors (in 0.8 μm CMOS technology).

With a larger scaling factor, the slope of the voltage transfer function of each buffer is smaller. This may potentially increase power consumption due to short-circuit currents when both the PMOS and NMOS branches may be simultaneously on for a longer period of time. However, hSPICE simulations indicated that the factor of short-circuit currents is negligible, consuming less than 5% of the total power.

10.5.1.2. Minimizing the occurrence of spurious switching

Spurious switching, which consumes power but does not generate correct output signals, was quoted to account for from 20% to 70% of total circuit switching power [32]. Spurious switching is caused by unmatched delays such as glitches in combinational circuits and input signals not arriving at the same time. Power-optimized logic synthesis and careful circuit design can limit spurious switching to a minimum.

In designing our library cells, extra care was taken to ensure that unwanted switching is never allowed to drive a buffer chain, as the power consumed is further amplified by the large capacitive load in successive buffer stages. To prevent spurious switching from driving a buffer chain, the input signal to the first buffer is either latched or gated. The former limits the switching to occur only when the latch is on, holding the signal constant when it is off. The latter is used if a completion signal can be identified from the logic module that drives a buffer chain to gate its outputs.

10.5.1.3. Equalizing the delays between logic paths

In our design, we tried to equalize the delays of all logic paths so that there is no single critical path. The reason is that if the delays of different logic paths are not matched, some logic paths will operate faster than others. As a result, energy will be wasted in those paths with shorter delays by delivering a current larger than necessary. Since switching speed is a function of transistor sizing, which determines switching currents and capacitive loads, we can effectively "slow down" the faster paths by using smaller transistors and fewer buffer stages to reduce both driving currents and transistor capacitive loads.

10.5.2. Design Example: the Adder

We use the design of an adder to illustrate the performance trade-off between low-power and high-speed designs. The choice of an adder design is determined by the area, speed and power budget available. Carry-propagate adders are compact and useful for adding numbers of short wordlengths (fewer than 4 bits), but too slow for adding numbers of wider wordlengths. On the other hand, tree adders perform both carry and summation computations in parallel, each carry taking only $\log_2 n$ gate delays, where n is the wordlength. The area needed for a full tree adder, however, is relatively large, requiring long interconnects and large capacitive loads. Hence, tree adders only yield significant advantage for adders with more than 32 bits. As our system requires adders of a wordlength between 10 and 20 bits, the adders to be considered are carry-select and carry-lookahead adders. By paying close attention to layout and the loading on input signals that an adder introduces, we can identify the most energy-efficient adder class meeting a give speed requirement.

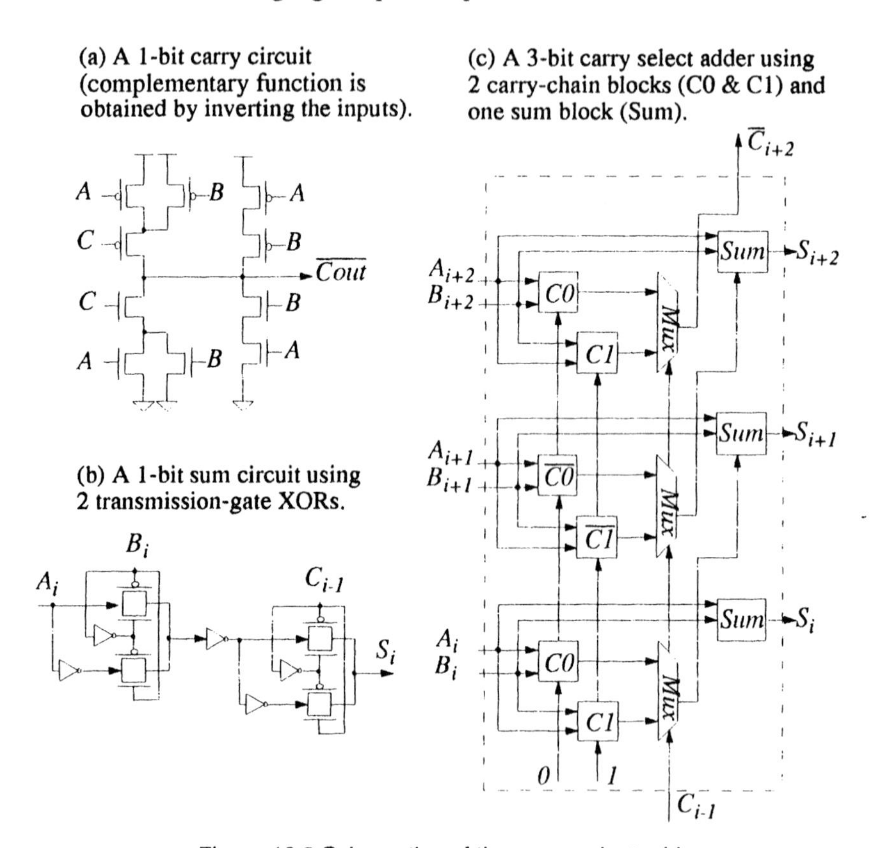

Figure 10.9 Schematics of the carry-select adder.

The adder used in our decoder design, shown in Figure 10.9, is a variant of the standard carry-select adder. Each adder block has two speed-optimized ripple-carry chains, one with a zero carry-in, the other with a one carry-in. The output carry bit is chosen by the actual carry-in to the adder block, then used in the final XOR to produce the sum bit. The transistor sizing in each adder block was designed to hide the delay of ripple-carry chains, guaranteeing that all output carry bits will be ready by the time the carry-in reaches the adder block, eliminating the probability of spurious switching mentioned earlier. By simulation, the different adder block sizes are chosen to be 2, 4, 5, 6 and 8 bits respectively. HSPICE simulations of a 17-bit adder indicated a worst case delay of 35 ns at 1.5 V, with an energy consumption of only 19 pJ per add operation.

A comparative study of various adder classes is given in Table 10.2, illustrating the types of adders compared and the sizes of their layouts in MAGIC. The performance comparisons were obtained by simulations of the extracted layouts using hSPICE in 0.8 μm CMOS technology (the BSIM model). Figure 10.10 (a) and (b) compare the delays of the sum and carry circuits respectively, while Figure 13 (c) graphs the energy consumed per add operation. Figure 10.10 (d), which shows the energy-delay product, is a measure of the efficiency of the various adders.

Name	Number of Bits	Description of Adders	Layout Area (in 0.8 μ CMOS)
Cp16	16	16-bit carry-propagate adder.	$890\lambda \times 190\lambda$
Man16	16	16-bit static Manchester carry adder, in 4-bit blocks with 4-bit carry-bypass.	$1284\lambda \times 265\lambda$
Tree16	16	16-bit tree adder (Brent-Kung).	$1292\lambda \times 545\lambda$
Cla16	16	16-bit carry-lookahead adder, in 4-bit blocks.	$1260\lambda \times 484\lambda$
Csf17	17	17-bit carry-select adder (standard), selecting sum, in 2-4-5-6 bit blocks.	$1236\lambda \times 440\lambda$
Cs17	17	Our implementation of a 17-bit carry-select adder, selecting carry, in 2-4-5-6 bit blocks.	$1236\lambda \times 332\lambda$

Table 10.2 Various adder classes in comparison.

Our carry-select adder (Cs17) differs from the traditional carry-select adder (Csf17) by selecting carry bits instead of sum bits. By eliminating one sum gate, which is equivalent to 2 transmission-gate XORs, our design consumes less power and is faster. By not computing both sums from carry 0 and carry 1, the only additional delay for the whole adder is a final XOR gate, negligible compared to the critical path of the carry-select circuit. Furthermore, removing one sum gate reduces area overhead and lowers capacitive loads on both the inputs and output carry bits, resulting in faster switching.

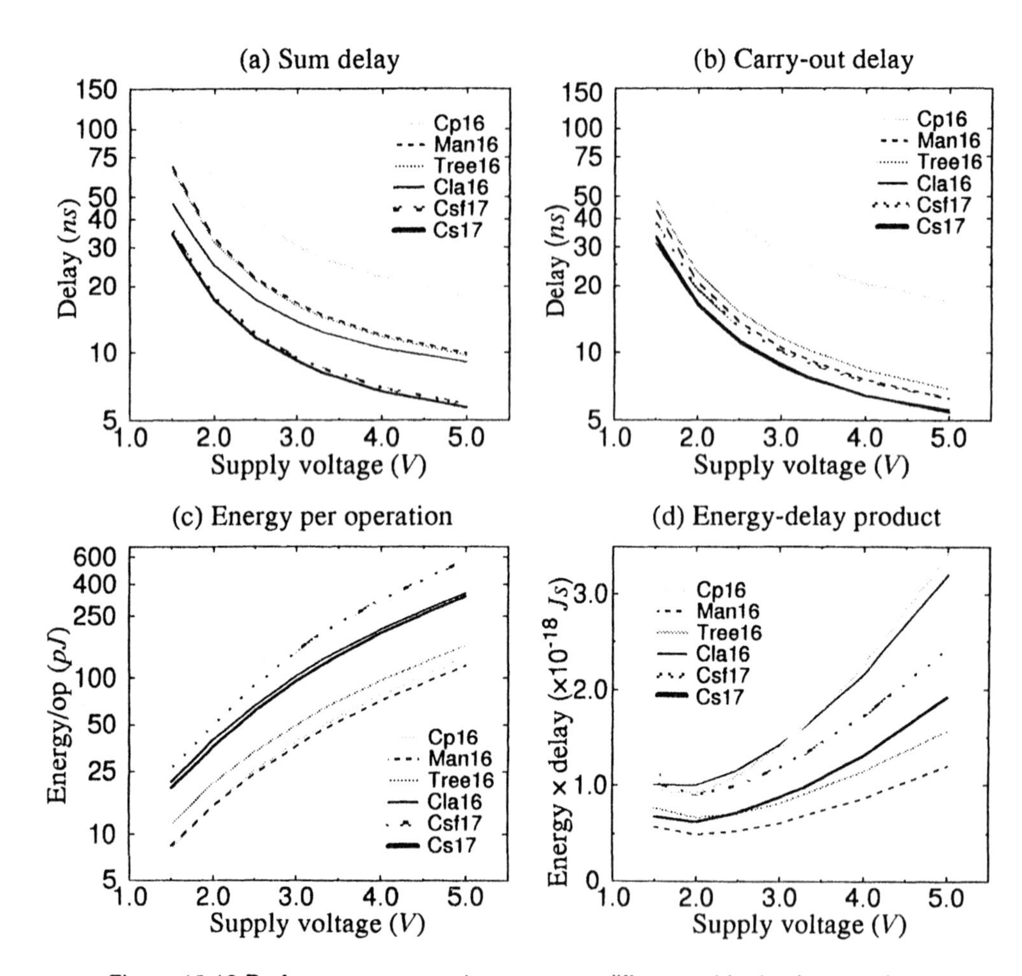

Figure 10.10 Performance comparisons among different adder implementations.

From the performance comparisons shown in Figure 10.10, it can be seen that our carry-select adder is the fastest among the various adders to compute a 16- or 17-bit sum. The carry-lookahead adder has a slightly faster carry-out delay because it is optimized for generating fast carries. Its sum delay is larger, however, and the layout is asymmetric because of the non-uniform carry function across the bits. The tree adder delivers slower performance, as its design is more suited for adding numbers of wider wordlengths.

Our design is not the most energy-efficient one. The reason for not choosing the more efficient Manchester carry-chain adder is to meet the timing constraint of a 50 ns clock cycle at 1.5 V. This choice is a trade-off between maintaining the throughput required by the system and the energy needed to achieve that throughput. A slower adder implies not only that more adders need to be implemented to meet a throughput requirement, but also that other faster logic blocks will be idle waiting for some addition to complete, resulting in unmatched delays in logic paths.

10.5.3. Low-Power Memory Design

Within a large system, the memory can potentially be the largest power-consuming element, thus requiring special attention to its design. In addition to the power dissipation, the memory speed is also important, as the memory access time may be in the critical path, determining overall system performance. The first design criterion is of course to guarantee correct circuit operation, especially with regard to the sense amps, at supply voltages down to 1 V. As with most memory design, this requires extensive modelling and simulation.

For low-power memory design, only those memory cells actually being accessed should be activated. This can be accomplished by using a local word line that activates only the bank of memory being used. The width of a memory bank should be selected to match the bandwidth requirement of the overall system, delivering multiple words per access, thus saving on access power overhead dissipated by global word line activation. Next, the power consumed in the path from memory cells to their input/output ports needs to be minimized. A self-timed circuit can be used to control the amount of voltage swing on the bit lines from memory cells, ensuring that the energy consumed in the bit lines will remain constant independent of operation frequency. Further savings can be achieved by using a small swing differential bus that connects all memory banks together [34]. This bus, which runs the entire width of the memory and connects to every memory bank, can have significant capacitance -- reducing the swing of this differential bus is vital to keeping the power consumption low. The required power consumption, however, is data dependent, as the activity of a full-swing single-ended bus will depend upon how often the data value changes.

From our decoder design, measurements on the actual memory design were taken. For 8 banks of memory each of 128×20 bits operated at a supply voltage of

1.5 V, our memory design delivers an access time of approximately 36 ns and consumes 17 pJ per bit for a write followed by a read operation.

10.5.4. Performance and Power Consumption of Library Cells

Table 10.3 summaries the hSPICE simulations of the energy consumed per operation in our cell library. The energy per operation numbers shown in Table 10.3 give us a guideline for eliminating some operations in favor of others in designing our decoder architecture. For example, a SRAM write access consumes 9 times, a chip I/O access 10 times, and a multiplication approximately 4 times more energy than a simple add operation. Given that it is possible to implement an algorithm in many different ways, the motivation is to replace memory accesses with computation, as computation dissipates less energy per operation. This trade-off will be crucial on architectural design, as will be discussed in the next section.

Module/Operation	Word Size	Energy/op (*pJ*)	Normalized to adder
Carry-select adder	16 bits	18	1
Multiplier	16 bits	64	3.6
Latch	16 bits	4	0.22
8×128×16 SRAM (Read)	16 bits	80	4.4
8×128×16 SRAM (Write)	16 bits	160	9
External I/O access[a]	16 bits	180	10

a. Estimate of the off-chip access energy is based on a total capacitance of 20 pF per pin. 180 pJ was obtained by multiplying the 720 pJ (CV2) by an activity factor of 1/4. This activity factor is based on a uniform distribution of 1's and 0's. The capacitance values were estimated from [33].

Table 10.3 Energy per operation at a 1.5 *V* supply in 0.8 μ*m* CMOS technology.

The relatively large energy dissipated by off-chip I/O accesses is the motivation for minimizing external data communication. In our design, we eliminated external frame buffers and placed memory on-chip whenever possible. As high-quality video implies high pixel bandwidth, with at least millions of pixel operations per second, external frame buffers lead to huge energy waste and overwhelm all other low-power improvements made. Requiring no off-chip memory support is one of the main factors enabling our portable decoder to deliver high-quality video at a minimal power level.

10.6. Low-Power Decoder Architectures

The largest savings in power can be achieved through careful algorithm and architectural design. In this section, we will discuss the architectural trade-offs made in the design of our decoder chip set to reduce overall power consumption, in addition to the power savings already obtained through voltage scaling and low-power circuit design.

Figure 10.11 shows the block diagram of our portable video decoder. The subband/PVQ-encoded bit stream is transmitted through a wireless link with a bandwidth of 500 Kbps to 1 Mbps. The received bit stream is decoded by the PVQ decoder chip into subband coefficients. The subband decoder chip then filters these coefficients and reconstructs them into image pixels each represented by three digital RGB color components. The PVQ decoder chip and the subband decoder chip form our decoder chip set.

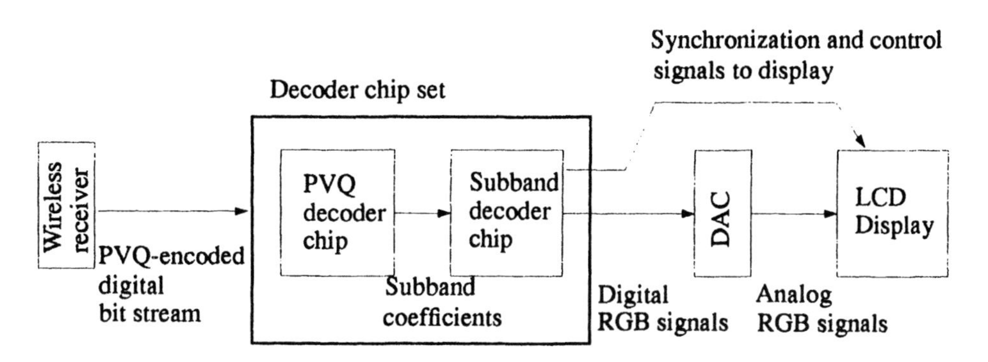

Figure 10.11 System organization of the portable decoder chip set.

The digital RGB color components are converted to an analog format by a D/A converter (DAC) suitable for a color LCD display. The video synchronization signals to control the LCD display are also generated by the subband decoder chip. For global synchronization, the entire portable video decoder runs on a common clock, with the subband decoder chip operated at half the rate of the PVQ decoder chip.

10.6.1. Low-Power VLSI Architecture for the Subband Decoder

In designing the subband decoder chip, we emphasized a low-power implementation without introducing noticeable degradation in decompressed video quality. As memory accessing is by far the most power-consuming operation, the main design strategy has been to eliminate memory accesses in favor of on-chip computation. Table 10.4 lists the major architectural decisions made for the implementation of a low-power subband decoder.

Technique	Description	Benefit
Filter choice	Use a 4-tap filter with coefficients (3,6,2,-1).	3× power saving vs. using a 9- tap filter.
External accesses	Reconstructed LL subband coefficients are kept on-chip, requiring no frame buffer. High-frequency subband coefficients are zero run-length encoded.	3× reduction in the number of external memory accesses.
Zero processing	Skip processing of zeros in high-frequency subbands.	1.5× reduction in processing cycles.
Word size	Rounding to 10 bits gives the same quantization error as truncation to 12 bits (from simulation).	1.2× reduction in memory and datapath width.
Internal memory	Interleave subband levels so that only one line of horiz. filter results and two lines of vert. filter results are stored.	2× reduction in the internal memory size.
Color conversion	Simplify coefficients to allow a multiplier-free implementation.	Over 5× reduction in converter area and power.

Table 10.4 Architecture strategies in designing a low-power subband decoder.

10.6.1.1. Filter selection and subband decomposition structure

One important design parameter for subband decoding is the subband filter that determines the numbers of multiplies and memory accesses required. To achieve low power consumption, the filter coefficients in our portable video-on-demand prototype were determined by extensive simulation, both to minimize the computation energy required and to deliver adequate compression performance. The coefficients [3, 6, 2, -1] for the low-pass filter and coefficients [1, 2, -6, 3] for the high-pass filter were chosen as a result [35]. The simple 4-tap filter requires only a single carry-save adder followed by a carry-select adder to implement one up-sampling and filtering operation. This reduces the energy and hardware by a factor of 3 when compared with an efficient shift-and-add implementation of a 9-tap QMF. If a full multiplier/accumulate implementation were used, the power saving would be another 20 times greater. Furthermore, the number of internal memory accesses and the amount of line-delay buffering required by a 4-tap filter are much less than those for a longer filter. For example, compared with a 4-tap filter which requires only 2 data samples from a line-delay buffer holding only one line of horizontally filtered data, a 9-tap filter would require 5 data samples from a line-delay buffer holding 5 lines of data. The on-chip implementation of the line-delay buffer for a 9-tap filter would be a major problem given the required pixel resolution.

The disadvantage of using a filter with a smaller number of coefficients is its lower compression performance due to a slower frequency roll-off. The asymmetric 4-tap filter, however, performs nearly as well as a symmetric filter of twice the length but with reduced memory requirements as mentioned above. Also, the 4-tap filter can easily handle exact reproduction at image edges which, when compared with the performance of approximated horizontal edges often used for longer filters, compensates for the overall coding loss. The decrease in coding performance is negligible, especially considering the large reduction in power, complexity, and size of the chip.

The subband decomposition structure used in our prototype decoder consists of four levels of subband decomposition into 13 different frequency bands, as shown in Figure 10.3. The compression performance achievable with 13 frequency bands is slightly less than that of using 31 bands. We made this compromise, however, to lower the total decoding power consumption, as a 31-band subband/PVQ decoder would require external memory given the available technology. The image quality delivered by our prototype decoder is still comparable to that of JPEG-compressed images [5].

10.6.1.2. External accesses

External data and memory accesses represent one major energy-consuming operation that need to be minimized. Intermediate reconstructed LL subband data are therefore stored internally and not written off chip. This reduces the number of external reads per pixel from 1.98 to 1.5 and the number of external writes per pixel from 3.48 to 3. Also, as the design does not require a frame buffer, an additional 3 external accesses per pixel are saved.

To reduce the communication bandwidth between the PVQ decoder and the subband decoder chips, zero run-length encoding is used to take advantage of the large number of zeros in the three highest-frequency subbands. In these high-frequency subbands, the PVQ uses a 32-dimensional horizontal vector which, when decoded, will contain only 3 non-zero coefficients. The PVQ decoder transmits to the subband decoder the values of non-zero coefficients and a zero length to the next non-zero coefficient. Consequently power is saved by reducing the size of the PVQ output buffer and the number of external accesses. By accepting zero lengths between non-zero coefficients, the number of external reads per pixel for the subband decoder chip is further reduced from 1.5 to 0.57. Without considering external access reductions, the total power of the subband chip would have been 40% more, not counting the power dissipated in accessing large external memory.

10.6.1.3. Power-down cycles

Because the output of the subband decoder chip directly controls the display device, long pauses during the horizontal and vertical synchronization periods required by the display operation may cause the output buffer to overflow. To prevent the output buffer from overflowing, the chip stalls and the clock is suppressed except in the video controller, thus saving global clock power dissipation. Additionally, run-length coding of zero coefficients in high-frequency subbands allows the subband decoder to skip processing of zero coefficients until a non-zero coefficient is present in the input stream. This effectively reduces the number of processing cycles per pixel from 1.98 to 1.33 cycles per pixel, saving clock power and reducing the number of arithmetic operations and internal memory accesses accordingly.

Application of the above power saving techniques, however, tremendously increases the level of complexity in the decoder's controller design, which consists of four finite-state machines controlling zero run-length decoding, idle-cycle stalls, levels of subband decomposition, and generating video synchronization signals to the display device.

10.6.1.4. Data word size

In digital signal processing design, the word size of internal data representations is an important parameter determining the accuracy of the final output. As video signals at a SNR higher than 46 dB usually appear nearly perfect, a word size of 12 bits is usually used. Our simulations indicated that, when using rounding instead of truncation for quantizing internal results, a word size two bits smaller will achieve a similar output SNR. This allows a word size of only 10 bits in our internal data representation, reducing the power and area of all datapath circuits, including the internal memory which uses the most silicon area. This 10-bit rounding strategy provides a power reduction of nearly 20% over the 12-bit one and a power reduction of 40% over the standard 16-bit approach with no perceptive loss in video quality.

10.6.1.5. Dataflow

The datapath of the subband operation is shown in Figure 10.12. The 10-bit input data from the four subbands of a given level are read horizontally and stored in a two-word input buffer. The data are then run through the high- and low-pass filters, combined, and stored in a line-delay buffer. The line-delay buffer outputs are filtered vertically producing reconstructed image pixels which are stored in the intermediate or the final result buffer.

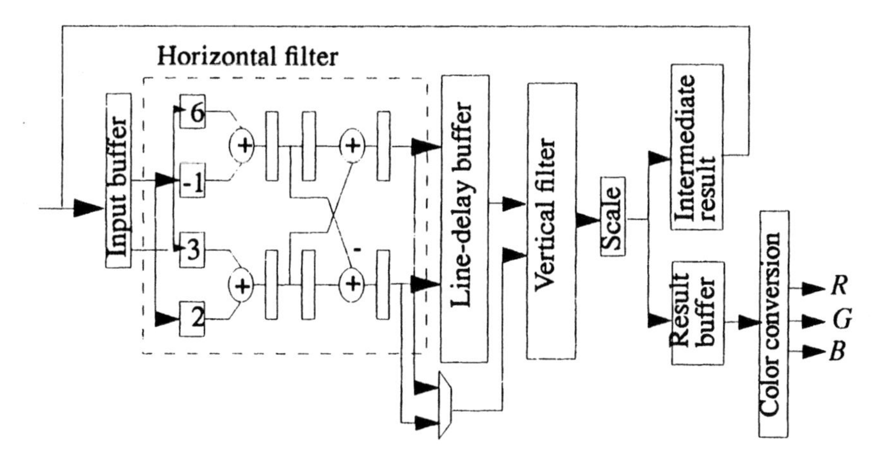

Figure 10.12 Subband decoder architectural organization

The datapath organization is derived from the natural flow of the filtering processing. Because the input data needs to be up-sampled before filtering, the even and odd samples are processed concurrently, implicitly implementing the up-sampling operation and generating two filtered samples per clock cycle. To save power, short interconnects are used between processing elements to replace a large high-capacitive, high-frequency global bus. Finally, all resources are fully utilized to achieve a 100% hardware efficiency. In our design, the datapath consumes approximately 30% of the total chip power, dissipating 0.34 mW out of the total 1.2 mW in processing 1.27 million pixels per second at a 1 V supply voltage.

10.6.1.6. Internal memory

The data ordering across the four levels of subbands is designed to minimize the amount of internal memory required, reducing both chip area and power. One line of data from a given level is processed producing two lines of outputs for the next level. At the next level, instead of processing both lines to produce four lines of outputs for yet another level, only one line is initially used thereby requiring only two lines of buffering for the result buffer and one line of buffering for the line-delay buffer. Processing of the second line is delayed until after the first line's outputs have been processed. This results in an interleaved sequencing of subband levels, but needs only half the memory space that would have been required if interleaving were not used.

The final output buffer consists of four horizontal scan lines to prevent buffer underflow. Since the output data is buffered and arranged in a raster scan order, it can be sent directly to a DAC which feeds the decoded analog signal to a display device without the need for an external frame buffer.

10.6.1.7. Color conversion

Another consideration is the power consumed by color conversion. Because human visual system is less sensitive to quantization error caused by course quantization of chrominance components (color) than of luminance components (intensity), a more efficient color representation than the standard RGB representation is to decompose the RGB components to the luminance component Y and the chrominance components U and V. Our compression scheme uses the YUV color representation in compression/decompression and therefore the subband decoder has to perform YUV to RGB color conversion before sending the data to a color display. A brute force approach requires 4 multiplies and 6 adds per pixel which would consume a significant amount of power. Because quantization errors introduced in the YUV to RGB color conversion are usually not visually perceptible, the conversion coefficients can be rounded to only two bits. This provides an efficient implementation requiring only 5 carry-select adds and 1 carry-save add per pixel per RGB component. Our color conversion circuit consumes only 90 mW at a 1 V supply for three components at 1.27 Mpixels/sec.

10.6.1.8. Edge handling

A final concern is the handling of image edges and different image resolutions. For our 4-tap filter, the first coefficient from a line is saved and reused with the last coefficient of the line to correctly reproduce the edges. This does introduce a stall cycle into the pipeline which is handled by the control logic. To decode higher-resolution images, multiple chips are used each processing a vertical slice of up to 256 pixels wide. To prevent artificial lines appearing down the center of an image, our design allows the subband chips that handle the middle slices of an image to access two additional coefficients from the next slice instead of using coefficients in the same slice, thus realizing a display-resolution independent design.

10.6.1.9. Total power consumption

The subband decoder chip, implemented in 0.8μm CMOS technology, occupies a chip area of 9.5 × 8.7 mm^2 (with an active area of 44 mm^2) and contains 415K transistors [36]. At a 1 V supply voltage, the chip runs at a maximum frequency of 4 MHz and delivers one output pixel of three color components at every two clock cycles. To satisfy the throughput requirement of 1.27 Mpixels/sec (176 × 240 pixels/frame at 30 frames/sec), the subband decoder chip only needs to run at 3.2 MHz. This fairly low operating frequency is an indication of the efficiency of our compression algorithm. Even for video of the standard SIF format (352 × 240 pixels/frame at 30 frames/sec), the subband decoder chip only needs to run at 6.4 MHz, while most JPEG decoding chips are required to run at a frequency between 30 MHz and 50 MHz.

Figure 10.13 illustrates the power dissipation of the subband decoder chip at 3.2 MHz over various supply voltages. At our target supply voltage of 1 V, the subband decoder chip consumes only 1.2 mW to perform real-time video-rate decoding. At a slightly higher supply voltage, for example at 1.5 V, the subband decoder chip would consume 2.64 mW.

Figure 10.14 plots the power consumption versus maximum operating frequency of the subband decoder chip at various supply voltages. The performance at 3.5 V achieves a 90 MHz clock frequency, generating 45 Mpixels/sec of three RGB components and dissipating 0.38 W.

Table 10.5 displays the breakdown of total power consumption of the subband decoder chip among the different sections of the chip, where the operations included are arithmetic computation, clocking, internal memory accesses, and external accesses. The power consumed on global clock lines is only a small fraction of the total power, but when combined with the power consumed on local clock lines in the datapath and control latches, the total clock power represents a modest percentage (approximately 25%) of the total power. Power consumption

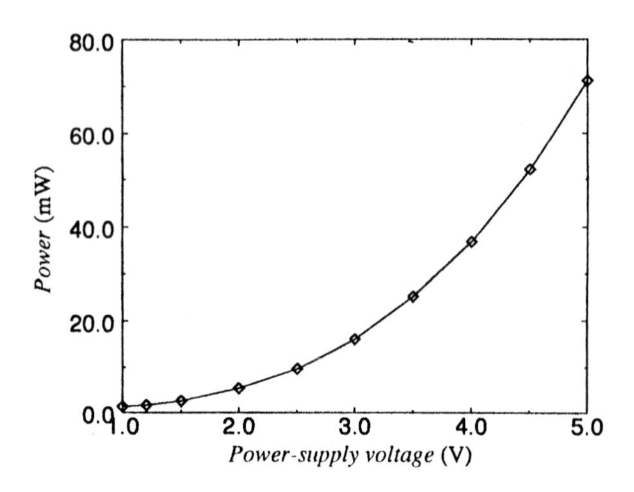

Figure 10.13 Total power consumption of the subband decoder chip at 3.2 *MHz* over various supply voltages.

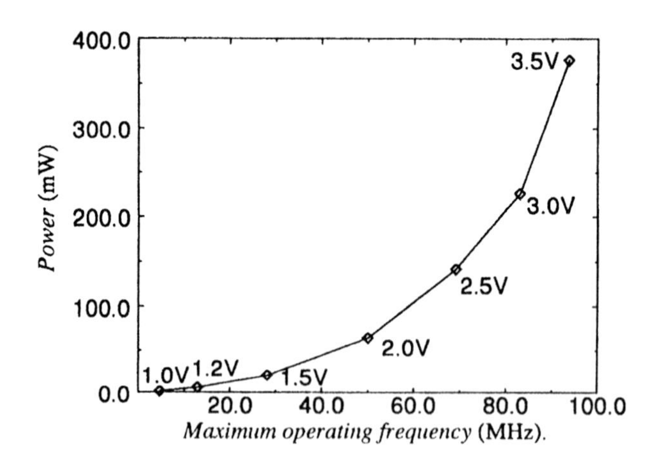

Figure 10.14 Power consumption and the maximum operating frequency of the subband decoder chip at various supply voltages.

in the control section remains a small percentage (approximately 15%) despite the increased complexity required to implement the power saving strategies applied to the memory and datapaths. This justifies the decision to concentrate the power reduction efforts on computation, memory, and external accesses.

Operation	Mops/sec	Energy/op (*pJ*)	Estimated Power (*mW*)	Measured Power (*mW*)
Total datapath			0.35	0.34
Carry-select add (16 bits)	17.7	7	0.12	
3-2 add (16 bits)	25	2	0.05	
Latching (16 bits) (including clock)	100	1.8	0.18	
Global clock	2.5	22	0.06	
Control	2.5	53	0.13	0.09
Internal memory			0.26	0.39
Internal read (16 bits)	2.4	36	0.09	
Internal write (16 bits)	2.4	71	0.17	
External access (16 bits)	2.7	80	0.22	0.38
Total			1.02 *mW*	1.20 *mW*

Table 10.5 Power breakdown of the subband decoder chip by operations at a 1 *V* supply (3.2 *MHz*).

This decoder chip was designed using a modular approach and therefore supports higher- resolution images as well. Table 10.6 illustrates the power dissipated at the required clock frequency when used for decompressing high-resolution images. The operating voltages were determined by the real-time computation requirements. For higher-resolution images, multiple chips would be cascaded, each operating on a slice of the image of up to 256 pixels wide. This additional parallelism keeps the operating frequency and thus the supply voltage low, resulting in extremely low power dissipation even for HDTV applications.

Format	Size	No. of chips	Clock freq. (*MHz*)	Power (*mW*)
Sharp LCD	176x240x30	1	3.2	1.2 (1 *V*)
CD-I (TV)	352x240x30	2	3.2	2.4 (1 *V*)
CCIR video	704x480x30	3	8.4	12.6 (1.2 *V*)
HDTV	1920x1035x30	8	19	122 (1.5 *V*)
Hires monitor	1024x768x75	4	37	192 (2.0 *V*)

Table 10.6 Power consumption of the subband decoder for various display resolutions.

10.6.2. Low-Power VLSI Architecture for the PVQ Decoder

The second custom chip designed for the portable video decoder is the PVQ decoder chip. It accepts an input bit stream encoded in a pre-arranged order, parses the bit stream into PVQ indices, decodes these indices into a vector, scales the vector, and stores the scaled vector elements in an output buffer for subsequent subband decoding. The overall block diagram in Figure 10.15 shows the general dataflow of the chip.

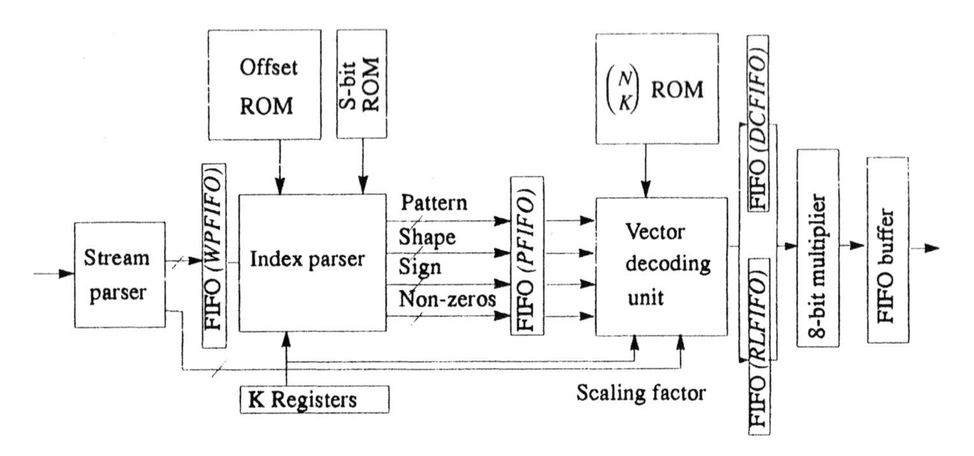

Figure 10.15 Block diagram of the PVQ decoder chip.

The PVQ decoder is divided into four main processing elements: (1) the stream parser, (2) the index parser, which parses a PVQ index into four intermediate fields that characterize the index's four properties (as described in Section 4.1), (3) the vector decoding unit, which translates the four properties to individual vector elements, and (4) a 6x8-bit pipelined multiplier, which performs the final scaling of each vector element. The bulk of the computations occur in the index parser and vector decoding unit, which mainly perform compares, subtracts, and shifts in a 16-bit datapath.

In addition to operating at a low supply voltage, the PVQ decoder chip employs several key architectural strategies to minimize its power consumption as listed in Table 10.7.

Technique	Description	Benefit
Independent processing elements (PE's)	Each PE operates independently, distributing processing cycles to match the non-deterministic nature of the decoding process.	2× reduction in total power by using gated clocks.
Fast block search in vector decoding	Fast block search (4 at a time) reduces processing cycles in the critical-path vector decoding unit.	2× reduction in processing cycles and the size of the output buffer.
Minimized external accesses	All combinatorial and indexing data, required for the decoding process, are stored on-chip.	2× reduction in the number of external accesses.
Run-length encoding	Decoder outputs of the high-frequency subbands are zero run-length encoded to compress the output bit stream.	10× reduction in the number of internal buffer accesses. 3× reduction in the size of output buffer storage.
Data reordering	Interleaved data between subband levels trades off additional control complexity for reduced overall system memory requirements.	Eliminates the off-chip frame buffer.

Table 10.7 Architecture strategies in designing a low-power PVQ decoder.

10.6.2.1. Independent processing elements

The PVQ decoding algorithm is non-deterministic in nature, i.e., the number of processing steps to decode a PVQ index depends on the vector data. As a result, the latency in the decoding process (the index parser and vector decoding unit) is non-deterministic. To reduce power consumption, each of the four processing elements operates independently, with its own local control, and enters a power-conserving stand-by mode when not being used. Each processing element is separated by FIFO's and only continues processing when either its input FIFO is not empty or its output FIFO is not full.

To minimize power in the stand-by mode, the global clock to each processing element is gated. Capacitance measurements indicate that more than half of the chip's total clock capacitance lies in gated clocks. With the exception of the vector decoding unit, whose activity rate is typically 90%, the other processing elements have a typical activity rate ranging from 40% to 50%. These relatively low activity rates lead to savings in total clock power dissipation by roughly a factor of 2.

10.6.2.2. Block search in the vector decoding unit

For typical image data, the vector decoding unit constitutes the critical path of the chip. Increasing the throughput of this unit directly increases chip throughput, reduces processing cycles, and reduces the amount of output buffering required to meet real-time throughput requirements. Direct implementation of the PVQ algorithm would require a linear search to locate correct index offset. For example, for $K = 30$, the average number of iterations required by a linear search will be 15 for uniformly distributed indices, which is unnecessarily slow and inefficient. Improved throughput was achieved by processing a block of 4 vectors at a time. This scheme significantly reduces the search time (speedup by a factor of 5 in the example above), and on the average results in a factor of 2 reduction in the number of processing cycles and the amount of output buffering.

10.6.2.3. External accesses and zero run-length encoding

Combinatorial calculations and index offsets, required for the vector decoding unit, were stored on-chip in low-power ROM's. In addition, parsing information for subband bit allocation was also stored in this ROM. This scheme essentially trades off greater programmability for less external accesses to off-chip memory.

As described in the previous subsection, zero run-length encoding of high-frequency subband coefficients reduces the amount of external accesses by a factor of 3. More significantly for the PVQ decoder chip, because zero run-length encoding compresses the representation of consecutive zero coefficients, it also reduces the amount of output buffering on the PVQ chip by a factor of 3 and the number of internal buffer accesses by up to a factor of 10. Here again, we traded off additional control to perform zero run-length encoding to lower I/O power and on-chip data buffering.

10.6.2.4. Data reordering

Interleaving data between subband levels, in fact, increased the control complexity of the PVQ decoder chip by about 15%, since each processing element needs to keep track of its location in the reordering scheme and set the appropriate PVQ parameters (e.g. the bit rate and vector dimension). However, this trade-off was made to eliminate the overall system requirement for an off-chip frame buffer.

10.6.2.5. Throughput

The PVQ decoder chip meets the throughput requirement of real-time video decoding at 176×240 pixels/frame at 30 frames/sec. From simulations of buffer occupancy of different image sequences, the worst-case sizes of various FIFO buffers between processing elements were found, each of a different size between 4 and 192. The different FIFO sizes indicate the relative speeds of production and consumption of data between different processing elements. By ensuring that all buffers are never full nor empty while there is incoming data to the PVQ decoder, the system is guaranteed to operate at maximum performance. From this performance figure, we calculated the maximum delay allowed to meet the video-rate throughput requirement at the lowest power consumption. A clock frequency of 6.4 MHz was found to be adequate to satisfy this requirement.

10.6.2.6. Total power consumption

The PVQ decoder chip occupies a chip area of 9.7×13 mm^2 (with an active area of 74.9 mm^2). Because the PVQ decoder chip will operate at a frequency twice of that of the subband decoder chip, to meet the video-rate throughput at 1.27 Mpixels/sec, the PVQ decoder chip needs to run at 6.4 MHz, powered by a 1.5 V supply. At this rate, the total power consumption of the PVQ decoder chip was measured to be 6.63 mW. The breakdown of the total chip power is shown in Table 10.8, where the operations included are arithmetic computation, control, FIFOs, clocking, internal ROM accesses, and external accesses.

Operations	Mops/sec	Energy/op (pJ)	Estimated Power (mW)	Measured Power (mW)
Total datapath:			0.88	
CSA (16 bits)	3.1	16.1	0.05	
Shift (16 bits)	1.5	8.3	0.01	
Multiply (8×6 bits)	0.9	26.0	0.02	
Registers (16 bits) (including clock)	52.3	15.3	0.80	
Control	0.56	177.1	0.99	

Operations	Mops/sec	Energy/op (pJ)	Estimated Power (mW)	Measured Power (mW)
Total FIFOs:			1.75	
Local clock	6.4	153.3	0.98	
Registers	10.1	15.3	0.16	
Control	10.1	60.4	0.61	
Clock			0.94	
Global clock	6.4	107.4	0.68	
Global half-clock	3.2	21.3	0.07	
Clock buffer	4.4	44.3	0.19	
ROM access (16 bits)	2.0	98.0	0.20	
External access (16 bits)	0.94	180.0	0.17	
Total			4.94 *mW*	6.63 *mW*

Table 10.8 Power breakdown of the PVQ decoder chip at a 1.5 V supply (6.4 *Mhz*).

The PVQ decoder consumes roughly three times the power of the subband decoder chip. The doubled clock frequency of the PVQ decoder explains much of this difference. A third of the chip power is consumed by the register FIFO storage with approximately 55% of this power dissipated on local clock lines. Using a memory-based design instead of the register-based FIFO storage would have helped reduce this power as well as area. It is also interesting to note that the datapath power is roughly equal to the control power. There are several reasons for this: (1) the datapaths use local gated clocks, while the control sections do not -- a significant factor in that the chip may be idle up to 40% of the time, and (2) the control sections require greater complexity due to the non-deterministic nature of the PVQ decoding algorithm and the data reordering scheme.

Because of the relatively large design, extensive global clock routing results in significant power consumption. Overall the total power consumed by clocking, including global clocking and local clocking in the control, datapaths, and FIFOS, makes up approximately 50% of the chip power. Finally, the power consumed in external accesses is a relatively small fraction, because the design limits off-chip accesses to only compressed data, at both the input and output.

The lower power estimate compared with the measured data is primarily due to the DC power consumed in the ROM's, which is not accounted for in the dynamic power simulation. Although the ROM's consume no power when not in use, the static design consumes both DC and AC power when the ROM's are being accessed. Using a pre-charged ROM design would have eliminated this additional power component.

10.6.3. Portable Video-on-Demand Prototype

The complete video-on-demand system consists of a portable decoder with a color LCD display that decodes and displays compressed video sequences, a radio transmitter and receiver, and an encoding base station implemented by a DSP multiprocessor board. The wireless data transmission is provided by three pairs of direct-sequence spread spectrum radio transceivers manufactured by Proxim, delivering a raw data rate at 720 Kbits/sec. The decoding chip set on the portable decoder receives compressed video data and decompresses them to RGB color components, which are then converted to analog signals for the color display. The display is a 4" color thin-film-transistor active matrix display with a resolution of 160 pixels by 234 lines.

As shown in Figure 10.16, the video-on-demand system accepts video data from two sources, a video server with a compressed-video database and an NTSC camera. Video sequences stored in the video database are pre-compressed using our PVQ/subband encoding algorithm. The NTSC camera is connected to a real-time encoder to allow for live video sources such as a camcorder as an input device. The encoder consists of an NTSC decoder and multiple TMSC40s on a DSP multiprocessor board, which implements the PVQ/subband encoding algorithm in real time. As the compressed bit rate ranges from 0.5 Mbits/sec to 1 Mbits/sec, a simple bus interface between the video server (a SUN workstation) and the radio transmitter has been built to support this rate of data transfer. Framing information necessary for synchronization at the decoder end is also added in at this interface.

In the portable decoder the digital RGB color components output from the subband decoder chip must be converted to analog signals to be displayed. Our design is a current-switching DAC, based on a current-mirror configuration with a bias current of 5 mA. The transistors in the current mirror were sized for digital inputs at extremely low voltages, delivering a resolution of 6 bits per pixel, an adequate resolution for the 4" color display. The DAC generates analog current outputs that can be converted to the required dynamic range through gain-control op-amps. The maximum frequency of the DAC runs up to 10 MHz at a supply

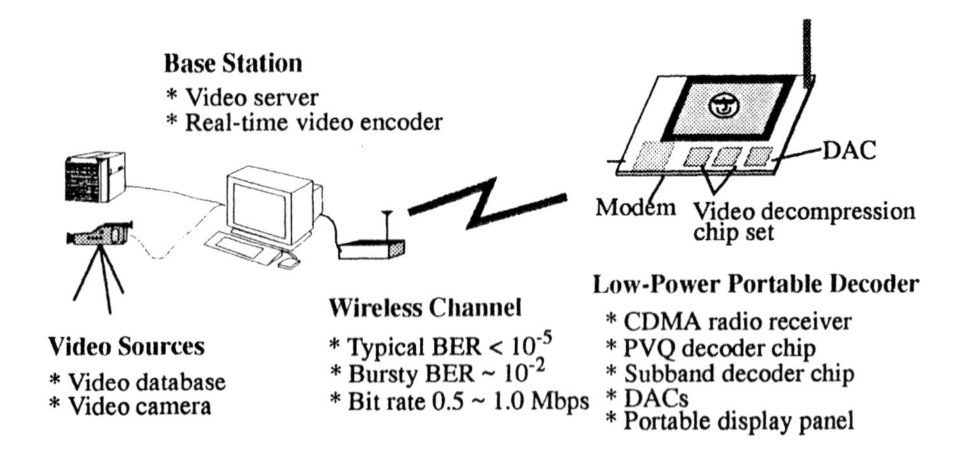

Figure 10.16 The portable video-on-demand system — Block diagram.

voltage of 1.5 V and each conversion operation consumes on average an energy of 42 pJ per color component.

10.7. Summary

The design of low-power electronics systems, especially portable systems, requires a vertical integration of the design process at all levels, from algorithm development, system architecture, to circuit layout. System performance needs not be sacrificed for lower power consumption, if the design of algorithms and hardware can be considered concurrently.

In our portable video decoder, more than a factor of 100 reduction in power consumption is demonstrated. We achieved this high level of power efficiency without degrading performance in video quality. In addition, by embedding error resiliency into our system development consideration, we were able to avoid the hardware complexity and bandwidth penalty incurred by the use of error-correcting codes, a procedure often considered necessary for wireless communication.

From designing this portable video-on-demand system, we learned that power reduction can be best attained through algorithm and architecture innovations, guided by the knowledge of underlying hardware and circuit properties. This hardware-driven algorithm design approach is key to the design of future portable systems under stringent power budgets. Without appropriate high-level synthesis tools, this vertical integration task is by no means trivial. The designer is required to be well-informed at all levels of the design process, willing to compromise and make sensible trade-offs in order to reach a globally optimal solution.

References

[1] W. B. Pennebaker and J. L. Mitchell, *JPEG Still Image Compression Standard,* Van Nostrand Reinhold, New York, 1993.

[2] D. LeGall, "MPEG: A video compression standard for multimedia applications," *Communications of the ACM,* vol. 34(4), pp. 37-48.

[3] M. Liou, "Overview of the Px64 Kbits/sec video coding standard," *Communications of the ACM,* vol. 34(4), pp. 60-63.

[4] N. Farvardin, "A study of vector quantization for noisy channels," *IEEE Trans. Inform. Theory,* vol. 36, pp. 799-809, July 1990.

[5] E. K. Tsern, A. C. Hung, T. H. Meng, "Video compression for portable communication using pyramid vector quantization of subband coefficients," Proc. *1993 IEEE Workshop on VLSI Signal Processing,* October 1993.

[6] A. C. Hung and T. H. Meng, "Error-resilient pyramid vector quantization for image compression," *Proc. International Conference on Image Compression,* Nov. 1994.

[7] K. C. Pohlmann, *The Compact Disc: a Handbook of Theory and Use,* A-R Editions, Madison, Wisconsin, 1989.

[8] N. Ahmed, T. Natarajan, K. R. Rao, "Discrete cosine transform," *IEEE Trans. on Computers,* vol. 23, pp. 90-93, January 1974.

[9] K. R. Rao and P. Yip, *Discrete Cosine Transform: Algorithms, Advantages, Applications,* Academic Press, Boston, 1990.

[10] D. A. Huffman, "A method for the construction of minimum-redundancy codes," *Proc. IRE,* vol. 40, pp. 1098-1101, September 1952.

[11] W. C. Lee, *Mobile Cellular Telecommunication Systems,* McGraw Hill, New York, 1989.

[12] J. Walker, Mobile Information Systems, Artech House, Boston, 1990.

[13] T. R. Fischer, "A pyramid vector quantizer," *IEEE Trans. Inform. Theory,* vol. 32, pp. 568-583, July 1986.

[14] E. K. Tsern and T. H. Meng, "Image coding using pyramid vector quantization of subband coefficients," *Proc. of IEEE ICASSP 1994,* pp. V-601-604, April 1994.

[15] S. Purcell, "C-Cube CL550 image processor," *Proc. HOT Chip Symposium,* 1990.

[16] J. W. Woods, editor, *Subband Image Coding,* Kluwer Academic Publishers, Boston, 1991.

[17] M. Barlaud, P. Sole, M. Antonini, and P. Mathieu, "A pyramidal scheme for lattice vector quantization of wavelet transform coefficients applied to image coding," *Proc. IEEE ICASSP 1992,* pp. 401-404, March 1992.

[18] S. K. Rao, M. Hatamian, et. al, "A real-time P¥64/MPEG video encoder chip," *Digest of Technical Papers. IEEE International Solid-State Circuits Conference*, February 1993.

[19] D. Taubman and A. Zakhor, "Multi-rate 3-D subband coding of video," *IEEE Trans. on Image Processing*, September 1994.

[20] T. Senoo and B. Girod, "Vector quantization for entropy coding of image subbands," *IEEE Trans. on Image Processing*, pp. 526-533, October 1992.

[21] A. Segall, "Bit allocation and encoding for vector sources," *IEEE Trans. on Information Theory*, Vol. 22, pp.162-169, March 1976.

[22] A. K. Jain, "Image data compression, a review," Proceedings of the IEEE, Vol. 69, pp. 349-389, March 1981.

[23] N. Chaddha and T. H. Meng, "Psycho-visual based distortion measure for monochrome image compression," *Proc. SPIE Visual Communications and Image Processing*, pp. 1680-1690, November 1993.

[24] H. C. Tseng and T. R. Fisher, "Transform and hybrid transform/DPCM coding of images using pyramid vector quantization," *IEEE Trans. on Comm.*, vol. 35, pp. 79-86, January 1987.

[25] A. Gersho and R. M. Gray, *Vector Quantization and Signal Compression*, Kluwer Academic Publishers, Boston, 1992.

[26] R. C. Reininger and J. D. Gibson, "Distributions of the two-dimensional DCT coefficients for images," *IEEE Trans. on Comm.*, Vol. COM-31, pp. 835, June 1983.

[27] G. C. Clark, Jr., and J. Bibb Cain. *Error-Correction Coding for Digital Communications*, Plenum Press, New York, 1981.

[28] S. Lin and D. J. Costello Jr., *Error Control Coding: Fundamentals and Applications*, Prentice Hall, Englewood Cliffs, 1983.

[29] A. Chandrakasan, S. Sheng, and R. W. Brodersen, "Low power CMOS digital design," *IEEE Journal of Solid-State Circuits*, pp. 685-691, April 1992.

[30] N. Weste and K. Eshragian, *Principles of CMOS VLSI Design: A Systems Perspective*, 2nd Ed., Addison-Wesley, MA, 1992.

[31] A. P. Chandrakasan and R. W. Brodersen, "Minimizing power consumption in digital CMOS circuits," *IEEE Proceedings*, April, 1995.

[32] A. Shen, A. Ghosh, S. Devadas, and K. Keutzer, "On average power dissipation and random pattern testability of CMOS combination logic networks," *IEEE ICCAD Digest of Technical Papers*, pp. 402-407, November 1992.

[33] H. B. Bakoglu, Circuits, *Interconnections, and Packaging for VLSI*, Addison-Wesley, 1990, pp. 423-425.

[34] B. Amrutur and M. Horowitz, "Techniques to reduce power in fast wide memories", *Proc. 1994 Symposium on Low-Power Electronics*, October 1994.

[35] B. M. Gordon and T. H. Meng, "A low power subband video decoder architecture," *Proc. IEEE ICASP 1994*, pp. II-409-412, April 1994.

[36] B. M. Gordon, T. H. Meng, and N. Chaddha, "A 1.2 mW video-rate 2D color subband decoder," *Digest of Technical Papers, 1995 IEEE International Solid-State Circuits Conference*, February 1995.

11

Algorithm and Architectural Level Methodologies for Low Power

Renu Mehra, David B. Lidsky, Arthur Abnous, Paul E. Landman and Jan M. Rabaey

11.1. Introduction

With ever increasing integration levels, power has become a critical design parameter. Consequently, a lot of effort has gone into achieving lower dissipation at all levels of the design process. It has been demonstrated by several researchers that algorithm and architecture level design decisions can have a dramatic impact on power consumption [25][27]. However, design automation techniques at this level of abstraction have received scant attention. In this chapter we explore some of the known synthesis, optimization and estimation techniques applicable at the algorithm and architectural levels. The techniques mentioned in this chapter are targeted for DSP applications but can readily be adapted for more general applications.

Two examples — a vector quantizer encoder and an FIR filter — are used throughout the chapter to illustrate how the methodologies may be applied. While the former is evaluated and optimized for ASIC design, the latter is targeted for a programmable processor.

11.2. Design Flow

A design environment oriented towards power minimization must embody optimization and estimation tools at all levels of the design flow. A top-down approach, annotated with examples of associated tools, is illustrated in Figure 11.1.

The most effective design decisions derive from choosing and optimizing algorithms at the highest levels. However, implementation details cannot be accurately modeled or estimated at this level of abstraction so relative metrics must be judiciously used in making design selections. More information is available at the architectural level, hence estimates are more accurate and the effectiveness of optimizations can be more accurately quantified.

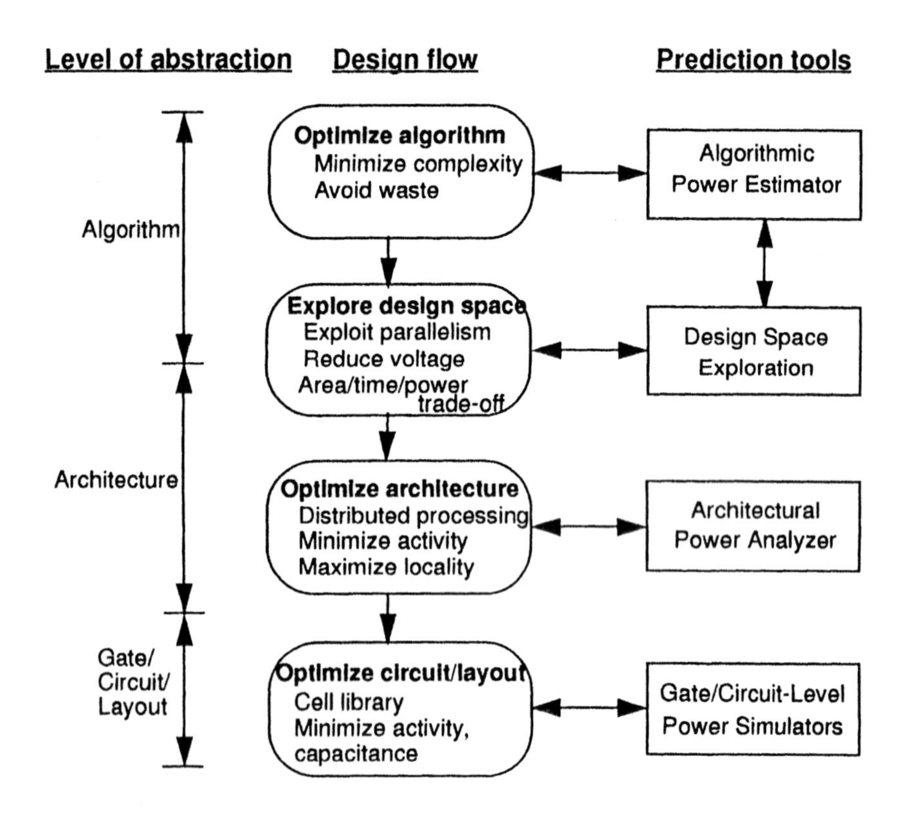

Figure 11.1 Design flow and some supporting tools for proposed low-power CAD framework

Design Example 1: Vector Quantization, Introduction

Throughout this chapter we will be using the design of a video vector quantizer to illustrate design flow at the different levels. Vector Quantization (VQ) is a data compression method used in voice recognition and video systems. This example implements 16-to-1 video compression. For further details on this example, readers are referred to [14].

In this approach to video quantization, an image is broken up into a sequence of 4×4 pixel images (Figure 11.2). Each pixel is represented by an 8-bit word indicating luminance. The 4×4 image, therefore, can be thought of as a vector of 16 words, each 8 bits in length. Each of these vectors are compared with a previously generated codebook of — in this case — 256 different vectors. This codebook is generated a priori with the intention of covering enough of the vector space to give a good representation of all probable vectors. The creation of the codebook is studied in greater detail in [5]. After compression, an 8 bit word is generated delineating the address of a codevector that approximates the original 4×4 vector image the best. This corresponds to a compression ratio of 16:1, since 16 8-bit words are now represented as a single word. Our design is directed toward a 240x128 pixel grey scale display. Processing the standard 30 frames/sec moving picture necessitates that one 4x4 pixel-vector be compressed every 17.3 µs.

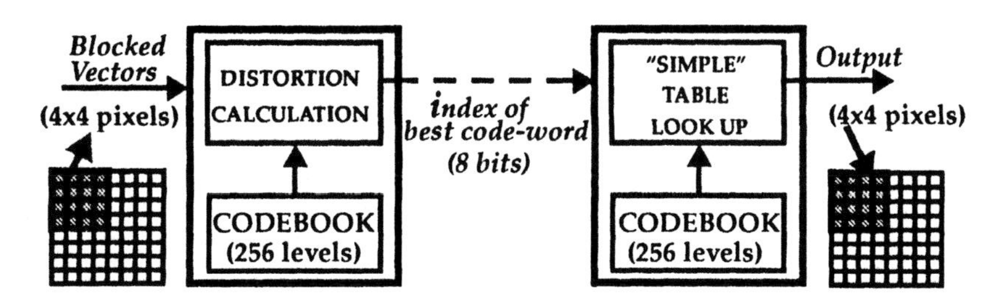

Figure 11.2 Vector Quantization, Encoding and Decoding

Design Example 2: FIR Filter, Introduction

The second example that will be used throughout this chapter is a 14 tap, low-pass Finite Impulse Response (FIR) filter. The algorithm will be optimized for targeted architectures and various implementations of the filter --using dedicated and programmable hardware-- will be analyzed in terms of their power consumption characteristics.

11.3. Algorithm level: Analysis and Optimization

Given a design specification, a designer is faced with several different choices such as selecting a particular algorithm, selecting and ordering transformations to be applied on it and determining various parameters such as clock period and supply voltage. The multi-dimensional space offers a large range of possible trade-off's, hence the need for fast prediction mechanisms. Section 11.3.1 explores estimation at the algorithm level, the techniques available and their limitations. In section 11.3.2, algorithm level approaches to power optimization are described.

11.3.1. Estimation (Analysis)

The sources of power consumption on a CMOS chip can be classified as dynamic power, short circuit currents and leakage. At the algorithm level, it makes sense to only consider the dynamic power. The contributions of short circuit currents and leakage are mostly determined at the circuit level and are only marginally effected by algorithm-level decisions. Using smart circuit techniques, we can reduce the dissipation due to short circuit currents to less than 15% of the entire power [28]. Also, lower operating voltages as advocated in this chapter tend to reduce the short circuit component. Hence, the power dissipated can be described by the following well-known equation for dynamic power:

$$Power = C_{eff}V^2f \tag{11.1}$$

where f is the frequency of operations, V is the supply voltage and C_{eff} is the effective capacitance switched. C_{eff} combines two factors — C, the physical capacitance being charged/discharged, and α the corresponding switching probability:

$$C_{eff} = \alpha C \tag{11.2}$$

Since algorithm-level power predictions are to be used for guiding design decisions and not for making any absolute claims about the power consumption, it is important that these predictions give accurate *relative* information. Since algorithm level decisions can result in orders of magnitude improvement in performance, the accuracy requirements on the tools can be relaxed, while still providing meaningful power predictions to guide high level decisions.

For the purpose of estimation, we can divide power dissipation into two components: *algorithm-inherent dissipation* and *implementation overhead*. The algorithm-inherent dissipation comprises the power of the execution units and memory. This is "inherent" in the sense that it is necessary to achieve the basic functionality of the algorithm, and cannot be avoided irrespective of the implementation. On the other hand, the implementation overhead includes control, interconnect and registers. The power consumed by this component depends largely on the choice of architecture/implementation.

Estimating the algorithm-inherent dissipation — The algorithm-inherent dissipation refers to the power consumed by the execution units and memory. This component is fundamental to a given algorithm and is the prime factor for comparisons between different algorithms as well as for quantifying the effect of algorithm level design decisions. Its dissipation can be estimated by *a weighted sum of the number of operations* in the algorithm. The weights used for the different operations must reflect the respective capacitances switched. Library-based approaches can use a high level model of the average capacitance switched by a hardware modules to determine the weighting factor [16]. In [26], a measurement based technique is proposed to determine the weighting factor for each operation in the Intel-486 instruction set.

To estimate the capacitance switched at this level it is necessary to ignore the statistics of the applied data and to employ a white-noise model, i.e. assume random data. Switching statistics are strongly influenced by the hardware architecture. The mapping of operators onto hardware resources effects the temporal correlations between signals. This information is not available until the architecture is finalized. Under this assumption, an algorithm with five additions consumes the same power (in the adders) irrespective of the amount of hardware sharing. Note that hardware sharing affects the power consumption units only in that it changes the switching activities at the inputs. It also introduces the extra overhead of multiplexors and tri-state buffers. In our power model, the latter factors are part of the implementation overhead.

For data-dependent algorithms such as certain sorting algorithms, the operation count and the number of memory accesses are input-dependent. These algorithms require advance profiling with typical input vectors to provide either average or worst case estimates.

Estimating the implementation overhead — The implementation overhead consists of the control, interconnect and implementation related memory/register power. The power consumed by these components depends on the specific architecture platform chosen and on the mapping of the algorithm onto the hardware. Since this overhead is not essential to the basic functionality of a given algorithm, several estimation tools ignore its effect for algorithm level comparisons [21]. However the power consumed by these components is often comparable if not greater than the algorithm-inherent dissipation. It is clear, therefore, that it important to get reasonable estimates of the implementation overhead for realistic comparisons between algorithms and to guide high level decisions. This is a formidable task without a complete architecture description. Fortunately, it is possible to produce first-order predictions of the overhead component given some properties of both the algorithm and the targeted hardware platform or architecture.

In fact, it has been determined that the implementation-dependent power consumption is strongly correlated to of a number of properties that a given algorithm may have [22]. An example of a structural property of an algorithm that may affect the amount of this overhead is the *locality of reference*. In terms of algorithm structure, *spatial locality* refers to the extent to which an algorithm can be partitioned into natural clusters based on connectivity and *temporal locality* refers to the average lifetimes of variables. A spatially-local algorithm renders itself more easily to efficient partitioning on hardware, allowing highly capacitive global buses to be used sparingly. A temporally-local algorithm tends to require less temporary storage and have small register files leading to lower capacitances. In terms of memory/register access, spatial locality refers to distance between the addresses of items referenced close together in time and temporal locality refers to the probability of future accesses to items referenced in the recent past. A spatially-local memory access pattern allows partitioning of memory into smaller blocks that require less power per access. The effect of structural properties on power is examined in [22].

Given a targeted hardware platform and a number of algorithm properties, techniques can be developed for early prediction of the implementation overhead.

Consider the interconnect power in an custom ASIC implementation. It has been established that in general the average length and, hence, the physical capacitance of the busses is proportional to the predicted die area [24]. In its turn, the active area is a function of algorithmic parameters such as the number of operations to be performed and their concurrency pattern. The switching activity can be derived from the number of bus accesses, which is proportional to the number of edges in the computational graph of the algorithm. The value of the weighting coefficients in the model can be determined from statistical studies. Such a study was performed in [16] for a custom ASIC hardware implementation platform based on the HYPER synthesis tool. Figure 11.3 demonstrates that such models can achieve reasonable accuracy levels. An important side-benefit of this approach is that it helps to understand the underlying dependencies and relationships of the various contributions to the power budget.

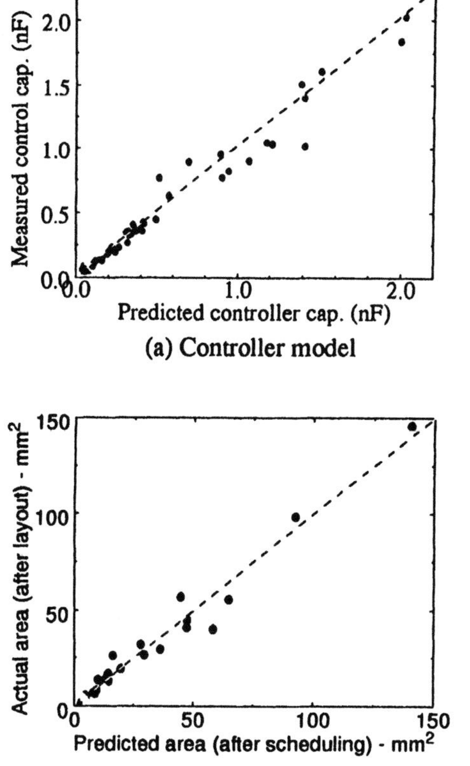

Figure 11.3 Controller and interconnect models: high-level prediction vs. chip measurements

Another method that has been used successfully is to solve a simplified version of the actual synthesis problem by relaxing some implementation constraints in order to obtain fast estimates [21]. The amount of relaxation/simplification determines the accuracy and speed of the estimation process.

Design Example 1: Vector Quantization, Algorithmic Estimation

We have established that at this high level of design description, there is no means to accurately estimate absolute power. However by using such metrics as operation count and first-order estimates of the critical path, design decisions can be made. To illustrate this level of estimation we use the most straightforward method of coding the vector, a full search through the entire codebook (FSVQ) combined with the standard Mean Square Error (MSE) distortion measure given by Eq. (11.3).

$$MSE = \sum_{i=0}^{15} (C_i - X_i)^2 \qquad (11.3)$$

where C is the codebook codevector, X is the original 4x4 vector representation, and i is the index of the individual pixel word.

The computational complexity per vector can be quantified by enumerating executions (e.g. memory accesses, multiplications, additions, etc.) required to search the codebook. This gives a reasonable first order approximation of relative power consumption. Computing the MSE between two vectors requires 16 memory accesses, 16 subtractions, 16 multiplies and 16 additions. In FSVQ, this is done for each of the 256 vectors in the codebook, and each of these vectors are compared with the leading MSE candidate at the time.

	Memory Access	Multiplications	Add/Subtract
Full Search	4096	4096	8448

Table 11.1 Full Search Vector Quantization Operation Count

Algorithm-inherent dissipation - Operation count can now be used to estimate the switching capacitance inherent to the algorithm if the targeted hardware library is known. Using black box capacitance models of the hardware [11], and making assumptions on the bit-widths of each operator, a first order estimate of capacitance can be made.

	Capacitance/ Operation	Operation Count/ Look-up	Capacitance Look-up
Memory Access	88.8 pF	4096	363.7 nF
Multiplications	16.2 pF	4096	65.91 nF
Add/Subtract	608 fF	8448	5.14 nF

Table 11.2 Full Search Vector Quantization Algorithmic Inherent Dissipation

Knowing that memory accesses and multiplications are power hungry, the first-order analysis produces the insight that these are the functions most in need of optimization. This picture can be refined by introducing some architectural constraints. Assume for instance that a single one-ported memory is used to store the codebook. This sets the maximum concurrency of the memory accesses to one or, equivalently, imposes a sequential execution of the algorithm. To meet the real time constraint of 17.3 μsec/block, this means that the memory access time has to be smaller than 4.2 nsec. For a power efficient implementation, it is obvious that either a more complex memory architecture, or a revised algorithm is necessary.

Design Example 2: FIR Filter, Algorithmic Exploration

We can go a step beyond a mere estimation of operation count if low-level design information (e.g., power consumption of the different modules in the hardware library) is available. Consider a direct-form structure (Figure 11.4) of the FIR filter and assume a throughput constraint of 3.125 MHz. As the voltage is reduced, it is necessary to choose faster hardware to meet the required time con-

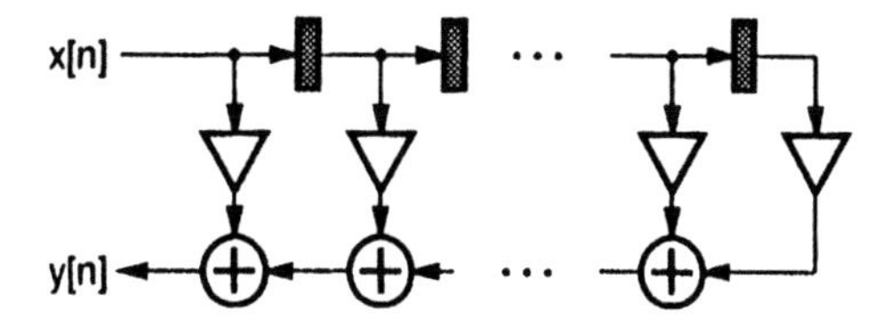

Figure 11.4 Direct-form structure of FIR filter

straint. Availability of power and area estimates of different modules in the cell-library allows us to evaluate power and area trade-off's involved in using different library cells. Though the ripple adder dissipates less power than the carry select adder (CSA), it fails to meet the required throughput below 5 V whereas the CSA continues to meet the throughput requirement down to 3 V. Table 11.3 summarizes the energy and area estimates obtained using the techniques described in Section 11.3.1. Using the carry select adder reduces the power consumption to a third of its original value with minimal area penalty.

Voltage (V)	Energy (nJ)	Area (mm^2)	Modules selected
5.00	24.00	16.75	ripple adder, carry save multiplier
3.00	8.51	16.86	carry select adder, carry save multiplier

Table 11.3 Area/energy predictions for direct form FIR filter.

Having looked at techniques of power estimation, we must now see how such a facility would be useful in an automated design environment. The primary importance of early prediction lies in the design guidance that it can provide. Therefore, estimation tools must be integrated with design space exploration and optimization tools in such a way as to provide an easy-to-use environment for the designer.

As design automation moves to higher levels of abstraction a large number of options are available that cannot be exhaustively explored manually and cannot be effectively automated into a single design tool. Design space exploration provides an interactive environment giving quick feedback to the designer about the effect of design choices on specified performance metrics and allows the user to make intelligent decisions. In [18], [1] and [16], several different case studies have shown that such design space exploration is useful in providing guidance for the selection of algorithms, as a cost function for transformations, and for hardware selection resulting in large power savings. In the next section, the FIR filter example illustrates how such an exploration tool may be used.

11.3.2. Power Minimization Techniques at the Algorithm Level

After examining methods for estimating power consumption at the algorithm level, the next logical step is to examine power minimization techniques at this level. We will start by mentioning some of the general approaches for power minimization and then look at specific techniques that can be used for minimization of both the algorithm-inherent dissipation and the implementation overhead.

The recurring theme in low power design at all levels of abstraction is *voltage reduction*. At the algorithm level, functional pipelining, retiming, algebraic transformations and loop transformations can be used to increase speed and allow lower voltages. Be aware that these approaches often translate into larger silicon area implementations, hence the approach has been termed *trading area for power* [1]. Estimation and exploration tools help us decide how much we can drop the voltage while still meeting the required performance constraints, as well as the associated area penalty.

Another technique for low power design is *avoiding wasteful activity*. At the algorithm level, the *size* and *complexity* of a given algorithm (e.g. operation counts, word lengths) determine the activity. If there are several algorithms for a given task, the one with least number of operations is generally preferable. This became clearly apparent in [13], where various filter topologies implementing an Avenhaus filter were compared with respect to power.

Reducing the algorithm-inherent dissipation — Important transformations in this category include operation and strength reduction [2]. *Operation reduction* includes common sub-expression elimination, algebraic transformations (e.g. reverse distributivity), dead code elimination. *Strength reduction* refers to replacing energy consuming operations by a combination of simpler operations. The most common in this category is expansion of multiplications by constants into shift and add operations. Though this transformation typically results in lower power, it may sometimes have the opposite effect if it results in an increase in critical path. Another drawback is that it introduces extra overhead in the form of registers and control. For transformations such as this one whose effect is not always obvious, estimation and exploration tools are indispensable in evaluating the effects. Another important component of the algorithm-inherent dissipation is the memory power. Algorithm level transformations for minimizing the power consumption of memories are presented in [3]. These include conversion of background memory to foreground register files and reduction of memory size using loop reordering and loop merging transformations.

Minimization of the implementation overhead — is a more challenging problem. In section 11.3.1, it was explained how certain algorithms have potentially less overhead than others as they possess certain structural properties such as locality and regularity. For selection of algorithms, therefore, we must be able to detect these properties. Optimizations on the algorithm level should enhance and preserve them. *Spatial locality* can be detected and used to guide partitioning. Regular algorithms typically require less control and interconnect overhead. Though there has been some effort to detect and quantify useful properties of algorithms [6],[7], transformations for enhancing these properties are largely unexplored.

One other way to reduce the implementation overhead is to reduce the chip area, as this typically translates into reduced bus capacitances. In [2], the authors demonstrate how retiming can be used to increase resource utilization by distributing operations more uniformly. Other area-reducing transformations can also be used for this purpose as well. Be aware however that a larger area can help to reduce the algorithm-inherent power. This shows again that power minimization is a subtle trade-off process. The availability of early estimation and exploration techniques is clearly indispensable.

Design Example 1: Vector Quantization, Algorithmic Optimization

Continuing with this design example, the properties of operation count and critical path will be used to aid in the choice and optimization of algorithms. Using an estimate of algorithm-inherent dissipation of Table 11.1, memory access has been identified as the main hurdle to achieving a low-power design. To achieve significant power savings, other algorithms are investigated.

Tree Search Vector Quantization (TSVQ) — Tree Search Vector Quantizer (TSVQ) encoding [5] requires far less computation. TSVQ performs a binary search of the vector space instead of a full search. As a result, the computational complexity is proportional to $\log_2 N$ rather than N, where N is the number of vectors in the codebook. Figure 11.5 diagrams the structure of the tree search. At each level of the tree, the input vector is compared with two codebook entries. If at level 1, for example, the input vector is closer to the left entry, then the right branch of the tree is analyzed no further and an index bit 0 is transmitted. This

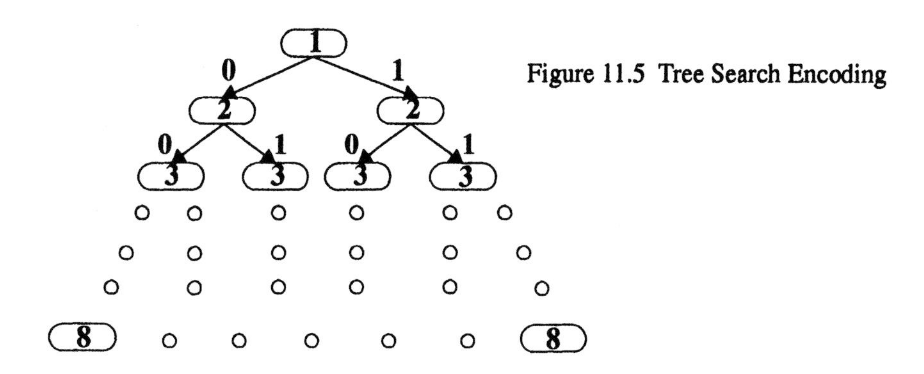

Figure 11.5 Tree Search Encoding

process is repeated until a leaf of the tree is reached. Hence only $2*\log_2(256) = 16$ distortion comparisons have to be made, compared to 256 distortion calculations in the FSVQ.

Mathematical Optimizations — In TSVQ, there is a large computational reduction available by mathematically rearranging the computation of the difference between the input vector X, and two codevectors C_a and C_b [4], originally given by Eq. (11.4).

$$MSE_{ab} = \sum_{i=0}^{15} (C_{ai} - X_i)^2 - \sum_{i=0}^{15} (C_{bi} - X_i)^2 \tag{11.4}$$

Since a given node in the comparison tree always compares the same two codevectors, the calculation of the errors can be combined under one summation. With the quadratics expanded, this yields

$$MSE_{ab} = \sum_{i=0}^{15} C_{ai}^2 - 2X_iC_{ai} + X_i^2 - \left(C_{bi}^2 - 2X_iC_{bi} + X_i^2\right) \tag{11.5}$$

which can be simplified and regrouped into Eq. (11.6).

$$MSE_{ab} = \sum_{i=0}^{15} \left(C_{ai}^2 - C_{bi}^2\right) + \sum_{i=0}^{15} 2X_i(C_{bi} - C_{ai}) \tag{11.6}$$

The first summation can be precomputed once the codebook is known and stored in a single memory location. The quantities $2(C_{bi} - C_{ai})$ may also be calculated and pre-stored.

Therefore, at each level of the tree the number of multiplications is reduced from 32 to 16, while memory accesses and add/subtracts go from 32 to 17, and from 33 to 17 respectively. The impact of the algorithm selection and the mathematical transformations is summarized in Table 11.4 for a 256 vector codebook. Observe that a side-effect of the transformations is that each memory word must be extended from 8 to 9 bits to preserve accuracy. Even though this results in an increased memory size (and capacitance per access), the substantial reduction in the number of memory accesses still translates into a sizable power reduction.

	Memory Access	Multiplications	Add/Subtract
Full Search	4096	4096	8448
Tree Search	256	256	520
Optimized Tree	136	128	136

Table 11.4 VQ Operation Count Summary

	Memory Access	Multiplications	Add/Subtract
Full Search	363.7 nF	66.3 nF	5.1 nF
Tree Search	22.7 nF	4.1 nF	155 pF
Optimized Tree	12.1 nF	2.1 nF	82.7 pF

Table 11.5 VQ Algorithmic Inherent Dissipation

Inherent dissipation values must be interpreted carefully. Since the timing requirements of the Full Search is much more stringent than for the Tree Search, the supply voltage and frequency will probably make the comparison even more favorable to a tree search. The two tree searches, however, have roughly the same timing requirements so a direct comparison of the capacitance numbers should be indicative of their final performance. The Optimized Search, therefore, is chosen as the favorable algorithm.

Design Example 2: FIR Filter, Algorithmic Optimization

The algorithmic transformations described in this section represent one of the most powerful and widely applicable class of optimization techniques. We revisit our FIR example to demonstrate their advantages. As mentioned before, the throughput required is 3.125 MHz. The direct form has 12 additions and 1 multiplication in the critical path and cannot meet the throughput constraint below 3 V for the given hardware library. We use retiming to reduce the critical

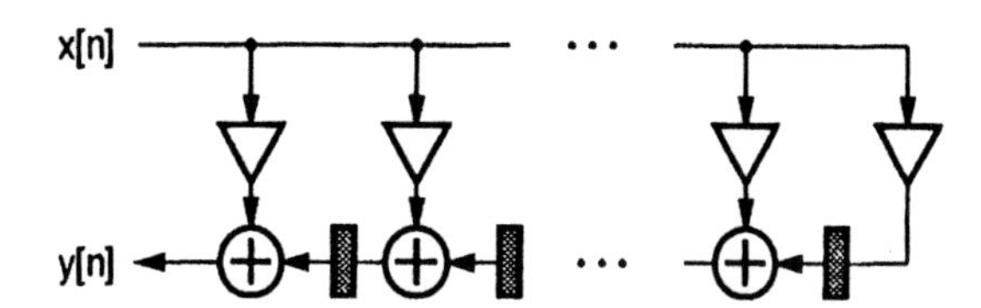

Figure 11.6 Structure of the retimed version of the FIR filter.

path. Figure 11.6 shows the structure of the retimed version. The critical path is now reduced to only 1 multiplication and 1 addition operation. This allow for a reduction in supply voltage below 3 V while maintaining the same throughput. The area-energy trade-off's for both the versions with the variation of the supply voltage, as generated by the algorithmic estimation tools, are shown in Figure11.7. The retimed version allows the voltage to be reduced to 1.5 V, thus reducing the power consumption drastically. However, the area penalty may be

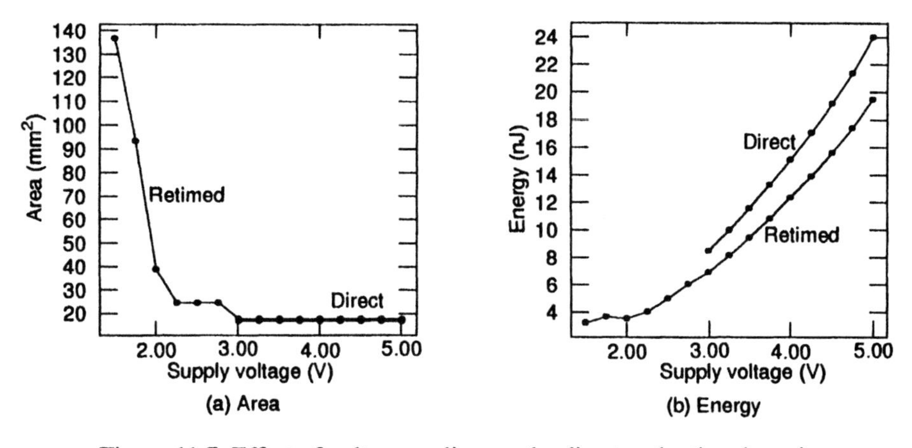

Figure 11.7 Effect of voltage scaling on the direct and retimed versions.

prohibitive. The designer can choose the voltage that best suited simultaneously taking into account the area, throughput and energy.

Design space exploration curves such as these clearly show the advantages of one design over another and also illustrate trade-off's. Similar exploration is possible over other design parameters such as the clock period.

11.4. Architecture level: Estimation and Synthesis

The lack of implementation-specific information limits the accuracy of estimation tools at the algorithmic level. More accurate results can be obtained at the architectural level once the datapath, memory, control and interconnect structures are fully defined. Section 11.4.1 describes some of the state-of-the-art architectural estimation techniques. Optimization at this level of abstraction means finding the best architectural composition for the given algorithm. The process of mapping a given algorithm onto a particular architecture style is called architecture synthesis. Section 11.4.2 discusses power optimizations possible during architecture synthesis targeted for an ASIC architecture style.

11.4.1. Architectural Estimation/Analysis

Estimating power at the architectural level can be more accurate for two reasons:

- More precise information can be obtained regarding the signal statistics, hence yielding more accurate models for the hardware operators and modules.
- The implementation overhead is now precisely defined in terms of controllers, memories and busses and can thus be estimated more accurately.

While obviously not of the same precision as power estimation at the circuit, switch or logic levels, architectural analysis has the advantage of being orders of magnitudes more efficient in terms of computation time. For complex circuits, this approach might be the only feasible one. Furthermore, it can be performed BEFORE the logical or physical design is completed, hence allowing for early design decisions.

Reliable power analysis at the module level requires two important entities: capacitance models for hardware modules and activity models for data or control signals.

Capacitance models for hardware modules — The capacitance of an RTL level module such as an adder, multiplier or memory can be expressed as a function of the complexity parameters of that module [19],[10],[15]. For example, the switching capacitance of a multiplier is proportional to the square of its input wordlength. The latter is a common complexity parameter for most modules. Some modules may require more parameters. For example, the capacitance model for a logarithmic shifter is given by

$$C_T = C_0N + C_1L + C_2NL + C_3N^2L + C_4MNL + C_5SNL \quad (11.7)$$

where N is the wordlength, S and M are the actual and maximum shift values, while $L = \lceil log_2(M+1) \rceil$ represents the number of shift stages [11].

Activity models for data signals: The average dissipation of a module is a strong function of the applied signals. Enumerating the capacitance model over all possible input patterns is clearly non-feasible. The Power Factor Approximation (PFA) technique [19],[20] uses a heuristically or experimentally determined weighting factor, called the *power factor,* to model the average power consumed by the given module over a range of designs. This approach works fine if the applied data resembles random white noise, but produces large errors for correlated data. A more accurate model for incorporating the effect of the switching activity is based on the realization that two's-complement data words can be divided into two regions based on their activity behavior [10],[11]. The activity in the higher order sign bits depends on the temporal correlation of the data, whereas the lower order bits behave similar to white noise data. The distinct activity behavior of the two bit types is depicted in Figure 11.8. The figure displays 0 -> 1 bit transition probabilities for data streams with different temporal correlations ρ. A module is now completely characterized by its capacitance models in the msb and the lsb regions. The break-points between the regions can be determined from the applied signal statistics, as obtained from simulation or theoretical analysis. This model, called the *Dual Bit Type* or DBT model, has been utilized to accurately determine datapath and memory power consumption [10]. Figure 11.9 shows a comparison between the power predicted by the DBT model and by a switch level estimator (IRSIM) for the logarithmic shifter mentioned previously. The overall rms error is only 9.9%.

Activity models for control signals — The DBT model is restricted in use to two's complement (or similar) signals and is, for instance, not applicable to control busses. For those signals, a simpler model has to be employed that uses

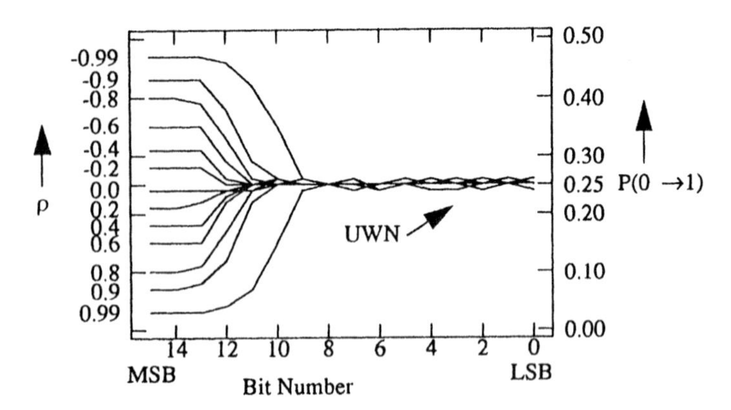

Figure 11.8 Transition activity versus bit for typical data streams

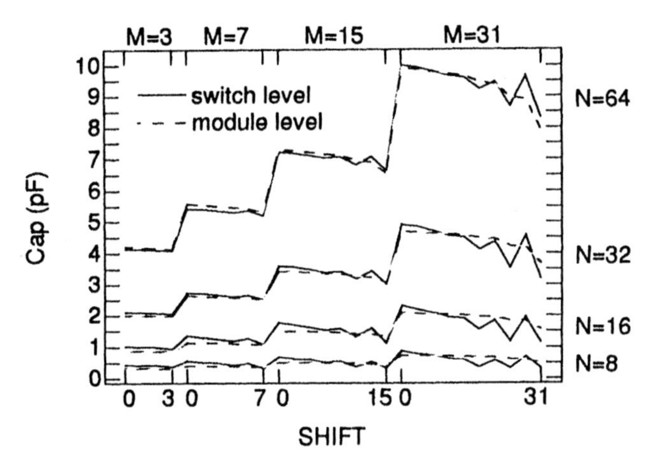

Figure 11.9 Log shifter: switch level power vs. module level power model

the average number of bits switching per event as the main statistical parameter. This so-called *Activity-Based Control* (ABC) model can be effectively used to estimate the power consumption of random-logic controllers [12]. The power dissipation of such a module is, once again, a function of a number of complexity parameters — N_I, the total number of inputs (primary and current state) to the combinational logic implementing the controller; N_O, the total number of outputs (primary and next state) from the logic block; and N_S, an estimate of the number of min-terms for the given logic function. The capacitive weighting coefficients are a function of the switching parameters α_I and α_O, that represent the average number of bits switching on the in- and output busses. Figure 11.10 compares the

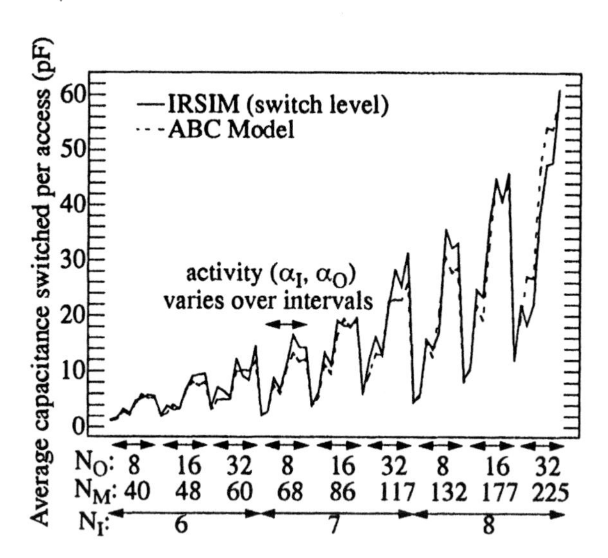

Figure 11.10 Random logic controller: IRSIM-CAP vs. ABC model.

power consumption estimated by the ABC model to that of a switch-level simulator (IRSIM) for a standard cell implementation. For the more predictable ROM and PLA implementations, estimation error is within 10%.

Uncertainties of the final placement and routing make it difficult to estimate interconnect power consumption at the architecture level. Possible solutions to this dilemma include using interconnect estimations based on derivatives of Rent's Rule [9] or back-annotation after early floorplanning.

Design Example 2: FIR Filter, Analysis at the Architectural Level

The techniques discussed in the above sections are utilized here to analyze the differences between the direct and retimed versions (Figure 11.4 and 11.6 respectively) of the FIR filter. The algorithmic level exploration curves (Figure 11.7) show that the retimed version has lower power consumption at a given voltage. A more accurate analysis is possible at the architectural level. SPA (Stochastic Power Analyzer) [12] is an architecture-level power estimation program that relies on the DBT and ABC models to provide fairly accurate estimates for the power consumption. A VHDL simulation is run over a set of typical input vectors to obtain the signal statistics. Table 11.6 shows a comparison of the two different

Component Class	Direct-Form: Capacitance (pF)	Retimed Version: Capacitance (pF)	Difference (%)
Memory	131666	106814	-18.9
Control	53275	43802	-17.8
Clock	18881	15483	-18.0
EXU	45249	32031	-29.2
Registers	13357	10740	-19.6
Total	293563	234637	-20.1

Table 11.6 Architecture level power analysis of direct-form and retimed versions of the FIR filter

FIR programs running on the datapath shown in Figure 11.11. The difference between the two is that the retimed version does not require any explicit memory transfers to implement the delay line. Even though the two programs perform the same computations, the multiplier consumes significantly less power for the retimed version because one of its inputs is fixed at the value of the input sample for the entire sample period. This is not the case for the direct form.

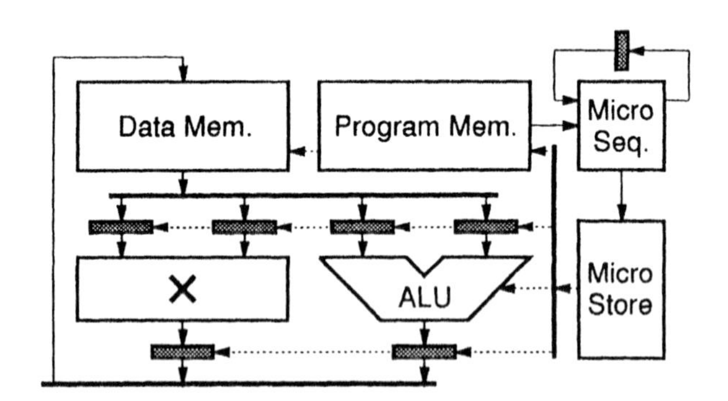

Figure 11.11 Processor datapath implementing the FIR filters

11.4.2. Architecture Optimizations/Synthesis

The power consumption of the final implementation of a given algorithm (especially the implementation-overhead component) depends on the quality of its mapping onto the architecture. The mapping/synthesis process must exploit the relevant properties of a given algorithm. In this section we briefly discuss a sub-set of the potential optimization techniques available during architecture synthesis.

Spatial locality inherent in the algorithm can be utilized during the binding of operations to hardware units operations. Resource sharing should be done only between operations that are strongly connected. This partitions the architecture according to the clusters inherent in the algorithm which reduces the number of "highly capacitive" data transfers and preserves data correlations. Figure 11.12 shows a parallel-form, sixth-order IIR filter. The structure consists of three distinct clusters. In the first case, the algorithm was mapped onto a single large unpartitioned chip with resource sharing between units in different clusters. In the second case, resource sharing was only allowed between operations in the same cluster. A 35% saving in power was obtained without any area penalty [22].

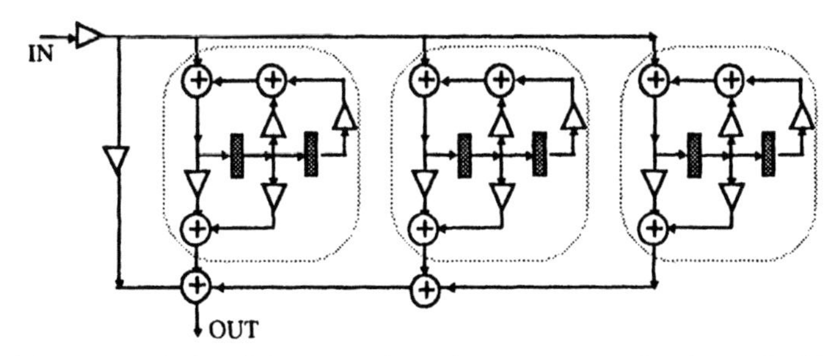

Figure 11.12 Spatial locality in a sixth order parallel form IIR filter.

The assignment process can further reduce the switching activity by mapping those operations with correlated input signals (or with the least number of bit transitions) to the same hardware unit [23]. This can be used effectively in the example of the FIR filter by assigning coefficients with small numbers of bit transitions to the same multiplier. Finally, *temporal locality* can be exploited during scheduling to reduce the size and hence the access capacitance of register files.

Another structural property of algorithms that can be exploited during architecture synthesis for low power is *regularity*. Regularity refers to existence of repeated patterns of computation in a graph. For example, a single pattern of computation occurs three times in Figure 11.12 (as highlighted by the gray circles). A dedicated hardware module constructed corresponding to this template (or a specialized instruction, so to speak) can be reused three times while simplifying interconnect (busses, multiplexers and buffers) considerably. Preserving regularity is therefore an important method to reduce the interconnect-overhead power component.

The DBT model suggests that the correct choice of the *number representation system* can have an important impact on the switching activity. Though common and easy to use, two's complement representation is not always the best for power purposes. A comparison between switching activity of two's complement and sign magnitude data streams is given in [2]. The latter is typically prone to less switching transitions, as only the MSB switches during a zero crossing transition, while all sign bits are toggled in the 2's complement representation.

Finally, *power-down of unused modules* is a commonly employed approach to power reduction. Since all power-down approaches incur some overhead, its is important to cluster related operations and compact the active time-slots of a unit into consecutive intervals. For example, spatial locality of memory references can be detected and exploited during memory assignment to induce effective power down of large blocks of memory.

Design Example 1: Vector Quantization, Architectural Optimizations

To illustrate the effectiveness of using algorithmic properties to guide architectural design, we take a final look at the Vector Quantizer example. The methodology is to start with a simple implementation and then, using algorithmic properties as a guide, improve upon it.

Single Memory/Processor — In one approach, all the operations are time-multiplexed onto a single data path as shown in Figure 11.13. Processing one node of the search tree requires 17 memory accesses, 16 multiply/accumulate operations, and a final add to produce the comparison bit indicating the location of the next node of the tree. A total of eighteen clock cycles are required per comparison: the first cycle to fetch the first word to be compared, then 16 operations for the multiply/accumulate, and the 18th cycle to produce the next location.

The same number of calculations is required for each node of the tree. Processing of a single block, therefore, requires 8x18=146 clock cycles to be executed every 17.3 μs. This necessitates a clocking frequency of 8.3 MHz and a memory access time less than 115 ns, which in turn puts a lower bound on the supply voltage.

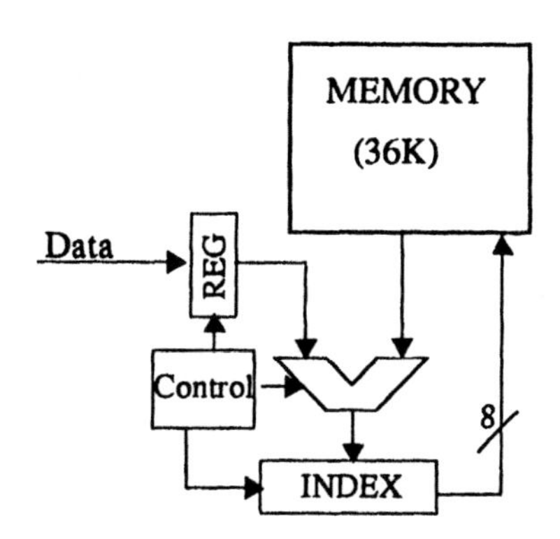

Figure 11.13 Centralized memory architecture.

The usual methods for shortening the critical path, and hence the operating voltage and clocking frequencies, are difficult to apply using the single memory configuration. The memory must be accessed 16 of the 18 clock cycles making it impossible to process more than one vector at a time. Therefore methods to reduce the critical path such as pipelining and parallelism can not be applied without duplicating the entire memory.

However, if the memory is partitioned and a distributed memory approach is used, pipeline stages can be introduced enabling reductions in voltage, clocking frequency, and switching capacitance.

Distributed Memory — In TSVQ each level of the search tree has specific codevectors C_i associated to it. Also, these codevectors are found only at that specific level. In other words, all the memory required for a single level of the codebook is referenced at one and only one level of the tree. This *locality of reference*, indicated by the dashed lines in Figure 11.14, enables the partitioning of the memory into smaller memories, each associated to a single level of the tree [8]. Associated to each memory are identical processing elements and controllers. This architecture can utilize a pipelined structure that greatly reduces the critical path (Figure 11.15).

The architecture can process eight vectors simultaneously with only a negligible increase in latency. The frequency of the clock can be reduced by a factor of eight, and since the execution units need only to process at 960 ns, the supply voltage can be dropped from 3.0 V to 1.1 V.

A less obvious benefit associated with a distributive architecture is a reduction in switching capacitance. A large fraction of the power on both the single and

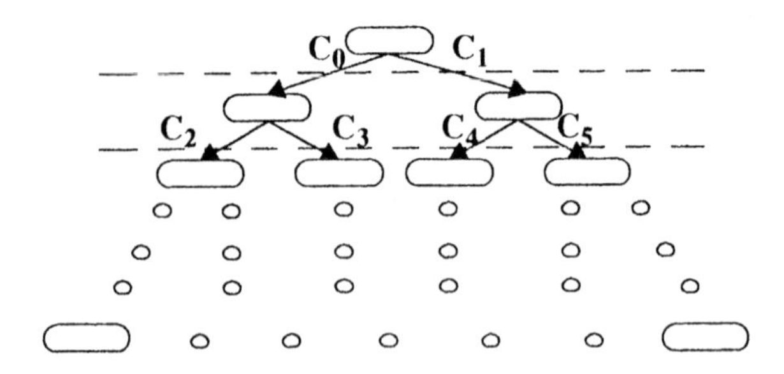

Figure 11.14 Locality of Reference Identification

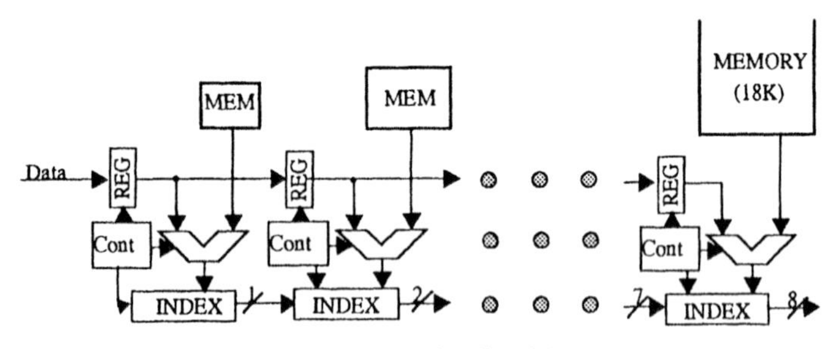

Figure 11.15 Distributive Memory

distributive memory architectures is dissipated by memory accesses. Compared to the centralized case, the distributive architecture switches less capacitance reading codevectors since there is less overhead in reading from smaller memories. Although there are now eight controllers and eight processors clocking on chip, they are clocked at one-eighth the frequency, so the *capacitance switched per vector* by these elements is unchanged. The effect of the interconnect between blocks is negligible since it is clocked at one-eighteenth of the clock frequency.

The final implementations of the two designs showed a savings in power from 4.28 mW for the centralized case to 412 μW for the distributed implementation.

Design Example 2: FIR Filter, Architectural Optimizations.

In the FIR example, the multiply-add pattern of computation is repeated several times. This regularity can be exploited when deciding the specific architecture to be used. Here we compare two different datapaths used to implement

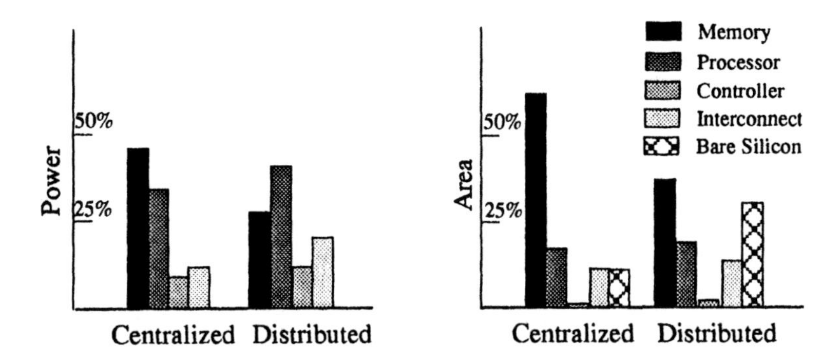

Figure 11.16 Power and Area Breakdown

the direct-form FIR filter. In the first case the regularity is ignored and the datapath of Figure 11.11 is used. In the second case, the datapath has been modified to allow for multiply-accumulate (MAC) instructions (Figure 11.17). This reduces the number of accesses to the data and instruction memories. Table 11.7 shows a comparison of the power consumption of the two different processor architectures that implement the FIR filter example. A 27% reduction in power is obtained by exploiting regularity.

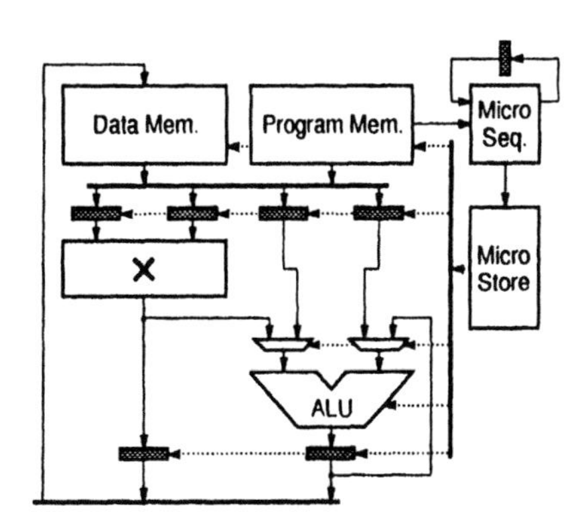

Figure 11.17 Processor datapath with MAC support

11.5. Summary

In this chapter, we have described techniques for power estimation and minimizations at the algorithm and the architecture levels. Figure 11.18 shows an

Component Class	Normal datapath Capacitance(pF)	Specialized datapath Capacitance(pF)	Difference (%)
Memory	131666	72424	-45.0
Control	53275	45553	-14.5
Clock	18881	14485	-23.3
EXU	45249	48648	7.5
Registers	13357	8634	-35.4
Total	293563	214432	-27.0

Table 11.7 Exploiting regularity in the direct-form FIR filter

overview of the essential tasks in an algorithm and architecture level low power design environment and the interplay between the optimization, synthesis and analysis tools.

At the algorithm level, the total power dissipated in a chip can be divided into two main components: the algorithm-inherent components and the imple-

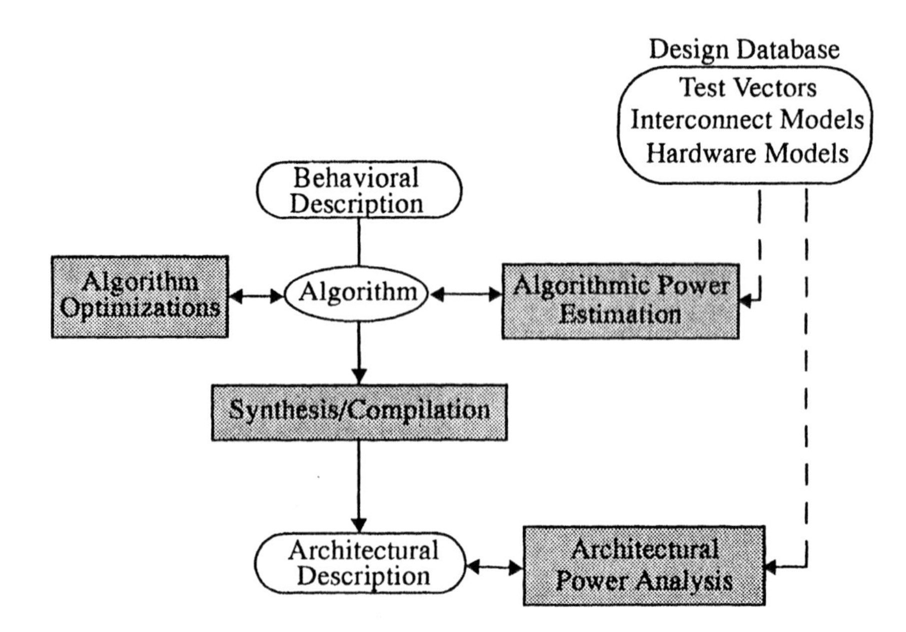

Figure 11.18 Algorithm and architecture level synthesis optimization and estimation environment.

mentation overhead. Due to lack of implementation details at this level, estimating the implementation overhead is difficult and requires stochastic modeling. At the architecture level, signals statistics can be used to estimate the power more accurately. Methods to analyze signal statistics for both data and control signals were discussed.

Optimization techniques discussed in this paper enable voltage reduction, operation reduction and switching activity minimization in the architectural implementation. This is achieved through transformations and through the exploitation of spatial locality, regularity and other structural properties of algorithms during the synthesis of the architecture.

References

[1] A. Chandrakasan, M. Potkonjak, R. Mehra, J. Rabaey, and R. Brodersen, "Optimizing Power Using Transformations," *IEEE Transactions on Computer-Aided Design*, Vol. 14, No. 1, pp. 12-31, Jan. 1995.

[2] A. Chandrakasan, R. Allmon, A. Stratakos and R. Brodersen, "Design of Portable Systems," *Proc. CICC Conference*, May 1994.

[3] F. Cathoor, F. Franssen, S. Wuytack, L. Nachtergaele and H. De Man, "Global Communications and Memory Optimizing Transformations for Low Power Signal Processing Systems", *Proc. of IEEE Workshop on VLSI Signal Processing*, San Diego, pp. 178-187, Oct. 1994.

[4] W. Fang et. al., "A Systolic Tree-Searched Vector Quantizer for Real-Time Image Compression", *Proc. VLSI Signal Processing IV*, IEEE Press, NY, 1990.

[5] A. Gersho and R. Gray, "Vector Quantization and Signal Compression", *Kluwer Academic Publishers*, Boston, MA, 1992.

[6] L. Guerra and J. Rabaey, "System Level Design Guidance using Algorithm Properties", *Proc. of IEEE Workshop on VLSI Signal Processing*, San Diego, pp. 73-82, October 1994.

[7] L. Jamieson, "Characterizing Parallel Algorithms," in *The Characteristics of Parallel Algorithms*, L. Jamieson, D. Gannon, R. Douglass (editors), MIT Press, Cambridge, Mass., 1987.

[8] R. Kolagotla *et al*, "VLSI Implementation of a Tree Searched Vector Quantizer," in *IEEE Trans. on Signal Proc.*, vol. 41, no2, February 1993.

[9] F. Kurdahi and A. Parker, "Techniques for Area Estimation of VLSI Layouts," *IEEE Transactions on Computer-Aided Design*, Vol. 8, No. 1, pp. 81-92, Jan. 1989.

[10] P. Landman and J. Rabaey, "Power Estimation for High-Level Synthesis," *Proc. of European Design Automation Conference '93*, pp. 361-366, Paris, February 1993.

[11] P. Landman and J. Rabaey, "Black Box Capacitance models for Architectural Power Analysis," *Proc. of the International Workshop on Low Power Design*, pp. 165-170, April 1994.

[12] P. Landman, "Low Power Architectural Design Methodologies", Ph. D. thesis, UC. Berkeley, August 1994.

[13] P. Landman, R. Mehra and J. Rabaey "An Integrated CAD Environment for Low Power Design," to be published, *IEEE Design and Test*, 1995.

[14] D. Lidsky and J. Rabaey, "Low Power Design of Memory Intensive Functions, Case Study: Vector Quatization", *Proc. of IEEE Workshop on VLSI Signal Processing*, San Diego, pp. 378-379, October 1994.

[15] D. Liu, and C. Svensson, "Power Consumption Estimation in CMOS VLSI Chips," *IEEE Journal of Solid State Circuits*. Vol. 29, No.6, June 1994, pp. 663-670.

[16] R. Mehra and J. Rabaey, "Behavioral Level Power Estimation and Exploration," *Proc. of the International Workshop on Low Power Design*, pp. 197-202, April 1994.

[17] M. Potkonjak and J. Rabaey, "Exploring the Algorithmic Design Space using High Level Synthesis," *Proc. of IEEE Workshop on VLSI Signal Processing VI*, pp. 123-131, 1993.

[18] M. Potkonjak and J. Rabaey, "Power Minimization in DSP Application Specific Systems Using Algorithm Selection", *Proc. of the ICASSP, 95.*

[19] S. Powell and P. Chau, "Estimating Power Dissipation of VLSI Signal Processing Chips: The PFA Technique," *Proc. of IEEE Workshop on VLSI Signal Processing IV*, pp. 250-259, 1990.

[20] S. Powell and P. Chau, "A Model for Estimating Power Dissipation in a Class of DSP VLSI Chips," *IEEE Transactions on Circuits and Systems*, pp. 646-650, June 1991.

[21] J. Rabaey and M. Potkonjak, "Estimating Implementation Bounds for Real Time DSP Application Specific Circuits", *IEEE Trans. on CAD*, pp. 669-683, June 1994.

[22] J. Rabaey, L. Guerra and R. Mehra, "Design Guidance in the Power Dimension," *Proc. of the ICASSP*, 1995.

[23] A. Ragunathan and N. Jha, "Behavioral Synthesis for Low Power", *Proc. of the ICCD*, pp. 318-322, 1994.

[24] G. Sorkin, "Asymptotically Perfect Trivial Global Routing: A Stochastic Analysis," *IEEE Trans. on Computer-Aided Design*, vol. CAD-6, pp. 820, 1987.

[25] S. Sheng, A. Chandrakasan and R. Brodersen, "A Portable Multimedia Terminal," *IEEE Commumications Magazine*, pp. 64-75, December 1992.

[26] V. Tiwari, S. Malik and A. Wolfe, "Power Analysis of Embedded Software: A First Step Towards Software Power Minimization", IEEE Transactions on VLSI Systems, No. $, Vol. 2, pp. 437-445, Dec. 1994.

[27] E. Tsern, T. Meng, and A. Hung, "Video Compression for Portable Communications Using Pyramid Vector Quantization of Subband Coefficients," *Proc. of IEEE Workshop on VLSI Signal Processing*, Veldhoven, pp. 444-452, October 1993.

[28] H. J. M. Veendrick, "Short-Circuit Dissipation of Static CMOS Circuitry and its Impact on the Design of Buffer Circuits," *IEEE Journal of Solid-State Circuits*, pp. 468-473, August 1984.

INDEX

P

Q

R

S

T

U

V

W

Y

Z

Printed in the United States
41209LVS00002BA/4

9 780792 396307